Fire in the Sky

Comets and meteors are spectacular and awe-inspiring natural phenomena, and are among nature's most compelling icons. Since the beginning of recorded time, they have mesmerized people, not least among them artists and astronomers. Britain during the eighteenth and nineteenth centuries produced a larger number and greater variety of representations of comets and meteors than any other country. The development of new technologies, and the burgeoning interest of the general public in science and art, dovetailed with the inherent British interest in nature and a strong literary tradition of comet and meteor symbolism. This beautifully illustrated book examines the link between these works, and the achievements of British science in the wake of Newton and Halley. This book will be of great interest to anyone interested in the art and culture of the period and/or the astronomy of comets.

ROBERTA J. M. OLSON is professor of Art History and Mary L. Heuser Chair in the Arts at Wheaton College, Norton, Massachusetts. Her specialities, in which she has published widely, are Italian Renaissance art, nineteenth-century art, and drawings. She has written many articles on various representations of comets in art and a book on the subject.

JAY M. PASACHOFF is Field Memorial Professor of Astronomy and Director of the Hopkins Observatory at Williams College, Williamstown, Massachusetts. He has twice been Chair of the Astronomy Division of the American Association for the Advancement of Science, and represents the United States on the Commission on the Teaching of Astronomy of the International Astronomical Union.

Professors Olson and Pasachoff have jointly written a number of articles on comet images. They participated in scientific symposia on the results of the 1986 appearance of Halley's Comet. That project led to this book, which has been supported by two grants from the National Endowment for the Humanities and two major Fellowships from the Getty Grant Program.

Fire in the Sky

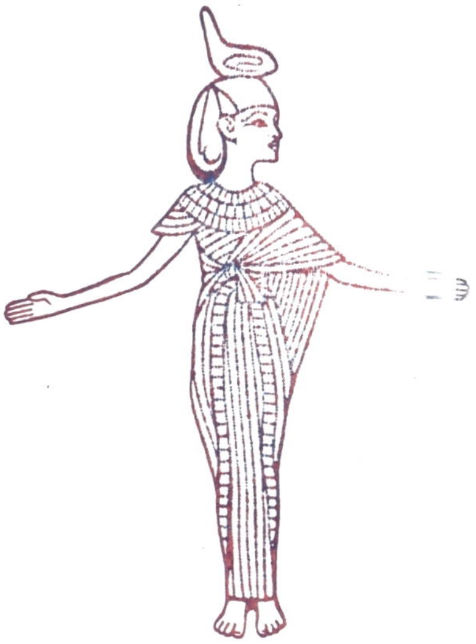

Anonymous, *Donati's Comet Over Christopher Wren's St. Paul's Cathedral 1858*, Collection of Gerald H. Morris, London (see Fig. 124).

Fire in the Sky

Comets and Meteors, the Decisive Centuries, in British Art and Science

Roberta J.M. Olson and Jay M. Pasachoff

Epilogue by Colin Pillinger

CAMBRIDGE
UNIVERSITY PRESS

Published with the support of Gresham College, London.

PUBLISHED BY THE PRESS SYNDICATE OF THE UNIVERSITY OF CAMBRIDGE
The Pitt Building, Trumpington Street, Cambridge, United Kingdom

CAMBRIDGE UNIVERSITY PRESS
The Edinburgh Building, Cambridge CB2 2RU, UK www.cup.cam.ac.uk
40 West 20th Street, New York, NY 10011–4211, USA www.cup.org
10 Stamford Road, Oakleigh, Melbourne 3166, Australia
Ruiz de Alarcón 13, 28014 Madrid, Spain

First published 1998
First paperback edition 1999

Printed in the United Kingdom at the University Press, Cambridge

Typeface Minion 11/15 pt *System* QuarkXPress® [DS]

A catalogue record for this book is available from the British Library

ISBN 0 521 63060 6 hardback
ISBN 0 521 66359 8 paperback

Gresham College was founded in 1597 as an independent institution in the City of London, supported by the Corporation of London and the Worshipful Company of Mercers.

Contents

Acknowledgments

This book was written with the generous support of the Getty Grant Program, which awarded us Senior Research Grants during 1994-95. We are extremely grateful for this wonderful opportunity to investigate thoroughly a subject we had long wanted to pursue. We thank all the people associated with the Getty, especially Gwen Walden and Joan Weinstein, who facilitated the implementation of this research project. Moreover, we thank Gresham College, which contributed a grant toward the publication of this volume, and Colin T. Pillinger, who believed in our project. In fact, the desire for the "New Learning" discussed in the early chapters of our study was anticipated at the incredibly early date of 1579 by the will of Sir Thomas Gresham and put into practice by the founding of Gresham College, which bears his name, in 1597. We also want to acknowledge the "seed" grants from the National Endowment for the Humanities, two Travel to Collections Grants, for the early phase of research on the topic, and to thank the NEH for their support.

Furthermore, we would like to express our personal gratitude to a galaxy of individuals, who made invaluable and varied contributions to different phases of the project, and without whose help, it would not have been possible. Prominent among them are a constellation of comet-hunters and meteor-spotters, including Alexander B.V. Johnson, Peter Hingley, Angus Macdonald, Wendy Sheridan, and Brian Marsden. Further thanks are extended to: William J.H. Andrewes; Elliot Aylwin; Henry Bedingfeld; James Bennett; Annette Blood; Mark Brady; John C. Brandt; Ron Bristow; C.N. Brown; Joan Linnell Burton; Christine Butler; Mungo Campbell; Mark Chalfont; Mary Chibnall; Gloria C. Clifton;

Melva Croal; Travis Crosby; Sandra Cumming; Cara Denison; Carrie Fake; Ronnie Fein; Celina A. Fox; Miranda Fyfe; John Gage; Owen Gingerich; Margaret Glover; John Goddier; George Goldner; David Goodwin; Susan Grayer; Antony Griffiths; Roger Griffon; Sandy Grindlay; William Griswold; Veronique Gunner; Mimi Harpur; Martha Hazen; Mary Henderson; Mike Hickox; Jeffrey Hill; Theo Hodges; Dorrit Hoffleit; David Hughes; Valerie Hunter; E.D.H. Johnson; Kevin Johnson; Susan Kaufman; Gary W. Kronk; Susan Lawson-Bell; Eric Lee; Benita Lehmann; Nigel and Lady Lindsay-Finn; Kristen Lippincott; Rupert Maas; Barbara Madden; Sir Malcolm Innes of Edingight, KCVO, The Lord Lyon King of Arms; Rosalind Marshall; Marie McCabe; Joslyn McDiarmid; Shona McEachern; Lindsay McFarland; Sally McIntosh; John H. Milburn; Hillis Miller; Duncan Mirylees; Jacqueline Mitton; Alison D. Morrison-Low; Patrick Noon; Anthony North; Wayne Orchiston; Ursula Overbury; Brian Overell; Alex Soojung-Kim Pang; Deborah Pasachoff; Eloise Pasachoff; Naomi Pasachoff; Robert L. Patten; Margaret Patterson; Adam Perkins; Diane Perkins; Clare Philips; Kevin Pluta; Roger Quarm; Martin Rees, Astronomer Royal; Christopher Ricks; Beatrice Riehl; The Hon. Jane Roberts; Pamela Roberts; Duncan Robinson; William Robinson; Caitlin Rooney; Judy Rudoe; Mary Sampson; Polly Sartori; Marty Schlabach; Richard E. Schmidt; Gail Schuster; Linda Seckelson; Marjorie Shelley; Joseph M. Silva; Robert Simon; Don Simpson; Peyton Skipwith; Kim Sloan; Ken Soehner; Anthony Spink; Lindsay Stainton; Miriam Stewart; John Sunderland; Ted Swystum; Albert Torres; Sue Tritton; J.K.H. Wales; Diane Walker; John Walker; Jessica Wapner; Brian Warner; Ian Warrell; Beth Carver Wees; Dustin Wees; Jane Wess; Rebecca Willetts; Andrew Wilton; Andrew Wyld; Martha Yellig; Donald K. Yeomans; and Kathy Yvinskas. We thank Brian Marsden, David Levy, Donald Yeomans, and Dorrit Hoffleit for their comments on the entire manuscript. In addition, we are most grateful to several anonymous readers of the manuscript who made valuable suggestions, and especially to our editor, Simon Mitton whose faith in our project, deft expedition, and judicious opinions are most appreciated. A special note of thanks to Christie's, in London and New York, and Sotheby's, in London and New York, and their staffs for facilitating orders of photographs and for assisting us in tracing works

of art that were sold through their auction houses. Finally, we want to acknowledge and thank the regular, science, and interlibrary loan librarians at Williams College and Wheaton College; the astronomy librarians at the Wolbach Library of the Harvard-Smithsonian Center for Astrophysics and at the Astrophysical Library of Princeton University; and the staffs of The Frick Art Reference Library, The New York Public Library, and The Thomas J. Watson Library of the Metropolitan Museum of Art. Many other individuals contributed to the realization of this book, some of whom are mentioned in specific footnotes, and we are most grateful to each and every one of them.

Illustration Credits

The authors wish to thank the following persons and collections for their kind permission to reproduce the following illustrations in the text.

© 1975 (revised 1997) The New York Academy of Sciences: Fig. 1

The Pierpont Morgan Library, New York. The Peel Collection: Figs. 2, 21, 68, 70-72

Rare Book and Manuscript Library, Columbia University: Fig. 3

Private Collection, London: Fig. 4

The Earl of Haddington: Fig. 5

President and Fellows of Corpus Christi College, Oxford: Fig. 6

By permission of the Syndics of Cambridge University Library: Fig. 7

Private Collection, Williamstown, MA: Fig. 8

Copyright ©The British Museum: Figs. 9, 20, 32, 34, 36, 37, 40, 41, 44-50, 61, 63, 64, 69, 73-76, 79, 81-83, 85, 86, 92, 108, 119, 121

By permission of the President and Council of the Royal Society: Figs. 10, 22-28, 30, 31

Birmingham Museums and Art Gallery: Figs. 11, 149

Science Museum London/Science and Society Picture Library: Figs. 12, 15, 16, 18, 93, 101, 104, 120

© Museum of the History of Science, Oxford University: Fig. 13

Courtesy of the Collection of Historical Instruments, Harvard University: Fig. 14

Private Collection, New York: Figs. 17, 77, 102, 109, 128, 148

By courtesy of the Trustees of Sir John Soane's Museum: Fig. 19

© Collection of the New-York Historical Society: Fig. 29

Royal Astronomical Society Library: Figs. 33, 51, 65, 97-99, 111, 129, 142-146

The Royal Collection © 1997 Her Majesty Queen Elizabeth II: Figs. 35, 110

Courtesy of Spink and Son Limited and the Yale Center for British Art, Paul Mellon Collection: Fig. 38

Private Collection, England: Fig. 39

Yale Center for British Art, Paul Mellon Collection: Figs. 42, 43, 56-58, 67, 80, 90, 105, 106, 126

The Pierpont Morgan Library, New York: Fig. 52

Rare Books Division, The New York Public Library, Astor, Lenox and Tilden Foundations: Figs. 53-55

© Tate Gallery, London 1997: Figs. 59, 60, 133, 136

The Fitzwilliam Museum, Cambridge: Fig. 62

Collection of Gerald H. Morris, London: Frontispiece, Figs. 66, 124, 137

Courtesy H. Schickman Gallery: Fig. 78

© Sterling and Francine Clark Art Institute, Williamstown, Massachusetts, USA: Fig. 84

Courtesy Christie's Images: Figs. 87, 94, 95, 152-154

Museum of London: Fig. 88

National Maritime Museum London: Figs. 89, 113, 114, 157

Reproduction of *An Attempt to Illustrate the Opening of the Sixth Seal* by Francis Danby courtesy of the National Gallery of Ireland: Fig. 91

By permission of the Syndics of Cambridge University Library and the Director of the Royal Observatories: Fig. 96

Courtesy of R.W. Lee, *A History of Valentines*, New York, 1952: Fig. 100

Courtesy of Brian Harpur: Fig. 103

© Indianapolis Museum of Art, Gift in memory of Dr. and Mrs. Hugo O. Pantzer: Fig. 107

Don Africana Library, Durban Metropolitan Library: Fig. 112

Reproduced by permission of the Royal Society of Edinburgh from the RSE Piazzi Smyth archive, held on deposit at the Royal Observatory Edinburgh: Figs. 115, 116

National Gallery of Scotland: Figs. 117, 118

The Victoria and Albert Museum, London: Figs. 122, 151

The Crawford Collection, Royal Observatory Edinburgh: Figs. 123, 132

Williams College Library: Fig. 125

Courtesy of Agnew's, London: Fig. 127

Private Collection, London: Fig. 130

The Maas Gallery, London: Fig. 131

Harvard College Observatory: Fig. 134

© 1983 Royal Observatory Edinburgh: Fig. 135

Courtesy of the Fogg Art Museum, Harvard University Art Museums, Bequest of Grenville L. Winthrop: Fig. 138

© Manchester City Art Galleries: Figs. 139, 147

Courtesy Satis House Collection, Tuxedo Park: Fig. 140

By permission of the Trustees of the Watts Gallery: Fig. 141

Courtesy of O. von Schleinitz, *Walter Crane*, Bielefeld and Leipzig, 1902: Fig. 150

Courtesy of Sotheby's: Fig. 155

Courtesy of Sotheby's, Inc.: Fig. 156

© 1986 Royal Observatory Edinburgh from a UK Schmidt plate. Processed by B.W. Hadley: Fig. 158

U.S. Naval Research Laboratory, courtesy Guenter Brueckner: Fig. 159

© 1986 copyright Max-Planck-Institut für Astronomie, Lindau/Harz, FRG, courtesy of H.U. Keller: Fig. 160

Courtesy of Jay M. Pasachoff: Fig. 161

Courtesy of Dennis di Cicco: Fig. 162

INTRODUCTION

Fire in the Sky: Comets and Meteors, the Decisive Centuries, in British Art and Science

> Now we know
> the sharply veering ways of comets, once
> A source of dread, no longer do we quail
> Beneath the appearance of bearded stars.
> Edmond Halley, dedicatory ode to Isaac Newton's *Principia Mathematica*

Comets and meteors are spectacular and awe-inspiring natural phenomena, which are among nature's most compelling icons. Since the beginning of recorded time, they have mesmerized people, not least among them artists and astronomers. Their fascination stemmed not only from the fact that their appearances seemed to violate what was perceived as the proper, elegant order of the universe but also from their sheer physical beauty. In the course of our previous work, we realized that there were a larger number and greater variety of extraordinary comet and meteor images produced in England during the eighteenth and nineteenth centuries than in other countries of the western world. Our subsequent findings suggest a correlation between these images and the achievements of British science in the wake of the momentous discoveries of Isaac Newton, Edmond Halley, and William Herschel, all of whose work involved comets and the nature of the universe. The scientific investigation of comets and meteors, which first occurred in earnest during the period in question, dovetailed with the inherent British interest in nature and a strong literary tradition of comet/meteor symbolism. Moreover, these trends were linked to the burgeoning interest of the general populace in science and art as well as to the development of new technologies, leading to a veritable explosion of captivating representations of these celestial apparitions between 1780 and 1860.

In the following chapters we examine these comet and meteor images more or less chronologically within the broader context of the unique British scientific, artistic, social and cultural developments that encouraged their genesis. While these representations can be grouped into three categories – realistic, satirical, and imaginative or symbolic – we relate each to larger pertinent themes and concerns, and establish the nature of their creator's intellectual engagement with comets and meteors. We believe that this investigation illuminates a corner of the scientific and cultural environment and elucidates yet another aspect of the rich spectrum of British history. Because this fascinating subject touches the boundaries of many disciplines and has complex implications, it leads to additional questions, bibliography, and methodological controversies that branch out from this study.[1]

Throughout we keep in mind the achievements of other nationalities, as well as the larger context of European art and astronomy, but focus on British contributions.[2] The treatment of comets and meteors in western art is more expansively surveyed in Olson's *Fire and Ice: A History of Comets in Art*, a book that accompanied an exhibition of the same title at the Smithsonian Institution, the National Air and Space Museum, Washington, D.C., during the apparition of Halley's Comet in 1985-86.[3]

The period spanned by the book – including the Enlightenment and the Georgian and Victorian periods – could be termed the Golden Age of both British astronomy and art.[4] In order to set the stage, we open with a discussion of the significant scientific developments and the tradition of comet and meteor representations during the seventeenth century and proceed into the Age of the Enlightenment. At this time, the "arts" and the "sciences" were not separated as they are today, and fields of knowledge were certainly less vast and more interconnected. With the increasing eighteenth-century interest in technological innovation and scientific investigation, which would evolve gradually into the full-blown Industrial Revolution, came a general curiosity about science as well as the public's involvement with scientific pursuits as entertainment. A knowledge of science and astronomy became one of the hallmarks of a British gentleman, just as sketching with artistic dexterity, together with art appreciation, was expected of a cultured individual of either sex. These endeavors coincided with the popularization of science and art for a more mass audience.

During this time, there was an unusually high level of amateur involvement in both British art and astronomy. In addition, the advent of renowned scientific societies, like the Royal Society, mixed individuals of all fields in an intellectual atmosphere fired by exciting discoveries and changes. (Not coincidentally, comets and meteors since time immemorial have functioned as symbols of change.) It was also a time when artistic societies for the organization and exhibition of works of art made the visual arts more accessible to a larger audience – for example, the Royal Academy and the Old Water-Colour Society. During the Victorian period, the age of the massive public exposition was inaugurated, including the first international Great Exhibition of 1851, in which some of the earliest astronomical photographs were exhibited in Joseph Paxton's Crystal Palace in London.

The birth of journalism in the eighteenth century spawned inventions and technological innovations in printing techniques that resulted in publications and prints which disseminated visual images as well as scientific information about comets and meteors. These influenced the opinions and tastes of the British and inflamed their interest in heavenly phenomena. A case in point involves the political and social satires and caricatures produced after 1760, in which the Tory-Whig struggles are occasionally illuminated by comet imagery. These prints continued during the crescendoing social discord of subsequent decades as Britons lived through political radicalism, conflicts between Parliament and the King, popular riots, revolt, war, colonial loss, urbanization, cultural crises and leveling, fiscal problems, and the spread of crime from London to the country – not to mention regional conflicts. During this time of flux, a very self-conscious interest in the present moment was reflected in both art and science. It was manifested in sweeping thematic changes and accompanied by the growth of periodical publishing. Our study deals with a variety of these publications, ranging from the popular lampooning *The Meteor* and *Punch* to the more information-oriented *Illustrated London News*, the latter used by artists (for example, Samuel Palmer) for information about comets. Published popular accounts of comets and meteors begin as early as 1737 in *The Gentleman's Magazine*, from which it is apparent that a learned person's knowledge extends to the observation of celestial phenomena. We also explore the theories and astronomical information published about comets and meteors in the *Philosophical Transactions of the Royal Society* and the *Monthly Notices* as well as the *Memoirs* of the Royal Astronomical Society.

Our study investigates larger trends, while focusing on the contributions of individuals. Especially interesting are the links between artists and scientists involved with specific astronomical apparitions, such as the association of the artist brothers Paul and Thomas Sandby with the scientist Tiberius Cavallo who observed and then collaborated on discussions and depictions of the meteor of 18 August 1783. This pivotal event caused much scientific ferment and elicited many reports and illustrations. Later artists, like David Scott and Joseph Mallord William Turner, were also associated with prominent scientists and astronomers. Still other individuals worked as both astronomers and artists, for example, Charles Piazzi Smyth, Astronomer Royal for Scotland, and John Brett, a member of the Royal Astronomical Society and associated with the Pre-Raphaelite group of artists.

Dating as far back as Elizabethan times, British literature featured vivid comet and meteor symbolism. This tradition played a not inconsiderable role in comet and meteor iconography. Moreover, the investigatory spirit of the period, allied with its archaeological and scientific quests and its interest in emotions and the imagination, led to profound explorations of these literary sources culminating in innovative illustrations. Our study examines the strong British illustrative tradition and selected illustrators who employed comet and meteor images. For example, the artist-poet William Blake produced the greatest number of them. Then, there is the case of Thomas Hardy, who not only used comet and meteor imagery in his writings, but also illustrated a poem with a huge comet. As one might imagine, both the rampant historicism and positivism of the nineteenth century also encouraged the use of these eminently vivid natural symbols. In both poetry and prose, comet and meteor symbolism, as well as an integral British sympathy to landscape and nature, is evident.[5] These tendencies appear in both centuries, although the interest in nature experiences a denouement at the end of the nineteenth century.

The magnetic attraction to stunning celestial occurrences was fed by the Romantic philosophical, literary, and artistic cult of Nature, in which natural phenomena provided inspiration and symbolic content for literature and elicited emotional and psychological responses in viewers. Both comets and meteors fit into the doctrine of the Sublime as articulated by Edmund Burke in *A Philosophical Inquiry into the Origin of our Ideas of the Sublime and Beautiful* (1757), wherein the grandest aspects of nature may excite awe or dread that is effectively an aesthetic experience.

Moreover, the moist maritime climate of the United Kingdom and the particular beauties of its atmospheric effects predisposed British artists to an interest in recording the sky, which in turn encouraged them to include comets and meteors in their works. Artists who concentrated on landscape focused on meteorology as well, and after 1700 utilized the watercolor medium in which these transitory effects are best captured. Since watercolors are portable, artists captured quickly what they observed outdoors rather than attempting to revisualize it in the studio in the more time-consuming, less-portable oil technique.[6]

During the eighteenth century, watercolor was elevated from its former status as a medium employed in preparatory studies to one that was valid for finished works of art. And, it was with watercolor – which before the advent of the camera was first used extensively by topographical draftsmen employed by antiquarians and nobility to render views encountered on trips – that British artists made many of their most significant contributions. Its portability and quick drying speed made it ideal for that purpose, while its transparent potentials could be strengthened by lines in pen and ink and by the addition of opaque body color or gouache. This change coincided with the development of pure landscape as a genre and as an important pictorial theme, liberated from the former requisite props of classical allusions. Although Thomas Gainsborough and others of his generation painted watercolor landscapes, it was Paul Sandby who really laid the foundations of landscape painting in England. And it is in the medium of watercolor that many of the most spectacular comet and meteor representations are recorded by Sandby and others.

Certainly the English predisposition to genre subjects, whose rise was accompanied by that of the commercial middle class to a dominant position in the social hierarchy, also created a context for the representation of comets and meteors. In Protestant Britain with its parliamentary system, official history painting on a grand scale, like that sponsored by the Catholic monarchies of the Continent, never took hold. Artists who, unlike many of their European contemporaries, were generally trained at home rather than abroad, responded to the taste of the general public. Although the reins of patronage remained in the hands of the aristocracy through the eighteenth century with its major outlet portraiture, the growth of the print-trade testifies to the demand for popular art forms to feed a middle-class audience and a growing interest in reportage. Just as in the case of their Dutch and Flemish predecessors, eighteenth-century British merchants looked to art not for exalted subjects inspiring noble thoughts but rather for a truthful reflection of their values and way of life. Since artists began to explore all fields of subject matter in their role as chroniclers of their time,[7] comets and meteors, which appeared frequently in the heavens during the period in question, were depicted with some regularity and even featured in witty satires of the time.

The period encompassed by the major part of our study ranges from the comets observed by Newton and Halley in 1680 and 1682 through the reappearance of the second of these – now called Halley's Comet – in 1910. It starts shortly after the discovery of the telescope and ends with the establishment of astronomical photography and the origins of astrophysics. We follow the effects of this technological revolution in astronomy on artists, the general public, and astronomers, who also drew comets and meteors as part of their observations. The development of astronomical photography, which we consider with an emphasis on comets and meteors, had a great impact on both astronomy and art. To ascertain the degree of accuracy in comet and meteor images, we have employed state-of-the-art computer programs to recreate the historical night skies and have consulted observational records and the drawings of astronomers.

Comet and Meteor Primer

The following brief explanation of comets and meteors will equip the reader to appreciate better their significance in Great Britain during the eighteenth and nineteenth centuries, when they attracted such great attention. The two phenomena will be discussed in alphabetical order, comets first and meteors second.[8]

Bright comets appear irregularly, and except for Halley's Comet, they cannot be predicted. They are one of nature's most wondrous features and have inspired awe, veneration, and often fear as well. Dramatically illuminating the dome of the night sky, these "blazing stars" or "hairy stars," as they have been called, seemed in the past to violate what was perceived to be the predictable order of the heavens. Unlike other spectacular celestial phenomena, such as eclipses, some comets are visible for months on end. The most dazzling have had tails that stretch far across the sky. It is understandable then that many observers throughout the centuries have attributed deep significances to their appearances, and that comets, therefore, have an enduring presence in Western art and literature.

Astronomers pinpoint the source of many comets to a vast cloud known as the "Oort cloud," proposed by Jan Oort in 1950. This spherical comet cloud, which is gravitationally part of the solar system, consists of the same primordial matter from which the sun and the planets were formed almost five billion years ago. A gravitational disturbance, such as one caused by a passing star, can occasionally dislodge one of these incipient comets and send it gliding toward the sun to become trapped in a new orbit. Most comets that have been observed from earth, including the most famous of all, Halley's Comet, travel in elongated elliptical orbits, which usually lie in planes that are highly tilted with respect to the orbits of the planets [Fig. 1]. "Long-period" comets, those with periods of two hundred years or longer, are plentiful. Some take thousands or hundreds of thousands of years for a complete circuit of their orbits (like Comet Kohoutek in 1973, with a period thought to be over 80,000 years). These comets seem nonperiodic to us since we never witness their return. Comets that round the sun in periods of two hundred years or less are called "short-period" comets. They are now thought mostly to come from the Kuiper belt, another comet reservoir beyond the orbit of

Neptune, proposed by Gerard Kuiper in 1951 and confirmed by recent observations from large ground-based telescopes. (This belt, which was earlier discussed by K.E. Edgeworth in Ireland in 1943 and published in the *Journal of the British Astronomical Association*, is often called the Edgeworth-Kuiper belt or the Kuiper-Edgeworth belt.) It is estimated that this belt contains at least 200 million comets, which have remained essentially unchanged since the birth of the solar system. Then, there are also nonperiodic comets, whose orbits have been modified by the gravity of a giant planet into open-ended paths in the shapes of hyperbolas. After rounding the sun, they glide off into interstellar space, never to return. Comet catastrophes include those comets that cease to be sungrazers and crash into the sun or those that crash into planets – for example, the large comet thought to have impacted earth 65 million years ago causing the extinction of the dinosaurs and other species; the chunk of a comet or asteroid that may have hit the earth in Tunguska, Siberia, in 1908; or Comet Shoemaker-Levy 9, which hit Jupiter in 1994.

Unlike meteors, comets do not move visibly across the sky as you watch, although their tails may give an impression of movement. However, over the course of time observers can see that a comet moves at a different rate than the nearby stars.

Comets are composed of ices of carbon dioxide, water, and other molecules, embedded with organic materials and dust. In

Figure 1. *A Comet and its Orbit through the Solar System*, updated from a drawing by Leon Tadrick in *The Sciences*, 15:8, 1975.

1950 Fred Whipple characterized a comet's nucleus as a dirty snowball several kilometers across, and his theory is generally accepted. The apparition of a reasonably bright periodic comet, Halley's Comet in 1985-86 (1P/1982 U1), which offered scientists a long time to plan experiments, led to unprecedentedly complete observations. Several spacecraft rendezvoused with the comet, most notably the European Space Agency's *Giotto*. This spacecraft provided electronic images beamed from only about 600 kilometers away from the comet's nucleus, which revealed a potato-shaped ice-ball about 16 kilometers long and 8 kilometers in each other dimension [Fig. 160]. The nucleus was black as soot, reflecting only about four percent of the light that hit it, and showed signs of hills and valleys. Several jets of dust on the sunward side reflected sunlight brightly. It is estimated that about 30 tons of water and 5 tons of dust per hour were lost from the comet's nucleus as it approached the sun. At this rate Halley's will last for hundreds of passes around the sun, although it is known to have appeared at least 24 times in recorded European history (Chinese documents record it 28 times).

There are a number of comet concepts and comet terms, which will be helpful in reading this book. The point of a comet's orbit closest to the sun is known as *perihelion* (*helios* meaning the sun in Greek). A comet's speed increases as it approaches the sun, following the second law of orbits discovered by Johannes Kepler in the early seventeenth century. The elliptical orbit of a comet can be disturbed by the gravitational pull of the planets as well as the sun, altering its course, speed, and brilliancy. During each passage around the sun, comets tend to lose a tiny fraction of their mass. Some comets, such as Biela's Comet in 1846 (3D-A and 3D-B) or comet Shoemaker-Levy 9 (D/1993 F2) freakishly break up into fragments; some disappear forever, and others continue but with noticeably less mass.

Comets are anatomically composed of three major parts. The icy *nucleus* – up to about 40 kilometers across (as with Hale-Bopp) or even larger (for objects related to comets like Chiron and some Kuiper-Edgeworth belt objects) and made up of frozen gases and chunks of matter – begins to sublime (change from a solid to a gas) as it nears the sun. Some of its frozen gases evaporate, releasing solid particles and creating a halo of diffuse material thousands of times the size of the nucleus called a *coma*. The coma, which reflects sunlight and fluoresces, sometimes has several layers referred to as *envelopes*. Together the nucleus and the coma make up the comet's *head*. The head is followed by the comet's *tail*, which is its most flashy feature and is responsible for the term "comet," derived from the Greek word "*kometes*," literally meaning "long-haired." Because comet tails can extend up to 250 million kilometers, greater than earth's distance from the sun, comets can be the largest objects in the solar system, but they have very little mass.

Comets have three types of tails, which can be exhibited simultaneously but look different. The most common is a dust tail, which is usually slightly curved. It is whitish in appearance since it reflects sunlight and trails behind a comet's orbit. The second type is a gas tail, also called an "ion" tail, which is blue because we see an emission of blue light from its molecules. Many comets, especially the brightest, display this kind of tail, which streams away at a faster rate and thus tends to be slimmer and straighter than the dust tail. A third type of tail, of neutral sodium, was discovered for Comet Hale-Bopp in 1997. A comet's gas tail always points away from the sun because it is formed by solar radiation and the outward flow of electrically charged particles (ions) from the sun. The "solar wind," an outward flow of ions, as an expansion of the solar corona, is responsible for the structure and rapid changes in the gas tail. The pressure of the radiation (that is light) controls the dust tail. Comet tails can split into parts, forming fanning sections, which in some cases contain as many as a dozen peacock-like rays. Some comets (usually the smaller ones) never develop tails, but rather look like fuzzy patches of luminescence in the sky.

Every comet is unique and its appearance fluctuates during each apparition. The appearance depends on several things: size and the shape of the nucleus, the comet's orbit, the sun's effect on the escaping material, and the angle and distance from which people see it. Prior to the late eighteenth century, Western records on comets are not entirely reliable. Chinese records, on the other hand, tend to be more accurate.[9] It was during the period of our study, with its obsession with comets, that their apparitions were carefully studied and first observed scientifically in the West.

There are several types of comet nomenclature. Since comets have long been named after their discoverers, each of the comets first discovered by Caroline Herschel could be called Comet Herschel. When the discovery rate of comets exceeded one per year, comets first were assigned Roman numerals after the year in which they were seen to indicate their order of appearance; however, starting in 1846, these Roman numerals were later modified to indicate the order in which they passed perihelion. Thus, under this system Donati's Comet was 1858 VI, the sixth comet to pass perihelion in 1858. But starting in 1870, with some retroactivity to 1869, a new system was adopted. Comets were given letters in order of their discovery. Thus, the three comets of 1869 were 1869a, 1869b, and 1869c. When they passed perihelion, something that took many months to find out, they were assigned Roman numerals in order of that passage. The order did not always correspond to the order of their discovery.[10] Moreover, there were variations in the usage and many exceptions, which are described in detail by Marsden.[11] Some comets could not even be given numbers at all, since they were discovered years later on photographs and could not be interpolated into the Roman-numeral scheme!

The International Astronomical Union, which has taken over responsibility for comet names, has unofficially used a limit of three individual names, a distinction awarded to the first three independent discoverers, and has devised a new identification system. As the number of comet discoveries in a year passed the number of letters in the alphabet and as a few short-period comets became visible all around their orbits, the old systems grew unwieldy. In 1994, therefore, they promulgated the new system of nomenclature, in which a comet is designated by the year and an upper-case letter to indicate the halfmonth in which the comet was discovered.[12] These designations are prefixed by a letter to indicate the type of object.[13]

The I.A.U. system no longer recognizes routine reappearances of periodic comets with very short periods. Thus Encke's Comet, with its three-year period, is now merely 2P. Allowable notations include 2P/Encke or 2P (Encke). For comets with periods longer than 30 years, letters are usually assigned to each recovery, such as 1P/1682 Q1 (Halley).[14]

Meteors are related to comets. While meteor showers are children of comets, individual meteors seen in the sky are, rather, chips off asteroids. They are among the celestial phenomena seen by many. We now use the word "meteor" interchangeably with "shooting star" to describe an unexpected streak of light lasting a few seconds that occasionally crosses the heavens. However, as we shall see this scientific definition is less than 200 years old. When these small bits of interplanetary matter enter the earth's upper atmosphere, at speeds as high as 40 miles (64 km) per second, they burn up. Each meteor can be as minute as a grain of sand, yet its bright track is visible 150 miles (240 km) away. Exceptionally bright meteors are known as "fireballs." Many eye-witnesses have stated that they have heard a hissing noise simultaneously with viewing the trail of a meteor as it rushed through the upper atmosphere with a parabolic speed of about 26 miles (42 km) per second. When a fireball breaks up at the end of its path, it is called a "bolide." Anyone who has observed one will never forget its visual impact. When fireballs were seen in the seventeenth or eighteenth centuries, they were regarded as sensational and always worthy of reporting, as we see in several series of papers published in the *Philosophical Transactions of the Royal Society*, such as those for the fireballs of 20 September 1676[15] and 11 December 1741.[16] Wallis, in the former case, even concludes that the meteor was so high that "it might be some small comet."[17]

The terms for meteors and meteorites (the remnants of meteors that survive passage through the atmosphere) are now standardized by the International Astronomical Union, although they were not always known to be celestial phenomena.[18] In the seventeenth century, it was thought that meteors were vapors in our atmosphere that we saw when they exploded.[19] Even Halley in 1686, when he established that the earth's atmosphere extended upward about 45 miles (72 km), thought that the meteors were "exhalations" from the earth.[20] The word "meteor" commonly appeared, even in the *Philosophical Transactions of the Royal Society*, for a wide variety of luminous celestial phenomena that today would not be called "meteors."[21] But ten years later, after a fireball passed over Italy, Halley concluded that the material "must be some Collection of Matter form'd in the Aether...and that the Earth met with it as it passed along in its Orb...."[22] And after he reported a fireball in 1719, he calculated that it was 74 miles (119 km) high and traveling at great speed, endorsing the idea that meteors were not terrestrial vapors.[23] Nevertheless, the confusion between what we call "meteors" and terrestrial phenomena continued.[24]

Already during the period under consideration in this book, Sir Joseph Banks, the autocratic President of the Royal Society in the eighteenth century, asked the young chemist Edward C. Howard to analyze some recovered meteoric stones. Howard examined several, both from Europe and Benares, India, and concluded that they were not terrestrial.[25]

Large meteors that survive their fall and hit the ground are known as "meteorites." The relation of meteors in the sky and meteorites, stones or iron chunks on the ground, has long been studied. The most famous early fireball associated with a stone on the ground took place at Ensisheim in Alsace in 1492 and is recorded in a number of works of art.[26] But the idea that such fireballs were associated with stones or iron chunks found on the ground grew to be ridiculed. The scientific establishment of the correct ideas about meteors is credited to Ernst F.F. Chladni in 1794.[27] His idea that small bodies existed in the solar system was endorsed by the discovery of the first asteroids in 1801 and 1802, which, in turn, validated the idea that meteorites could be extraterrestrial. Jean-Baptiste Biot in France is widely credited with conclusive proof of the extraterrestrial nature of meteorites from a meteor shower through an investigation of the phenomenon in L'Aigle, France, in 1803.[28] After the L'Aigle meteorites, only a few persons seriously hypothesized that meteorites were atmospheric condensations, although the general acceptance of the idea dates to the 1880s. The relation of meteorites with craters on the earth and the moon was fixed in 1906 with the work of Daniel Moreau Barringer in Arizona and proven with the space program in the 1960s.[29] Presently, there is an increased interest in meteorites hitting the earth because of the suggestion of Lewis and Walter Alvarez in 1980 that such a collision caused the extinction of the dinosaurs about 65 million years ago.[30]

There are two main kinds of meteors. The first kind comes in "showers" of meteor dust – debris of ancient but now disintegrated comets, whose matter has become distributed around their orbits – that appear at a given time every year. This phenomenon was satisfactorily explained by Daniel Kirkwood in 1861.[31] The second kind is "sporadic" and at any random hour an observer with a clear sky can see about six sporadic meteors. In a shower like the Perseids that appear in mid-August each year, however, at least one meteor per minute may be visible. In the rare "meteor storms," such as occurred with the Leonid showers in November of 1833 and 1866, thousands of meteors per minute appeared! Since the period is about 33 years, and since a storm was visible in 1966, we await the Leonids of the next few years to see if such a meteor storm will again occur.

Popular interest in meteor storms seems to date from the Leonid storm in 1799, though over a dozen November meteor storms had been previously reported, dating back hundreds of years.[32] Only in 1863 did H.A. Newton argue that these storms represented, as we now believe, the intersection of particles in space with earth's orbit.[33] He predicted the next storm for 1866, when the earth was again to intersect the densest part of this particle cloud in space, a prediction that was borne out. Reports came from several scientists whose names appear in later parts of this book, including John Russell Hind, John Herschel, and Charles Piazzi Smyth.[34]

Let us now discuss, mostly chronologically, comets and meteors and how they were observed and portrayed.

Roberta J.M. Olson
Department of Art
Wheaton College
Norton, MA 02766 USA

Jay M. Pasachoff
Hopkins Observatory
Williams College
Williamstown, MA 01267 USA

Notes

Without compromising our theme, we cannot resist adding results from comets of the late twentieth century, such as Comet C/1995 O1 (Hale-Bopp). In the epilogue we also look into the future via the British involvement with the European Space Agency mission intending to land on Comet 46P/Wirtanen in 2011.

1 Provenances of the works discussed often can be located in the bibliography cited. Exh. cat. = exhibition catalogue.

2 R. Rosenblum, "British Art and the Continent, 1760-1860," in F. Cummings and A. Staley, *Romantic Art in Britain: Paintings and Drawings 1760-1860*, exh. cat., Philadelphia, 1968, 11-16, demonstrates that British art of the period, while distinctive, was not as insular as some have thought. The cross currents are also evident in studies such as H. Honour et al., *The Age of Neoclassicism*, exh. cat., London, 1972.

3 R.J.M. Olson, *Fire and Ice: A History of Comets in Art*, New York, 1985. See also I. Baroni, "Note Scientifiche: Le Comete. Storia – Fantasia – Realtà," *Emporia*, 31, 1910, 212-304. S.Y. Edgerton, Jr., *The Heritage of Giotto's Geometry*, Ithaca and London, 1991, addresses the intersection of art, the sciences, and history.

4 For an introduction to art of the period see D. Irwin, *English Neoclassical Art: Studies in Inspiration and Taste*, London, 1966; R. Rosenblum, *Transformations in Late Eighteenth Century Art*, Princeton, 1967; F. Cummings, "Romanticism in Britain, 1760-1860," in F. Cummings and A. Staley, *Romantic Art in Britain: Paintings and Drawings 1760-1860*, exh. cat., Philadelphia, 1968, 17-24, who stresses the romantic need to experience everything empirically and connects this artistic impulse with techniques in the sciences; Rosenblum, 1968. John Locke and other empirical British philosophers had prepared the intellectual climate during the seventeenth century. It is clear that the overwhelming curiosity which motivated both scientists and artists during the period in question engendered a great demand for visual evidence as well.

5 See, for example, E.D.H. Johnson, *The Poetry of Earth: A Collection of English Nature Writings*, New York, 1966; A. Staley, "British Landscape Painting, 1760-1860," in F. Cummings and A. Staley, *Romantic Art in Britain: Paintings and Drawings 1760-1860*, exh. cat., Philadelphia, 1968, 25-30; C. White, *English Landscape 1630-1850: Drawings, Prints and Books from the Paul Mellon Collection*, exh. cat., New Haven, 1977.

6 E. Waterhouse, *Painting in Britain 1530 to 1790*, Harmondsworth, Middlesex, 1962, 225-226, believes that the true discovery of watercolor was not made until the romantic obsession with nature became pronounced, at the time of the early poems of William Wordsworth and Samuel Coleridge. For the Scottish school, see J. Halsby, *Scottish Watercolours 1740-1940*, London, 1986. See also M. Hardie, *Water-colour Painting in Britain*, 1: *The Eighteenth Century*, eds. D. Snelgrove, J. Mayne, and B. Taylor, London, 1967, 3-43.

7 E.D.H. Johnson, *Paintings of the British Social Scene from Hogarth to Sickert*, New York, 1986, 12-14, who further notes that only with the 1855 Universelle Exposition in Paris was international recognition accorded to the British school of painting. Its British section had a preponderance of domestic genre that excited critics like Théophile Gautier.

8 See also J.M. Pasachoff, *Astronomy: From the Earth to the Universe*, Philadelphia, 1998, 272-291 (fifth edition).

9 See Ho Peng Yoke, "Ancient and Medieval Observations of Comets and Novae in Chinese Sources," *Vistas in Astronomy*, 5, 1962, 127-225.

10 Thus Comets 1871 I, 1871 II, 1871 III, 1871 IV, and 1871 V were the names assigned for Comets 1871a, 1871b, 1871d, 1871e, and 1871c, respectively, and Comet 1885d became Comet 1886 I when it was the first comet to pass perihelion in 1886.

11 B.G. Marsden, "New Designations for Old," *International Comet Quarterly*, 17, 1995, 3-4.

12 A=Jan 1-15, B=January 16-31, C=February 1-15, D=February 16-28/9, E=March 1-15, and so on. Note that I is omitted, as is traditional in astronomical nomenclature, to avoid confusion with J. See B.G. Marsden and G.V. Williams, *Catalogue of Cometary Orbits 1996*, Cambridge, MA, 1996 (eleventh edition), 5-8; B.G. Marsden, "Resolution of International Astronomical Union Commission 20," Minor Planet Circular Nos. 23803-23804; *idem*, 1995, 3-6.

13 In particular, such prefixes should be applied in cases where comets have possibly been misdesignated as minor planets, or vice versa. If necessary, the prefix A/ would precede a comet designation that actually refers to a minor planet (or asteroid). Short-period comets are prefixed by "P/," where "short" is defined as less than 200 years. They are numbered in historic order of the determination that they are periodic; over 116 comets are so designated. Other comets are usually prefixed by "C/." A few "defunct" or lost comets that have not returned when expected or whose orbits are poorly known are assigned "D/." If a comet is observed to return (or have its periodicity established through aphelion), the P/ (or D/) shall be preceded by an official sequential number assigned to each (e.g., 1P/1682 Q1 = Halley). "X/" comets are questionable cases and for which in any case no orbit has been able to be computed. The names of the discoverers can still be given in parentheses. Thus Donati's Comet, formerly 1858 VI, is now C/1858 L1 (Donati).

14 The best sources for listings of historical comets begin with the eighteenth-century *Cométographie* by Alexandre-Guy Pingré (Paris, 1783-84). It lists comets and other celestial objects identified as comets back through antiquity, although its accuracy by today's standards is flawed. Many books on comets appeared during the centuries of this study, a large number written in English. But only with S.K. Vsekhsvyatskii's *Physical Characteristics of Comets*, published in Moscow in 1958 and translated into English in 1964, was such an extensive cataloguing attempted again. Though inaccurate in many ways, Vsekhsvyatskii's volume serves a similar role of identifying historic comets and comments on their characteristics and their observers in an ambitious manner. More recently, Gary Kronk has taken up this task in *Comets: A Descriptive Catalogue* (1984 and updated in a forthcoming publication). Both can be used with the definitive work, issued approximately annually on the orbits of comets, Brian G. Marsden and Gareth V. Williams's *Catalog of Cometary Orbits*, published by the International Astronomical Union's Central Bureau of Astronomical Telegrams, a concise listing of orbital data for comets whose orbits have been determined. (D.K. Yeomans, Comets: *A Chronological History of Observation, Science, Myth, and Folklore*, New York, 1991, in a long appendix, pp. 361-424, compiles naked eye comets prior to A.D. 1700, including evaluations of discussions by Ho, Pingré, and others.)

15 J. Wallis, "An Extract of Two Letters... concerning a considerable Meteor seen in many distant places of England at the same time," *Philosophical Transactions*, 12, no. 135, 1677, 863-864; *idem.*, "A Third Letter from the same hand, concerning the same Meteor," *Philosophical Transactions*, 12, no. 135, 1677, 864-866. J.B. Burke, *Cosmic Debris. Meteorites in History*, Berkeley, Los Angeles, and London, 1986, 17, writes that during the eighteenth century there were 189 reports of fireballs and that two thirds of these appeared after 1750.

16 Lord Beauchamp, "An Account of the Fire-ball seen in the Air, and of the Explosion heard Dec. 11, 1741, near London," *Philosophical Transactions*, 41, 1739-41, 870-871. J. Fuller, "In Sussex," *Philosophical Transactions*, 41, 1739-41, 871-872. W. Gostling, "In Kent," *Philosophical Transactions*, 41, 1739-41, 872-873. C. Mason, "Extract of a letter...concerning a Fire-ball seen in the Air, and a great Explosion heard, Dec. 11, 1741," *Philosophical Transactions*, 42, no. 462, 1742, 1. B. Cooke, "Giving an Account of the Fire-ball seen Dec. 11, 1741," *Philosophical Transactions*, 42, no. 462, 1742-43, 25-27. W. Gordon, "A Letter from Capt. William Gordon to Capt. Samuel Mead, F.R.S. inclosing an Account of the Fire-ball seen Dec. 11, 1741," *Philosophical Transactions*, 42, no. 463, 1742-43, 58-60. W. Gostling, "Concerning the Fire-ball seen Dec. 11 last, and the Mock-Suns seen the 19th of the same Month," *Philosophical Transactions*, 42, no. 463, 1742-3, 60-61. T. Milner, "An Account of a meteor seen at Peckham, Dec. 11, 1741," *Philosophical Transactions*, 42, no. 464, 1742-43, 138-139.

17 K. Mark, *Meteorite Craters*, Tucson, AZ, 1987, 23, quoting Wallis.

18 M. Beech and R. Youngblood, "That Which We Call a Meteorite," *The Observatory*, 114, 1994, 312. M. Gingrich, "Great Comets, Novae, and Lady Luck," *Sky and Telescope*, 89:6, 1995, 87, characterizes them as the sky's nickel slot machines because of their small, frequent payoffs.

19 D.W. Hughes, "The History of Meteors and Meteor Showers," *Vistas in Astronomy*, 26, 1982, 325-345.

20 *Ibid.*, 325; Mark, 3-7.

21 For example, R. Vans, listed as "Extract of a Letter," indexed as "An Account of an Extraordinary Meteor, or kind of Dew resembling Butter, that fell last Winter and Spring," *Philosophical Transactions*, 19, 1696, 224. Burke, 6, quotes the eighteenth-century definition of meteors from the *Lexicon Technicum* (1704) of John Harris that follows Aristotle's ideas:

> METEORS...are various impressions made upon the Elements, exhibiting them in different forms,...for the most part, they appear high in the Air, and they are either Fiery, Airy, or Watery...

See also W. Whiston, *An Account of a Surprising Meteor Seen in the Air March 19. 1718/19*, London, 1719. By mid-century scientists had eliminated two types of fiery "meteors," the aurora borealis and lightning.

22 Hughes, 1982, 326.

23 E. Halley, "An Account of the Extraordinary Meteor seen all over England, on the 19th of March $171^8/_9$. With a Demonstration of the uncommon Height thereof," *Philosophical Transactions*, 30, no. 360, 1717-19, 978-990.

24 For example, T. Barker, "An Account of an extraordinary Meteor seen in the County of Rutland, which resembled a Water-Spout," *Philosophical Transactions*, 46, 1749-50, 248-249.

25 E. Howard, "Experiments and Observations on certain stony and metalline Substances, which at different Times are said to have fallen on the Earth; also on various Kinds of native Iron," *Philosophical Transactions*, 92, 1802, 168-212, which includes the following indexed articles: J.L. Williams, "Account of the Explosion of a Meteor, near Benares, in the East Indies; and the falling of some Stones at the same Time, about 14 Miles from that City," *Philosophical Transactions*, 92, 1802, 175-179; E. Howard, indexed as "Of one seen in [South] America," *Philosophical Transactions*, 92, 1802, 202. See also Mark, 12-13.

Perhaps the most famous story about the relation of meteorites and meteors came from the fireball of 14 December 1807, seen over New England. Stones were found on the ground in Connecticut and analyzed at Yale. The story is widespread that President Thomas Jefferson, when told about them, reputedly said that he "would prefer to think that two Yankee professors would lie than that stones could fall from the sky." The incident has been studied and this quotation is often said to be apocryphal (Marvin, 145-146). Instead Jefferson replied: "We are not to deny whatever we cannot account for.... It may be very difficult to explain how the stone you possess came into the position in which it was found. But is it easier to explain how it got into the clouds from whence it is supposed to have fallen? The actual fact, however, is the thing to be established, and this I hope will be done by those whose situations and qualifications enable them to do it." (*Ibid.*, 146.) Jefferson's statement seems overcautious, in view of the fact that by 1807 the scientific journals were filled with articles on meteorites. D. Hoffleit, "Stones From Heaven," *Memoirs Connecticut Academy of Sciences*, 23, 1992, 15-19, and personal communication, 1996, also evaluates the event and thinks that the quotation may actually be essentially real.

26 See J.T. Wasson, *Meteorites*, New York, 1985, 4; U.B. Marvin, "Meteorites, the Moon and the History of Geology," *Journal of Geological Education*, 34, 1986, 141-142; Mark, 6-10; Yeomans, 189; C.T. Pillinger and J.M. Pillinger, "The Wold Cottage meteorite: Not just any ordinary chondrite," *Meteoritics and Planetary Science*, 31:5, 1996, 589-605.

27 E.F.F. Chladni, *Ueber den Ursprung der von Pallas gefundenen und anderer ihr ahnlicher Eisenmasse und über einig damit in Verbindung stehende Naturerscheinungen*, Riga, 1794. See also Burke, 7, as well as R. Westrum, "Science and Social Intelligence about Anomalies: The Case of Meteorites," *Social Studies of Science*, 8, London and Beverly Hills, 1978, 461-493.

28 Wasson, 4-5; Mark, 13.

29 H.J. Melosh, *Impact Cratering: A Geologic Process*, New York and Oxford, 1989, 3.

30 L.W. Alvarez, W. Alvarez, F. Asaro, and H.V. Michel, "Extraterrestrial Cause for the Cretaceous-Tertiary Extinction," *Science*, 208, 1980, 1095; W. Alvarez, E.G. Kauffman, F. Surlyk, L.W. Alvarez, F. Asaro, H. Michel, "Impact Theory of Mass Extinctions and the Invertebrate Fossil Record," *Science*, 223, 1984, 1135-1141; L.W. Alvarez, "Mass extinctions caused by large bolide impacts," *Physics Today*, 40, 1987, 24-33; W. Alvarez, *T. Rex and the Crater of Doom*, Princeton, 1997.

31 Kirkwood, 1861, quoted in Hughes, 1982, 335.

32 Hughes, 1982, 327-328.

33 See also Yeomans, 1991, 193-195.

34 Hughes, 1982, 333-335.

I

Prelude: The Beginning of Telescopic Astronomy and the Background of British Astronomy and Artistic Traditions

In order to understand the contributions of eighteenth- and nineteenth-century British astronomers and artists to comet/meteor studies, we must backtrack to the seventeenth century when the telescope made its debut with Galileo (1609-10). The telescope ushered in the opportunity to view comets and meteors in more detail, helping to continue the momentum initiated by Tycho Brahe's proof, based on the comet of 1577, that comets were distinct objects in space. Prior to that time, most people followed Aristotle's ideas that comets were threatening atmospheric phenomena and regarded them superstitiously as prescient signs from God. With the advent of telescopic astronomy, comets were approached with a more objective attitude. Their physical properties began to be investigated, initiating a trend that continued with ever-increasing sophistication.

This shift in attitude is apparent in a comparison of comet broadsides – single-sheet prints describing an event – from the schizophrenic sixteenth century with those from the seventeenth century. During the sixteenth century broadsides functioned simultaneously as the *National Enquirer* and the CNN of their time, spreading both superstition and real news of natural disasters and topical occurrences. They were especially popular in Germany, which was torn apart by religious and political upheavals and which possessed a legacy of excellent printmakers.[1] In the subsequent century they became less sensational.

An etching by Wenceslaus Hollar (1607-1677) of around 1647, *Map and View of the State of England, Scotland, and Ireland* [Fig. 2], reveals the impact of the Germanic broadside tradition on an expatriate artist working in England. It also embodies his

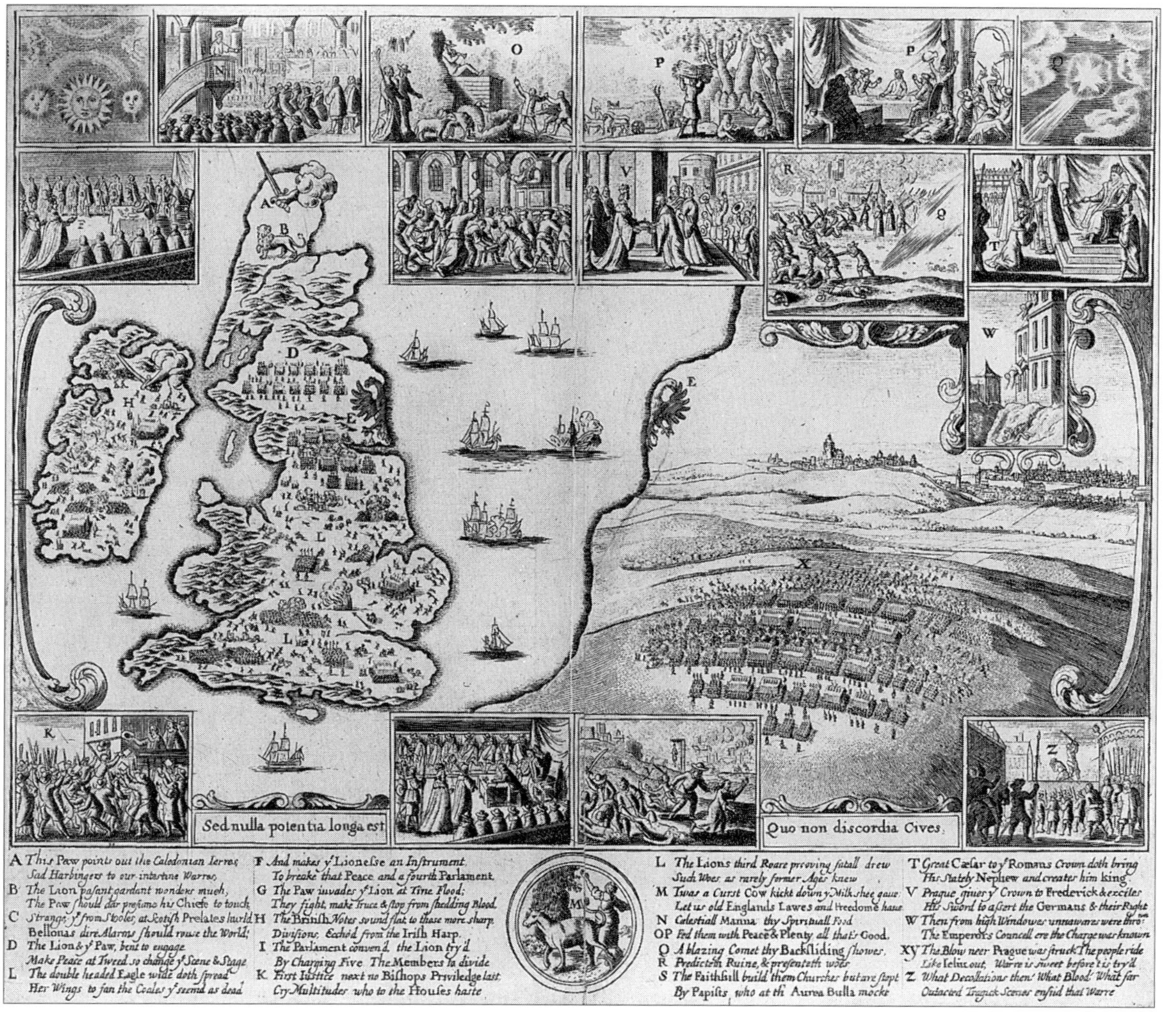

personal misfortune, in that he left the civil strife in Bohemia and the Continent for the peaceful haven of England, only to experience that country's descent into civil war. Hollar's two main images symbolize his dilemma: on the left a map of the British Isles and on the right a bird's eye view of Prague, each with various battles raging.[2] The sixteen subsidiary vignettes depict events from both conflicts that are explained by the verses in the lower margin. The one in the upper right designated "Q" represents a stylized comet, which is described as *A blazing Comet thy Backsliding showes.*

Figure 2. Wenceslaus Hollar, *Map and View of the State of England, Scotland, and Ireland,* after 1642, etching, 305 × 369 mm, The Pierpont Morgan Library, Peel Collection, New York.

That the English were not entirely immune to contagion from sensationalist Germanic broadsides is also evident on

Figure 3. Title page of John Gadbury's book *De Cometis*, 1665, Columbia University, Rare Book and Manuscript Library, New York.

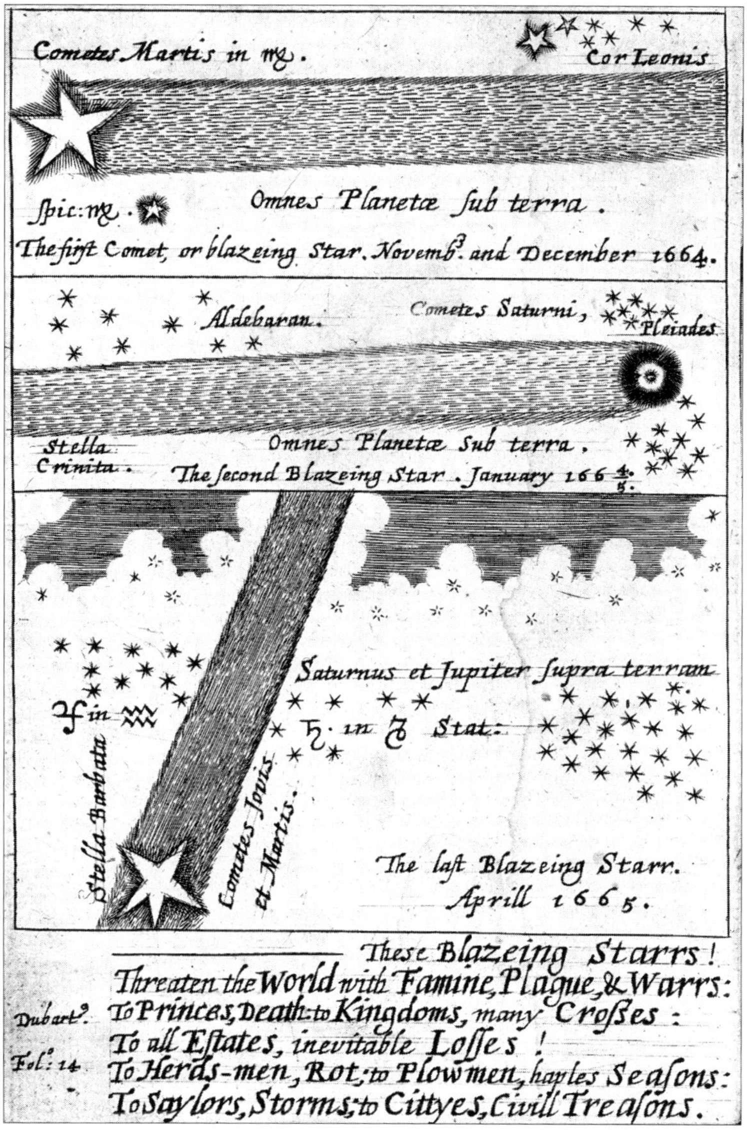

the title page [Fig. 3] of John Gadbury's book *De Cometis* (*A discourse on the nature and effects of comets... or blazing stars*). It was written about the comets of 1664 (C/1664 W1) and 1665 (C/1665 F1), associated in the minds of many with the plague and fire of the following years, respectively.[3] Its bold broadside quality is evident not only in its typography and text but also in its crude execution, anachronistic layout, and inflammatory sentiments. Unlike broadsides, however, Gadbury's illustration and rhetoric have been tamed by inclusion in a book where they are accompanied by the following poetical verse rather than by bald portentous prophecy:

—These Blazeing Starrs!
Threaten the World with Famine, Plague, & Warrs:
To Princes, Death: to Kingdoms, many Crosses:
To all Estates, inevitable Losses!
To Herds-men, Rot; to Plowmen, haples Seasons:
To Saylors, Storms; to Cittyes, Civill Treasons.

(*Guillaume de Salluste du Bartas,* La Semaine)

Gadbury's illustration wrongly cites the appearance of three comets or blazing stars, when, in fact, the first two show the same comet at different times (C/1664 W1). This comet was recorded from 17 November 1664 until 18 February 1665. The second (C/1665 F1), which was discovered on 27 March, appeared for less than a month but was naked-eye.[4]

None of the six comets that followed the spectacular comet of 1618 (C/1618 WI, formerly 1618 II)[5] struck terror in the hearts of Europeans until the Comet of 1665 lit up the sky. At that time England was primed for catastrophic changes (after the execution of Charles I in 1649, the ensuing the civil war and the Restoration in 1660, generally viewed as auspicious but not without its controversies). In 1665 the plague descended on the Continent and England with a fury that had not been seen for centuries. Soon thereafter, the city of London burned for four days in the great conflagration of 1666, destroying most of the wooden medieval metropolis. These two vast destructions, one following before the first had abated, were linked with comets. Subsequently, the ferocity of this plague and its comet portent were recorded by Daniel Defoe in his 1722 *Journal of the Plague Year 1665.*

The Comet of 1665 also precipitated comet treatises that are among the most lavish ever published. One of the most important was the continental scientist Johannes Hevelius's *Cometographia*, published in 1668. Hevelius, unlike Kepler, who believed comets travelled in straight lines, hypothesized more correctly that comets travel in a curved trajectory around the sun, although he mistakenly thought that they were condensed planetary exhalations.

Before this more scientifically oriented period, comets and meteors had been viewed as ambiguously symbolic portents, a fact borne out in the literary images of William Shakespeare and others. On the one hand they augured auspicious fortune, as well as births of rulers or important persons. On the other they were omens of disasters (literally, "bad stars"), calamities, and deaths of rulers or important persons.[6]

Although English sixteenth- and seventeenth-century literature is peppered with references to comets and meteors, the period's visual culture was not flooded in the same manner. Instead, it was dominated by the production of portraits. Nevertheless, portraiture and its related heraldry afforded a place for the insertion of comet representations. For example, a comet is included in a half-length portrait of a woman identified tentatively as Lady Arabella Stuart [Fig. 4 color]; it is by an anonymous English painter and dated around 1590 on the basis of the sitter's costume.[7] At the upper left is the sitter's personal emblematic device without a surrounding escutcheon. It is crowned by a large, yellow comet, as stylized as the sitter's intricate coiffure. Below the comet, a white banderole winds around a column resting on three steps; it is inscribed: *FIDENS NON CONFIDENS* (Faithful and Tenacious).[8] This device is positioned where armorial bearings frequently decorate Elizabethan portraits. Its symbolic content, including its comet, harmonizes with what is known about Arabella Stuart (1575-1615).[9] Certainly, her story of star-crossed love reads like one cursed by "bad stars." However, the conventional comet rendered in this portrait does not necessarily refer to the sitter's fate. Rather, it is part of the sitter's personal heraldry, an emblem that remains unidentified.

In contrast to Arabella's tragic destiny – and to the stylized comet in the above portrait – is the successful career of the first Earl of Haddington, Sir Thomas Hamilton – and the naturalistic comet painted in the upper right half of his portrait by an unknown painter [Fig. 5].[10] Sir Thomas (1563-1637) was created Earl of Melrose in 1619 and Earl of Haddington in 1627. He was friendly with James VI and managed the finances of Scotland for the monarch. In 1596 he was appointed the King's Advocate and also served as President of the Court of Session (1616-26).

From the inscription in the upper left of the panel, we know the large comet to the right of Haddington, who sat for this portrait when he was 55 years old, was the one that appeared in 1618. (It reads: *Aetatis suo 55/Anno 1618*, while the inscription at the upper

Ætatis suæ 55
Anno 1618
Eodem tempore
Cœlis Cometa
apparuit

Figure 5. British School, *Portrait of Sir Thomas Hamilton, First Earl of Haddington, King's Advocate, 1596, and The Comet of 1618*, 1618, oil on panel, 61 × 51.4 cm, Collection of the Earl of Haddington.

right reads: *Eodem tempore/Coelis Cometa apparuit.*). In this context, the comet is not tied to Haddington's heraldry but rather commemorates an historical comet. It probably depicts the second that year (C/1618 W1, formerly 1618 II). The first (C/1618 Q1, formerly 1618 I) was faint and only briefly naked-eye, whereas comet 1618 II, the first brilliant comet of the seventeenth century, appeared for a longer time with a tail up to 70° in length. The length of this tail is comparable to that of the comet depicted in the Haddington portrait, wherein the comet is partially obscured because of discolored varnishes. Because we do not know the motives for its inclusion in this portrait, we can only speculate on them. Did the sitter believe that the comet was a portent of his good fortune and/or his status, or did its appearance coincide with an occasion he deemed significant, or was he simply charmed by its astounding beauty? The directness of the sitter's gaze and the psychological openness of the half-length format complements the naturalistic representation of this famous comet. In fact, the relaxed informality of Haddington's portrait and the wonderfully observational comet with hindsight can be interpreted as foreshadowing the new world that British science was about to reveal. This world held a potential for great change, which led ultimately to new attitudes towards comets and meteors.

At times, however, the naturalism of seventeenth-century artists, writers, and natural philosophers was at loggerheads with the old, deeply ingrained, superstitious beliefs about comets. For example, England's poet laureate, John Dryden, confessed that he could not decide which of the many theories about comets to believe.[11] Yet despite the uncertainty about their nature before Newton's and Halley's pivotal theories in the later seventeenth and eighteenth centuries, comets were a popular feature in the century's literature. John Donne used a positive image in a song beginning "Goe and catch a falling star...," while his poem "To the Countess of Huntington" demonstrates a great command of the old traditions associated with comets:

> Who vagrant transitory Comet see,
> Wonders, because they are rare; But a new starre
> Whose motion with the firmament agrees,
> Is miracle; for there no new things are;

In woman so perchance milde innocence
A seldome comet is, but active good
A miracle, which reason scapes and sense,
For, Art and Nature this in them withstand

As such a starre, The Magi led to view....[12]

The Telescopic Comets of Newton and Halley

Newton's and Halley's observations of the comet of 1680 followed by only seven decades the first astronomical use of the telescope. Galileo's refracting instrument had revealed a panoply of new astronomical phenomena, such as the moons of Jupiter and the myriad stars of the Milky Way. Newton himself, incorrectly believing that the chromatic effects of lens systems could not be eliminated, improved on Galileo's telescope by constructing the first reflecting telescope that was an all-mirror system and reflected light of all colors equally. His design is still in use today; the type is called "Newtonian."

The first telescopic discovery of a comet, which occurred in 1680 (of Comet C/1680 V1), caused a flurry of astronomical activity that went hand-in-hand with the revival of old superstitions. With a tail measuring over 30° in November, growing to 70°, this daylight comet was first spotted accidentally before sunset on 14 November 1680 by Gottfried Kirch of Coburg, Germany. While astronomers coolly gazed at this apparition through telescopes, an irrational element in society grew hysterical about the gigantic red-tinted, naked-eye comet, one of the most brilliant ever recorded. It was observed by a wide variety of people, including the British astronomers and scientists Robert Hooke, Halley, and Sir John Flamsteed, then the Astronomer Royal. Newton was the last person to observe it on 19 March 1681, and eventually he concluded that comets are controlled by the sun's gravitation. (Earlier, like others, he had thought that comets travelled in straight lines.) Newton's gravitational theory was a direct attack on all superstitions connected with comets that had grown to encompass witchcraft in the seventeenth century.

Sir Isaac Newton (1642-1727), the natural philosopher and mathematician, was a pivotal figure in British astronomy and

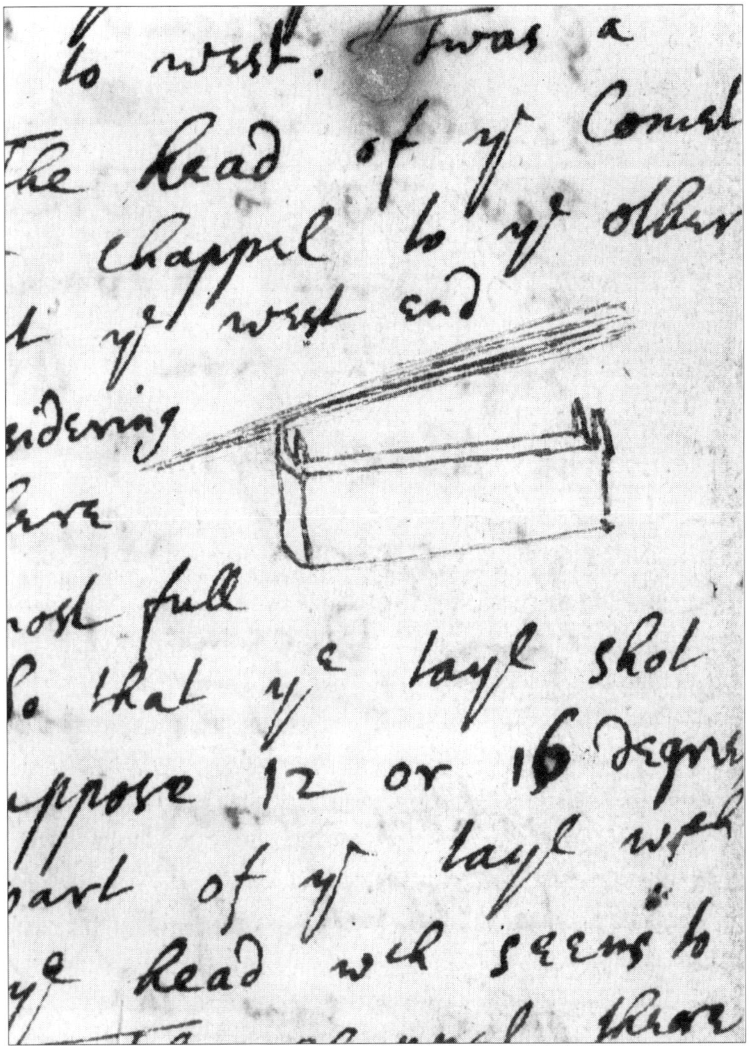
to west. Twas a
The head of ye Comet
chappel to ye other
ye west end
sidering
ost full
so that ye tayl shot
ppose 12 or 16 degree
part of ye tayl wch
ye head wch seems to

Figure 6. Sir Isaac Newton, *The Great Comet of 1680-81 over King's College Chapel (Ms. CCC 361, detail of page 8)*, pen and dark brown ink, 315 × 190 mm (letter sheet), Corpus Christi College Library, Oxford.

comet theory. Although his approach was often unorthodox, his brilliance was appreciated when in 1669 he became Lucasian Professor of Mathematics at Cambridge, where he had earned his degree. In the summer of 1665, as the plague menaced, the University closed for most of the next two years. Newton's resulting period of rustication at home in Woolsthorpe led to one of the greatest scientific revolutions the world has ever seen. Though he did not formulate his law of gravity in one fell swoop when he saw an apple falling from a tree, the incident sparked in Newton thoughts about gravity. (A remnant from this tree is still in the Royal Astronomical Society.) Traditionally, Newton is quoted as

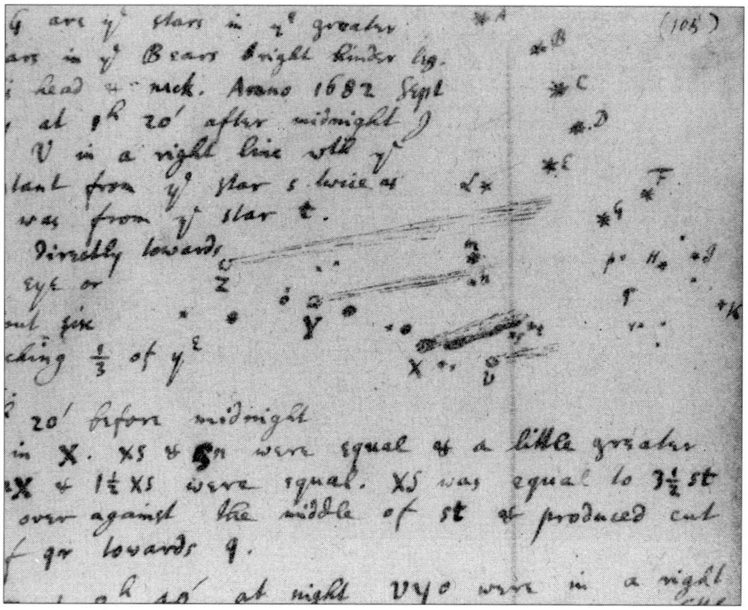

Figure 7. Sir Isaac Newton, *Sightings of Halley's Comet on Four Successive Nights in September 1682 (Ms. Add. 4004, detail of folio 105r)*, 1682, pen and dark brown ink, c. 262 × 209 mm (paper irregular), Cambridge University Library.

answering the question of how he discovered the law of gravitation, "I keep the subject constantly before me."[13]

In the winter of 1680-81 Newton observed the very bright comet with its tail filling almost half the sky. He even sketched it in the margin of a letter, noting its tail seen by Dr. Babington over King's College Chapel, Cambridge, in November [Fig. 6].[14]

The Great Comet of 1680-81 also piqued Edmond Halley's (1656-1742) interest. After undergraduate work at Oxford University, Halley – an explorer, naturalist and inventor as well as an astronomer – had been elected a Fellow of the Royal Society at the age of 22 years. Later, armed with Newton's theory of gravitation, he would gather data about many comets and calculate their correct orbits. While other astronomers had argued against Kepler's theory that comets travel in straight lines and in favor of different orbits, no one until Newton and Halley possessed the scientific ability afforded by Newton's new law to calculate it correctly.

After 1680, eight more comets crossed the heavens before the close of the century, including the apparition in 1682 (1P/1682 Q1) that was later to be called "Halley's Comet." Slightly over a year after Newton lost sight of the comet of 1680, Arthur Storer, an Englishman living in Maryland who corresponded with Newton and may have been a school friend, independently discovered this

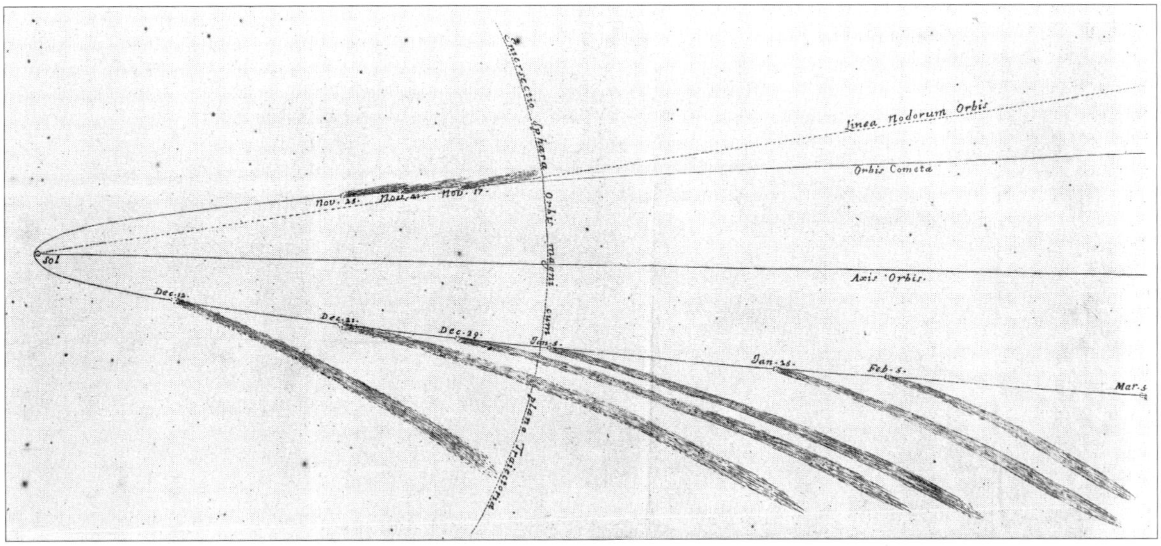

Figure 8. Sir Isaac Newton (after), *The Orbit of the Great Comet of 1680-81*, from the first edition of the *Principia*, 1687, engraving, 125 × 290 mm (image), Private Collection, Williamstown, MA.

new comet on 24 August 1682. Hevelius found it the next day and reported a tail of 12° in length, only about a week after it had been seen in England and France. It was again spotted in England by Flamsteed on 30 August. Halley first spied it on 5 September from his house in Islington, although he did not announce his momentous calculations of its orbit for another 23 years. It was visible for only 31 days in 1682, compared with four months for the comet of 1680.[15] Newton, too, jotted down its appearance on four successive nights in September [Fig. 7].[16]

A seminal meeting between Halley and Newton occurred in 1684 when, after a discussion at the Royal Society, Halley journeyed to Cambridge to ask Newton about planetary orbits. Subsequently, with Halley's prodding, the elder man prepared for publication the manuscript of his *magnum opus*, *Philosophiae Naturalis Principia Mathematica* (*The Mathematical Principles of Natural Philosophy*), submitted in 1686 and published in 1687 by the Royal Society. Although he was employed in only a minor post, as a clerk at the Royal Society, Halley believed so strongly in the genius of Newton's work that with great largesse he paid for its publication. This volume outlines Newton's theory of universal gravitation and in a few mathematical equations presents the laws governing the movements of the planets together with a view of the universe as an ordered, harmonious system with profound implications of a newly-perceived universal order. Naturally, this view affected people's attitude toward themselves, toward their place in the natural order, and ultimately, toward God, traditionally held to be the architect of the universe. Comets were also governed by these laws, and in his tome Newton hypothesized a parabolic orbit around the sun for comets, as demonstrated in the frontispiece of his *Principia* [Fig. 8]. In fact, he based much of his gravitational theory on the orbit of the Great Comet of 1680-81. He also ventured to describe the physical nature of comets, concluding that a comet's tail was formed by a flow of minute particles emanating from its head under the influence from solar rays – an explanation that agrees substantially with modern theories. Newton, who had become a Fellow of the Royal Society in 1672, was to be elected President in 1703, after which it became a prominent forum for exploring the most current ideas about comets and meteors.[17]

In 1695 Halley began to study comets intensively, building on the *Principia*, in which Newton had calculated a parabolic orbit for the comet of 1680-81. Halley discovered that Newton had mistranscribed one of the positions for the comet and that an ellipse fit these observations better. With the assistance of Newton, he obtained records of old observations, and with those in hand calculated the orbits of all the comets he could locate. He discovered that the orbits of the comets of 1531, 1607, and 1682 were similar. All were tilted with respect to the plane in which the planets orbit and all had the comet travelling backward with respect to the planets. Halley, therefore, suggested that these were actually the same comet. The only problem was that the intervals between the appearances were not quite the same: 76 years for the first interval, 75 for the second, and so forth. To solve the problem Halley used Newton's theory of gravitation to show that the gravitational pulls of the planets Jupiter and Saturn accounted for this variation.

In 1704 Halley was elected to a professorship at Oxford and took advantage of the opportunity to publish his work on comets (*Astronomiae Cometicae Synopsis*). This book of only about a dozen pages came out in 1705 in both Latin and English; the Latin version was reprinted in volume 247 of the *Philosophical Transactions of the Royal Society*.[18] Halley went on to predict that the comet would reappear in 1758 and later wrote (in his *Tabulae Astronomicae*, published posthumously in 1749): "Where, if according to what we have already said, it should return again about the year 1758, candid posterity will not refuse to acknowledge that this was first discovered by an Englishman."[19]

In 1719 Halley was appointed Astronomer Royal, a position he held until his death in 1742, sixteen years before the return of the comet he had predicted. All during 1758, European scientists tried to locate with revised calculations the comet Halley had predicted – in fact, the French writer Voltaire wrote that no astronomer slept that year! Then, on Christmas Eve, an amateur astronomer, the farmer Johann Georg Palitzsch near Dresden, found it with his homemade, seven-foot telescope. Thus, this comet (1P/1758 Y1, formerly 1759 I), which went through perihelion only 33 days before the prediction by Alexis-Claude Clairaut, which was updated from Halley's own prediction, marked a triumph for Newton's theory. Ever since, it has been identified as "Halley's Comet."[20]

The Royal Society's Link with Comets and Meteors

Both Halley and Newton, as well as the study of comets and meteors, were intimately linked to the early years of the Royal Society, whose formation and first meetings were held in Gresham College, London, at the time when Christopher Wren was the Gresham Professor of Astronomy (1657-1660). In 1660, various scientists and other individuals, who had met as early as 1645 in informal discussions about experimental philosophy, set up the Royal Society of London for the Promotion of Natural Knowledge. Two years later, the organization was chartered by Charles II. Despite its royal charter, the Royal Society, like the Royal Academy founded at a later time, was always very much of a private organization, unlike the governmentally subsidized academies on the Continent. Not only is it the oldest scientific society in the British Isles, but also it is one of the oldest in Europe and still one of the premier scientific organizations in the world.

Its purpose, as defined in 1664, was to form "a Corporation of a number of Ingenious and knowing persons, by ye Name of ye Royall Society of London for improving Narturall knowledge, whose dessein it is, by Observations and experiments to advance ye Contemplations of Nature and to Use and Practice."[21] While this definition limited the theoretical work discussed, one of its most important functions was the performance of experiments at meetings. Consequently, observational subjects like astronomy received less time. The Society also formed a museum, at one time the most famous in London; upon the Society's move to Somerset House in 1781, its contents were presented to the British Museum. Moreover, the Society provided advice on scientific matters to the government and from 1710 played a role in the Royal Observatory, Greenwich.

Among the early members of the Royal Society were individuals associated with excellence in different fields. They included: Robert Hooke, known for his work on microscopy and who observed the comet of 1664-65; Christopher Wren, known primarily as an architect; Halley; and Newton. All of them were present at a significant scientific meeting regarding Kepler's laws in January 1684. (In the early seventeenth century, Kepler had presented three laws showing how the planets moved in elliptical orbits around the sun, but these laws were only empirical). Hooke claimed that he could derive Kepler's laws theoretically, a feat at which Halley had already tried and failed. Wren raised a challenge, and while a prize was offered for a solution within two months, when the time arrived Hooke claimed that although he could prove the laws, he would keep his proof secret. Halley, still puzzled by the matter five months after the deadline, went up to Cambridge to consult Newton, who replied that he had the answer: the orbit of a planet would be an ellipse. Subsequently, Halley wrote an encomium to Newton for the preface of the *Principia* (see the first page of the Introdution of the present study).[22]

Although Kepler had formulated the law that the planets revolve around the sun in ellipses, he did not know that his laws also applied to transitory comets, which were generally thought to travel in straight lines.[23] Indeed, even Newton at first agreed and thought that the comet of 1680 that he saw going one way in November and the other in December was really two different comets. But by 1684, Newton had begun to think that comets might travel in elliptical orbits as well.[24]

But it was Halley, employing Newton's law of gravity, who first showed that comet orbits closely resembled parabolas, curves that represent the boundary between open paths and ellipses (charac-

Figure 9. William Hoare, *Emblematic Profile Portrait of Isaac Newton with a Flaming Comet*, 1734, red chalk, 302 × 260 mm, The British Museum, London.

terized by closed paths). By 1695, Halley had come to suspect that the comet of 1682 was really the return of the comet of 1607; he reported his conclusion to the Royal Society in 1696. But it was only in 1705 that Halley published his extensive calculations of the parabolic orbits for 24 comets.[25] With the second edition of the *Principia* in 1713, Newton seemed a little grudging in his acknowledgment of Halley's extension of Newton's cometary theory to include an elliptical orbit. According to Hall, it is not clear whether or not Newton now supposed *all* cometary orbits to be ellipses.[26]

With the *Principia*, Newton's fame was insured and he continued to be enshrined as one of the preeminent scientific figures, as demonstrated by a red chalk drawing by William Hoare (1707-1792) featuring a comet [Fig. 9]. The drawing was executed in preparation for an etching by the artist, dated 1734, in which the image is reversed.[27] Hoare, the leading portraitist in the fashionable spa of Bath until the arrival of Gainsborough, drew this profile portrait of Newton, linking his fame and fortune to the plunging five-pointed comet star with a flaming tail (probably the comet of 1680-81) behind his head. Hoare framed Newton's visage with the device of a serpent biting its tail, an ancient symbol of eternity. Thus, this emblematic portrait with a neoclassical veneer, in which Newton wears a classical coiffure similar to those in Roman copies of Greek sculptures, argues for Newton's immortality based on the implications of his theories involving the comet.

The Royal Society published many notices about comets and meteors during the period under consideration. A fair number were illustrated by plates lavish for their time, including elaborate foldouts. The Society's short-lived journal (1681-82), *Philosophical Collections*, edited by Hooke, featured three articles concerning comets, all by authors not from Great Britain.[28] The greatest number of articles appeared in the Society's long-lived, celebrated journal, the *Philosophical Transactions of the Royal Society*, which began publication in 1665. The lion's share of these were written by British authors and usually had been first read or outlined at meetings. It was in this periodical where correspondence with continental philosophers, which was steadily maintained, also appeared (as is the case with comet and meteor material). Up until about 1840, the *Philosophical Transactions* included a wide range of astronomical papers, with dozens addressing comets and meteors. (A complete list appears as Appendix I, although strangely, there is nothing about the comet of 1811.) Thereafter, articles concerning comets and meteors were published in the *Memoirs of the Royal Astronomical Society*, which began in 1822, and in the continental *Astronomisches Nachrichten*, which commenced in 1823. In addition, shorter, unillustrated notes appeared in the *Monthly Notices of the Royal Astronomical Society* from 1831. However, these mainly list positions and are shorn of the fascinating descriptions found in the accounts of the *Philosophical Transactions*. The *Proceedings of the Royal Society*, which started in 1832, did not publish comet/meteor articles.[29]

To gauge the scientific curiosity about comets and meteors in Britain, a survey of the articles published in the *Philosophical Transactions*, the premier scientific journal of the English-speaking world during the period of this study, reveals the following statistics. There were 147 articles addressing comets and meteors between the first volume (1665) and volume 130 (1840), an average of 0.84 articles per year over the 175-year span. Of these, 90 were about comets (0.51 per year) and 57 about meteors or meteorites (0.33 per year). These numbers correspond to an average of one article about comets every two years or one article about meteors or meteorites every three years. Articles frequently appeared in groups with several on individual comets or fireballs. Comet articles were more evenly distributed over the span than were articles on meteors. Graphing the number of articles about comets and meteors that appear annually yields an approximate bell curve that extends from 1655 to 1840, and reaches an apogee in 1759 with a dozen articles. The heaviest concentration is from 1737 through 1795. The peak years for comet articles are 1737 and 1759 (with six and seven articles respectively), while the peak years for meteor articles are 1739, 1742, 1759, 1764, and 1784.

One reason why the Royal Society's journals no longer featured comet and meteor articles was that the nature of observational astronomy was changing. Perhaps the gentlemanly tone and the manner of letters to the Royal Society, which mixed professionals and amateurs in a whole range of fields, went out of favor.[30] Perhaps the discursive exchanges about comets and meteors found in the earlier *Philosophical Transactions* articles were out of date as astronomical effort went into numerical measurements. In addition, the founding of the Royal Astronomical Society must have also played a role in the cessation of comet and meteor articles. Some astronomical topics, however, such as the construction of a giant telescope by the Earl of Rosse, together with their implications for the overall structure of the universe, continued to be covered in the *Philosophical Transactions*.

Besides Halley's comet paper of 1705, another still-famous comet paper was published in the *Philosophical Transactions*: William Herschel's "Of a Comet" in 1781. Herschel termed the celestial object a "comet" in his title, even though after thorough examination he was uncertain about its nature. Indeed, within several months, other individuals, especially Nevil Maskelyne, then the Astronomer Royal, had measured its position and calculated an orbit. Instead of a comet, it turned out to be a new major planet, the first to be discovered since antiquity. Though Herschel tried to name it *Georgium Sidus* after King George III, the new planet was eventually named Uranus (see chapter II).

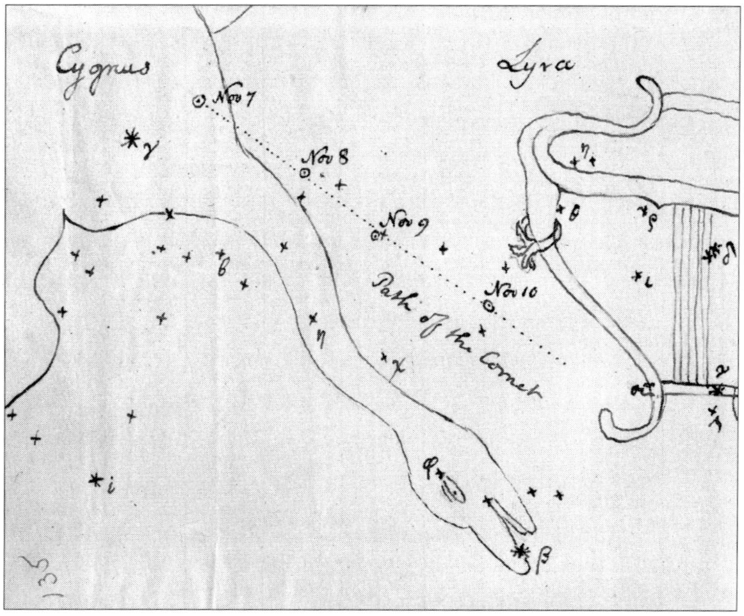

Figure 10. William Herschel, *Detail of a letter with Comet's Path*, No. L&P X. 135, p. 2, 1795, pen and brown ink, 225 × 185 mm (page), Archives of the Royal Society, London.

The handwritten manuscripts of the letters that Herschel wrote about it to the Royal Society, the substance of which was published in the *Philosophical Transactions*, are in the Archives of the Royal Society. In the first of these documents, Herschel drew an unpublished map of the "comet's" position.[31] Another unpublished drawing by Herschel in a letter from 1795 depicts another comet's path across the constellations Cygnus and Lyra, presumably 2P/1795 V1, Encke's Comet [Fig.10].[32] Herschel's observations of comets as well as his theorizing about them was extensive, and thus it is interesting to observe that he also recorded his observations in drawings.[33]

A wide variety of handwritten manuscripts submitted for publication in the *Philosophical Transactions* are still in the Archives of the Royal Society. Few from the earliest years exist and some periods are represented not by manuscript copy but by copybooks. Additional manuscripts about comets and meteors will be discussed in the following chapter, because the Royal Society continued to play an important role in the British history of comet/meteor studies. Many of these manuscripts feature illustrations that testify to the visual nature of astronomical phenomena and the need not only to report them but also to record them before the advent of astronomical photography.

Figure 4. British School, *Portrait of a Lady, possibly Arabella Stuart*, c. 1590, oil on canvas?, 90 × 72.5 cm, Private Collection, London.

Figure 11. Josiah Wedgwood, *Plaque of Isaac Newton*, before 1780, jasper, white relief on blue, 92 × 71 mm, City Museums and Art Gallery, Birmingham.

Figure 38. Paul (?) Sandby, *The Meteor of 18 August 1783 from the East Angle of the North Terrace, Windsor Castle*, 1783, watercolor with touches of pen and ink over graphite, 318 × 484 mm, Yale Center for British Art, Paul Mellon Collection, New Haven.

Figure 39. *(Left and detail below)* Formerly Samuel Scott, *The Meteor of 18 August 1783 Over the Thames*, oil on canvas, 81.9 × 111.8 cm, Private Collection, England.

Figure 60. *(Opposite)* William Blake, *The Ghost of a Flea*, 1819-20, tempera heightened with gold on panel, 21.4 × 16.2 cm, The Tate Gallery, London.

Price One Shilling Colour'd
Rowlandson Del
by Thos Tegg No 111 Cheapside
LOOKING AT THE COMET TILL YOU GET A CRIEK IN THE NECK.

Figure 77. *(Opposite)* Thomas Rowlandson, *Looking at the Comet till You Get a Criek in the Neck,* 1811, hand-colored etching, 345 × 247 mm, Private Collection, New York.

Figure 87. *(Above)* Thomas Rowlandson, *Viewing the Comet,* signed and dated 1821, pen and brown ink and watercolor over graphite, 330 × 270 mm, whereabouts unknown, formerly Christie's, New York.

Figure 89. *(Opposite)* Anonymous, *Comet Weathervane on the Old Royal Observatory,* c. 1899, gilded brass, 54 × 84 cm, National Maritime Museum, Greenwich.

Figure 101. *(Above)* Anonymous, *Watch Fob with a Gem Engraved with a Portrait of Edmond Halley and His Comet,* after 1835, carnelian in a silver mount, 10 × 10 × 20 mm, The Science Museum, London.

Figure 102. *(Right)* Anonymous, *Comet Pin,* c. 1759-1811, cut steel and brass, 114 mm, Private Collection, New York.

Figure 103. John J. Chalon, *Halley's Comet 1835*, brown, gray, and black watercolor wash over charcoal, 355 × 440 mm, National Maritime Museum, Greenwich.

Figure 107. J.M.W. Turner, *Napoleon's Abdication at Fontainebleau*, 1834-35, watercolor strengthened with pen and sepia ink, 174 × 148 mm, The Indianapolis Museum of Art.

Figure 109. Anonymous, *Letter Opener in the Shape of a Comet with the Moon*, c. 1830-40, brass, c. 25.4 cm, Private Collection, New York.

Figure 110. *(Above and detail opposite)* John Martin, *The Eve of the Deluge*, signed and dated 1840, oil on canvas, 142.9 × 218.4 cm, Collection of H.M. The Queen.

Figure 115. Charles Piazzi Smyth, *Coggia's Comet 12 July, 1874 at Midnight*, 1874, watercolor with pen and brown ink, 124 × 164 mm (image), RSE Piazzi Smyth Archive, held on deposit at the Royal Observatory Edinburgh.

Notes

1 For German comet broadsides see R.J.M. Olson and J.M. Pasachoff, "Historical Comets Over Bavaria: The *Nuremberg Chronicle* and Broadsides," in *Comets in the Post-Halley Era*, 2, eds. R.C. Newburn, Jr., et al., Dordrecht, Boston, and London, 1991, especially pp. 1326-1336. See also Olson, 1985b, pp. 41ff; S. Schaffer, "Comets and the world's end," in *Predicting the Future*, eds. L. Howe and A. Wain, Cambridge, 1993, 52-76.

2 Vol. 1, p. 10, no. 16, in the Peel Collection, The Pierpont Morgan Library, New York, etching measuring c. 305 × 369 mm. It is one of two versions; see R. Pennington, *Wenzel Hollar: Beschreibendes Verzeichnis seiner Kupferstiche*, Berlin, 1982, 85, no. 543; R.T. Godfrey, *Wenceslaus Hollar. A Bohemian Artist in England*, exh. cat., New Haven, 1994, 92, no. 58, ill. See also M.D. George, *Catalogue of Political and Personal Satires Preserved in the Department of Prints and Drawings in the British Museum*, 1, London, 1870-1954, 107-109, nos. 144, 145.

3 The title continues *De Cometis (A discourse of the nature and effects of comets, as they are philosophically, historically, and astrologically considered. With a brief (yet full) account of the III late comets, or blazing stars, visible to all Europe. And what (in a natural way of judicature) they portend...)*. It was published by L. Chapman in London in 1665. It follows earlier tracts, such John Bainbridge's *An astronomical description of the late comet from the 18 of November 1618* (1619).

4 See S.K. Vsekhsvyatskii, *Physical Characteristics of Comets*, Jerusalem, 1964, 116-117; G.W. Kronk, *Comets: A Descriptive Catalog*, Hillside, NJ, 1984, 10-11; Yeomans, 420-421; G.W. Kronk, *Cometography. A Catalogue of Comets*, 1, Cambridge and New York, forthcoming, 191-194, for C/1664 W1; 194-195, for C/1665 F1.

5 See Vsekhsyvatskii, 113; Kronk, 1984, 10; Yeomans, 418; Kronk, forthcoming, 185-186. 1618 II (C/1618 W1), whose tail was over 70 degrees and was observed splitting into numerous star-like pieces, was an important brilliant comet. It was observed by John Milton when he was a boy – he continued to be haunted by it since it was the only comet he would see before going blind – and he frequently alluded to comets in his writings, which were an influential source for comet/meteor representations in subsequent centuries. Thomas Harriot studied this comet, as well as Halley's Comet of 1607, as noted by A. Chapman, "The Astronomical Work of Thomas Harriot (1560-1621)," *Quarterly Journal of the Royal Astronomical Society*, 36, 1995, 97-107.

6 See Olson, 1985b. For an indication of British beliefs that comets were malevolent signs, see L. Digges the Elder, *A Prognostication Everlasting of Ryght Good Effecte*, London, 1556 (second impression).

7 In a private collection, London, in oil on canvas?, measuring 90 × 72.5 cm; see Sotheby's, London, sale cat., 16/XI/83, no. 24. We are grateful to Rosalind Marshall for for bringing to our attention photographs in the National Portrait Gallery of Scotland of portraits identified as Arabella Stuart. Two of these resemble the portrait under consideration: (1) in the North Carolina Museum of Art (inv. no. 52.9.72; Raleigh, North Carolina Museum of Art, *Catalogue of Paintings*, 2: *British Paintings to 1900*, Raleigh, 1969, 30, pl. 79); (2) in the National Portrait Gallery of Scotland, Edinburgh, attributed to Robert Peake and dated 1605 on the sitter's watch. For another in Hampton Court, see O. Millar, *The Tudor, Stuart and Early Georgian Pictures in the Collection of Her Majesty the Queen*, 1, Greenwich, Conn, 1963, 74-75, no. 87; C. Lloyd, *The Queen's Pictures. Royal Collectors Through the Centuries*, London, 1991, 64-65, no. 10, neither of whom believe the sitter is Arabella. For English portraiture, see D. Piper, *The English Face*, London, 1957.

8 Stars were more common in Scottish heraldry than comets. G.H. Johnston (*Scottish Heraldry Made Easy*, Edinburgh and London, 1904, 66), notes that comets also appear on some foreign coats of arms. In fact, a study of Italian heraldry reveals many comets in personal *imprese* and family coats of arms through the nineteenth century. The Lord Lyon informed us that *imprese* were not frequently used in Scotland and that four Scottish armorial bearings with comets, none allied to Arabella, are referred to in volume I of *The Public Register of All Arms and Bearings in Scotland* (commenced in 1672), pp. 288, 388, 389. In a letter of 6 February 1995 he also notes that in "Papworth's Armorial" three other families used a comet or flaming star in their armorial bearings.

9 Cousin of James VI of Scotland (later James I of England in 1603 after Queen Elizabeth I's death), by strict pedigree she ranked next to him in succession to the throne of England. Some individuals, however, believed that Arabella had a greater claim to the throne, since she was born and bred in England, whereas her cousin was Scottish by birth and upbringing. Moreover, she was judged by some to be more directly descended by blood line. Naturally she was regarded with suspicion by Elizabeth I, a posture later adopted by James, especially when Lady Arabella's name was linked to plots to prevent his succession in 1591 and 1597. After his accession to the throne, James also discouraged marriage proposals for her with anyone either English or foreign who was viewed as even remotely in line of succession to the British throne. Following several abortive attempts at romance, Arabella fell in love with William Seymour (later the Duke of Somerset), and they were married secretly. After intrigues and an ill-fated plan to escape, Arabella was captured in the Straits of Dover and imprisoned in the Tower of London, until she reputedly lost her sanity and died. G. Davies, *The Early Stuarts 1603-1660*, Oxford, 1949, 1ff.

10 In the Collection of the Earl of Haddington, in oil on panel measuring 61 × 51.4 cm. It was sold at Christie's, London, 10/VI/19, according to the photographic archive of the National Portrait Gallery of Scotland, which includes the note that O. Millar suggests that it may be an early work of Myrtens, and D. Piper noted that other portraits by the same hand with a comet are known.

11 For Dryden's use of "comets," "meteors," and "star," see G. Montgomery, ed., *Concordance to the Poetical Works of John Dryden*, Berkeley and Los Angeles, 1957.

12 For Donne's use of "comet" and "meteor," see H.C. Combs and Z.R. Sullens, eds., *A Concordance to the English Poems of John Donne*, Chicago, 1940, 55, 229.

13 R.S. Westfall, *Never at Rest: A Biography of Isaac Newton*, Cambridge, 1980, 174.

14 Ms. CCC 361, detail of p. 8, in the Corpus Christi College Library, Oxford, in pen and dark brown ink, measuring 315 × 190 mm. The sketch measures c. 30 × 65 mm. The letter of 28 February 1680 is addressed to Mr. Compton for Mr. Flamsteed's notice. Westfall, fig. 9.6 and p. 392, n. 154, incorrectly identifies it as from Ms. Add. 4004, fol. 44, and in the caption as showing the comet which appeared on 20 or 23 December four days after its first evening sighting. This caption corresponds to

Newton's drawing of the comet in 1682 (see n. 16). We thank Christine Butler for checking the reference. See also J.F.W. Herschel, *Outlines of Astronomy*, London, 1849, 348.

15 For the comets of 1680 and 1682, see Vsekhsyvatskii, 120-121; Kronk, 1984, 12; Yeomans, 422-423; Kronk, forthcoming, 199-202. See also Westfall, 395; for Storer, P. Broughton, "Arthur Storer of Maryland: His Astronomical Work and His Family Ties with Newton," *Journal for the History of Astronomy*, 19, 1988, 77-96, who believes that Storer's observations of the comet of 1682 were superior to those of Halley and Hevelius. Kronk has calculated that if Storer was using the Julian instead of the Gregorian calendar, his positions were accurate (Kronk, personal communication, 24 March 1995).

16 Ms. Add. 4004, fol. 105r, in the Cambridge University Library, in pen and dark brown ink measuring c. 262 × 209 mm (paper irregular).

17 For Newton, see Westfall; G. Christianson, *In the Presence of the Creator: Isaac Newton and His Times*, New York and London, 1984; P. Lancaster-Brown, *Halley's Comet and the Principia*, Aldeburgh, Suffolk, 1986; N.J.W. Thrower, ed., *Standing on the Shoulders of Giants: A Longer View of Newton and Halley*, Berkeley, Los Angeles, and Oxford, 1990; A.R. Hall, *Isaac Newton: Adventurer in Thought*, Oxford and Cambridge, MA, 1992. See also S.S. Genuth, "Comets, Teleology, and the Relationship of Chemistry to Cosmology in Newton's Thought," *Annali dell'Istituto e Museo di Storia della Scienza di Firenze*, 10:2, 1985, 31-65; *idem.*, "Newton and the Ongoing Teleological Role of Comets," in Thrower, 298-311.

18 It also appeared in "A Synopsis of the Astronomy of Comets," *Miscellanea Curiosa*, 2, London, 1723-26, 321-344.

19 Lancaster-Brown, 1986, 132. Halley in a letter to Newton of 28 September 1695 (Ms. Add. 3982, fol. 2, in the Cambridge University Library) reveals to Newton his great discovery (Halley's Comet). He writes: "I am more and more confirmed that we have seen that Comet three times, since ye Yeare 1531." Quoted by D. Dewhirst, "Christmas Stars and Celestial Portents," exh. labels, Cambridge University Library, 1985, who also noted that the February 1758 issue of *Gentleman's Magazine* published the first predicted return of Halley's Comet. See also Olson, 1985b, 15.

20 For Halley, see E. F. MacPike, *Hevelius, Flamsteed, and Halley*, London, 1937; Westfall; P. Lancaster-Brown, *Halley and His Comet*, Poole, Dorset, 1985; *idem*, 1986; Thrower. For the 1758-59 apparition (1759 I), see Vsekhsyvatskii, 128-129; Kronk, 1984, 228-233; Yeomans, 123-139; Kronk, forthcoming, 228-233. For the voluminous literature on P/Halley, see the bibliography in R. Freitag, *Halley's Comet: A Bibliography*, Washington, D.C., 1984; D.W. Hughes, "The history of Halley's Comet," *Philosophical Transactions*, 323, 1987, 349-367.

21 M.B. Hall, *Promoting Experimental Learning: Experiment and the Royal Society, 1660-1727*, Cambridge, 1991, 9.

22 M. Hunter, *Establishing the New Science: The Experience of the Early Royal Society*, Woodbridge, 1989; *idem., The Royal Society and Its Fellows, 1660-1700: The Morphology of an Early Scientific Institution*, Oxford, 1994.

23 J.A. Ruffner, "The Curved and the Straight: Cometary Theory from Kepler to Hevelius," *Journal for the History of Astronomy*, 2, 1971, 178-194; P. Barker, "The Optical Theory of Comets," *Physis (Firenze)*, 30, 1993, 1-25.

24 Westfall, 391-396; Yeomans, 101-103; A.R. Hall, 171. In 1684 an important anonymous treatise (attributed to John Edwards), *Cometomantia, A Discourse on comets; shewing their original substance, place, time, magnitude, motion, number, color, figure, kinds, names and more especially their prognosticks, signficiations and presages...*, was published in London. It concerned the great comets of 1680 and 1682. It is notable for its errata that the Lord Bishop of Sarum (Seth Ward, FRS) "asserteth that Comets are perpetual bodies, and that no new Comets appear, but the very same which were seen before."

25 See n. 18; Yeomans, 101-119.

26 A.R. Hall, 301-302.

27 Inv. no. 1894-12-19-1 in the British Museum, London, in red chalk measuring 302 × 260 mm. L. Binyon, *Catalogue of Drawings by British Artists and Artists of Foreign Origin Working in Great Britain Preserved in the Department of Prints and Drawings in the British Museum*, 2, London, 1900, 314. E. Croft-Murray, in an unpublished catalogue of the drawings by British artists (c. 1665-1715), noted that it may have originally been intended as a design for a memorial relief-medallion, perhaps by the artist's brother Prince Hoare I. In addition, the medal with a profile of Newton to the right, struck in the mid-nineteenth century and signed *L.B.*, also has a comet in the same position. Thanks to Kim Sloan for this information. For the artist, see E. Newby, *William Hoare of Bath. R.A. 1707-1792*, exh. cat., Bath, 1990.

28 One featured Hevelius's observations of the comet of 1680 (J. Hevelius, "The First Astronomical Observations made by Mr. Hevelius at Dantzick, Jan. 1, 1681. St. N. being an Occultation of the Bulls-Eye by the Moon. As also a short account of his Observations of the Late Comet," *Philosophical Collections*, 3, 1681, 65-67). Moreover, P.J. de Fontaney, "Observations of the comet of 1680 and 1681, made at the Colledg of Clermont," *Philosophical Collections*, 4, 1681/92, 106-113, includes an engraving with ten views of that comet (opp. p. 83).

29 London, Royal Society, *The record of the Royal Society of London for the Promotion of Natural Knowledge*, London, 1940, 178-179.

30 H.J. Cook, "The Practice of Science: A Behaviorist's View," *American Scientist*, 83, 1995, 271-273, comments on a possible difference between English scientific practice and elsewhere due to the "gentlemanly traditions of England." This amateur involvement was one of the strengths of British astronomy, resulting in a heavier cultural saturation of astronomical interests.

31 No. L&P VII.192, p. 5, in the Archives of the Royal Society, London, in pen and brown ink.

32 No. L&P X.135, p. 2, in the Archives of the Royal Society, London, in pen and brown ink measuring 225 × 185 mm (page).

33 See S. Schaffer, "The great laboratories of the universe: William Herschel on matter theory and planetary life," *Journal for the History of Astronomy*, 11, 1980, 81-111.

II

The Heavens on Fire:
The Eighteenth Century

The eighteenth century was an age motivated by optimism and a belief in progress. Comets and meteors, as well as science in general, became an obsession for some artists and scientists, and for the public at large. Because of the previous century's discoveries and the development of new technologies, people's insatiable curiosity about comets and meteors engendered more objective observations of these celestial fires, whose natures were controversial. Eventually, the flurry of scientific investigations led to the differentiation between them. While astronomers, like William Herschel, and artists, like Paul and Thomas Sandby, recorded these phenomena with elegant objectivity, other individuals, such as William Blake, retreated from the Enlightenment's rationalism to a private symbolic world where they expanded the symbolism associated with comets and meteors in fantastic poetic and visual images. In this chapter, we investigate the situation in Great Britain at this formative time when many connections existed between individuals involved in the sciences and the arts.

Attitudes toward Change and Progress

British attitudes toward comets and meteors, and their representations, were affected by general cultural and intellectual trends. At the beginning of the eighteenth century, the national temper was characterized by a vitality and pride in being English. This prevailing spirit of optimism had its sources in the newly gained political stability provided by parliamentary rule, the economic well-being resulting from burgeoning trade, and the ethical principles of Puritanism supporting freedom of conscience and individual enterprise.[1] Some of these characteristics are noted by the intellectual historian David Spadafora, who believes that the notion of progress is the crucial ideology of modernity.[2] Rampant during the eighteenth century, it was inherited by the Victorians, to whom it was a central tenet of self-definition. Though other interpretations of the period exist (including cyclical theories), several of Spadafora's points are pertinent to our study. Indeed, our study of comets and meteors shows that many types of progress were made during the era – in telescopes and other technology, in banishing superstition, and in scientific understanding.

He points out that eighteenth-century intellectuals had not yet determined a definitive way to categorize the various disciplines. The trend epitomized by the French *Encyclopédie* was to consider branches of knowledge as an interconnected whole, a tendency that gained strength from Newton's achievements as theoreticians in the various fields sought to become the "Newtons" of their own disciplines. As a result, "art" aspired to become "science," or at least, to adopt the scientific approach. In 1776, the English philosopher Jeremy Bentham stated that the same object that is called art, viewed in another light is called science, while Samuel Johnson listed the seven liberal arts under the category "Science." Further, Spadafora argues that the arts and the sciences were viewed as two sides of a coin. The word "science," from the Latin *scientia*, which translates into "knowledge," was not necessarily limited to what we now call "science." These attitudes help to explain the exchange of ideas among individuals involved in the arts and sciences.

While the notion of progress had been prevalent since the Renaissance, Spadafora believes that eighteenth-century Britain experienced a massive departure from earlier epochs.[3] No longer was there any doubt that the moderns excelled over their predecessors, a watershed from the past when the Ancients were worshipped. (Certainly, the prevalent Protestant orientation of Great Britain encouraged the break with tradition and Roman Catholicism.) The same was true in the practical realm, where inventions elicited much attention, such as innovative engraving techniques and new instruments – telescopes, the air pump, and orreries. Spadafora believes that the classicism of Josiah Wedgwood, as expressed in designs such as his reproduction of the

Portland Vase and his portraits of modern heroes like Newton in the manner of the ancient cameos was an attempt to exceed antiquity via pioneering techniques in the manufacture of pottery at his Etruscan Works factory. Like others, Spadafora traces the inception of the industrial revolution to this era, citing as an example the number of patents granted in the four decades after 1750 which were four times as many as in the three previous decades.

At the time, a new blossoming of the arts rescued British artists from a decline in patronage and a challenge by continental artists. Artists and writers – such as Robert Adam, Sir Joshua Reynolds, and Edward Young, who participated in it – had predicted the change. As a result, a whole range of academies and exhibition facilities for the arts developed. These British institutions – societies, museums, and botanical gardens – facilitated a cultural climate that encouraged the pursuit of knowledge by professionals and amateurs alike. Among the most noteworthy were: the Royal Society; the British Museum; the Royal Academy; the Society for the Promotion of Arts, Manufactures, and Commerce; the Royal Institution; and the London Society of the Arts (which sought to unite science or theory with art or practice). They instilled an optimistic view of the future, which harmonized with the Christian vision of history and eschatology (both Anglican and Catholic) and touched everything – from the inventiveness of the English language, to the attitudes of Dr. Samuel Johnson, to the idea of governmental improvement.[4] The reign of George II and victories in the Seven Years' War after 1758 reinforced it. Even pessimistic thinkers, such as Edward Gibbon (*Decline and Fall of the Roman Empire*, 1774, published in 1781), expressed some belief in the progress of civilization which provides a backdrop for our study of comets and meteors.

Wedgwood, The Lunar Society, and the Intersection of Art and Science

Since many significant eighteenth-century scientific developments build on the discoveries of Newton and Halley, it is fitting that two of the four oval plaques with Newton's visage credited to Josiah Wedgwood (1730-1795) include a comet [Fig. 11 color]. Most likely, they represent the comet 1680-81, which confirmed

Newton's gravitational theory.[5] Wedgwood, who was a Fellow of the Royal Society and a friend of artists like Reynolds, George Stubbs, John Flaxman, and Joseph Wright of Derby, belonged to the Lunar Society, which was a microcosm of the Enlightenment. The society was one of the significant regional intellectual English groups, in which philosophers and scientists met to discuss new ideas and to solve practical problems in a scientific manner.[6] It provided an environment in which philosophers and scientists could freely exchange radical ideas about scientific, religious, intellectual, philosophical, and literary thought and practice.

The Lunar Society derived its astronomical name from the fact that it scheduled meetings in the Midlands near Birmingham on the Monday nearest the full moon, beginning in 1764. Among its members were the eminent thinkers Erasmus Darwin, Joseph Priestley, and James Watt. In addition, Benjamin Franklin and Sir Joseph Banks, President of the Royal Society, occasionally attended. Another pivotal member of the Lunar Society and friend of Wedgwood was Richard Lovell Edgeworth, the mechanical genius. Edgeworth, who was also a Fellow of the Royal Society, wrote one of the articles on the 18 August 1783 meteor discussed below.

Yet another friend of Wedgwood and member of the Lunar Society was the painter Joseph Wright of Derby (1734-1797), who embodied some of the Society's scientific ideas about progress in two of his most famous paintings, *A Philosopher Lecturing with an Orrery* and *An Experiment on a Bird in the Air Pump*. These galvanizing canvases represent a complex synthesis of art, science, and philosophy, influenced by his contacts with the Lunar Society. Each painting captures the intense, almost mystical excitement generated by these experiments that implied a new universal order. For example, the artist's *Orrery* (1766) represents a cosmological theme under the guise of a public demonstration.[7] Wright's painting relates to the work of his friend John Whitehurst, whose practical experience as a clockmaker would have enabled him to advise Wright about the orrery's operation. Wright had the opportunity in 1762 to personally witness a lecture on an orrery, like the one in his painting, at a talk presented in Derby by Whitehurst's friend, the Scottish astronomer James Ferguson. Ferguson, who made scientific instruments in his London shop, published scientific treatises and travelled the country delivering lectures, using an orrery, an armillary sphere, an astronomical clock, and other equipment. Actually, he was only one of a number of instrument makers and philosophers to engage in such activities, which constituted a feature of polite society. Therefore, the scene depicted in Wright's painting was rather typical of the day. Although the contents of Ferguson's Derby lecture are not recorded, he advertised another course in Bath in 1767, which offers a clue. The latter was billed as intending: "To explain the Laws by which the Deity Regulates and Governs all the Motions of the Planets and the Comets by Machinery: To show the Courses... by means of an Orrery." Similarly, Wright's painting of the orrery, in addition to its overt scientific content, assumes a religious significance, because the people are contemplating not just the workings of the solar system but also the operations of God himself. Thus, the tenebristic light symbolizes knowledge and reason while doubling as the traditional metaphysical metaphor of divinity. Yet the primary subject of Wright's canvas is secular science, and the religious implications are secondary.

During the late eighteenth century, societies similar to the Lunar Society were all the rage in Great Britain. It was partly due to forums like these that English intellectual and cultural life flourished. For example, an offshoot of the Lunar Society to which Josiah Wedgwood also belonged was the Derby Philosophical Society, founded in 1782-83 by Erasmus Darwin, the physiologist, poet, and grandfather of Charles Darwin. In fact, the movement to disseminate scientific knowledge and to promote invention became even more consolidated at the local level, where individuals explored the consequences of the scientific and industrial revolutions into the nineteenth century.

The Popularization of Science, Comets, and Meteors

Like Ferguson, Benjamin Martin (1714-1782) advertised the wonders of science as a writer and as an itinerant lecturer. He came to London in 1756 to set up shop in Fleet Street, where he traded as an optical, mathematical, and philosophical instrument maker. Martin was savvy about the location of his shop, which was near the Royal Society at Gresham College before it moved to Somerset

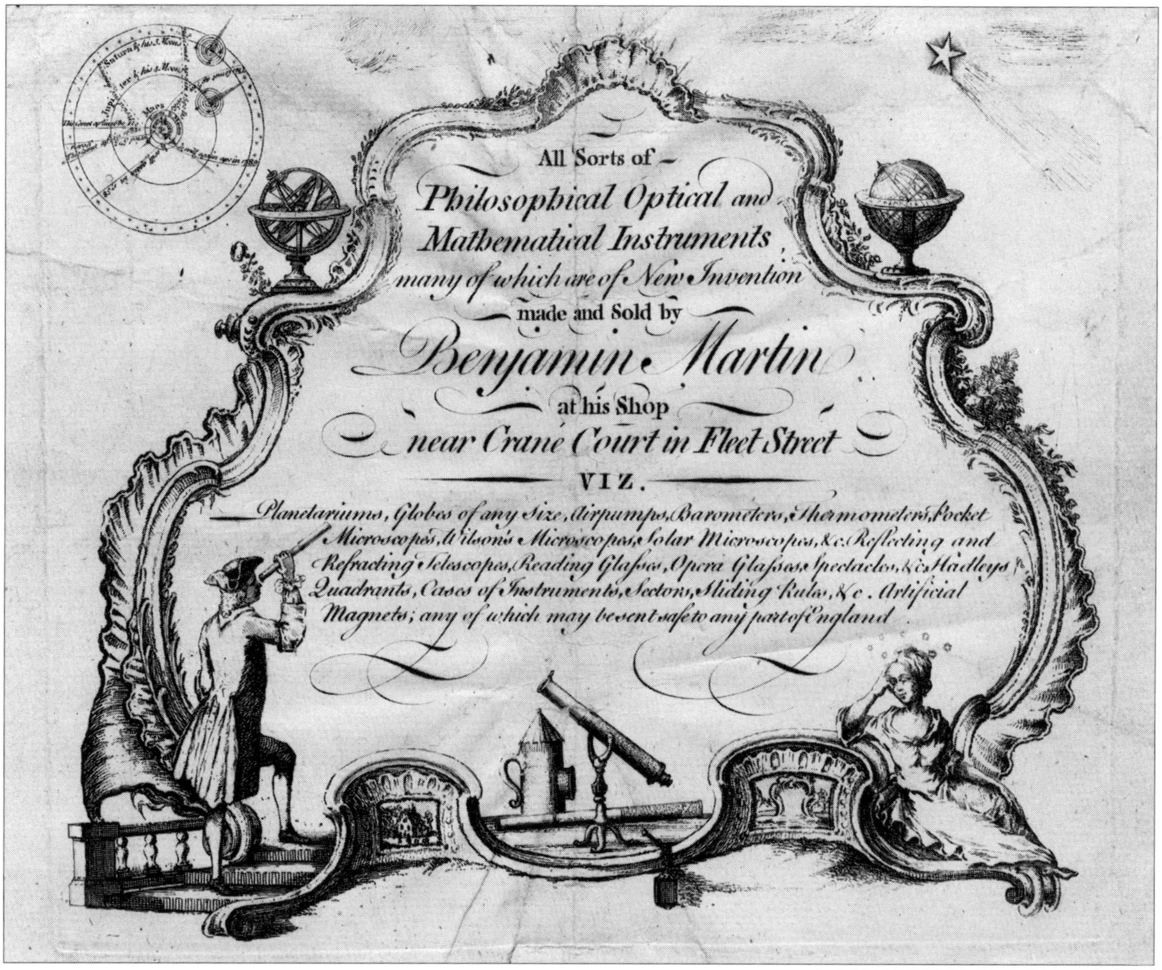

Figure 12. Benjamin Martin, *Trade Card*, c. 1750-60, engraving, 203 × 238 mm, The Science Museum, London.

House in 1780, although he himself failed in his attempts to become a Fellow of the Society. To spread notice of his business, Martin used a printed trade card [Fig. 12], which features a man looking at the heavens through a telescope.[8] Dominating the composition are: at the left a diagram of the solar system including comets in their elliptical orbits and at the right a large, five-pointed comet-star. Martin's trade card not only reveals his own involvement with comets but also suggests that they were topical attractions at the time.

Furthermore, in 1757 Martin issued *The Theory of Comets Illustrated* and a hands-on illustration of the orbit of Halley's Comet [Fig. 13].[9] A separate sheet of paper, attached horizontally at its mid-point, can be held out at the proper angle to the ecliptic

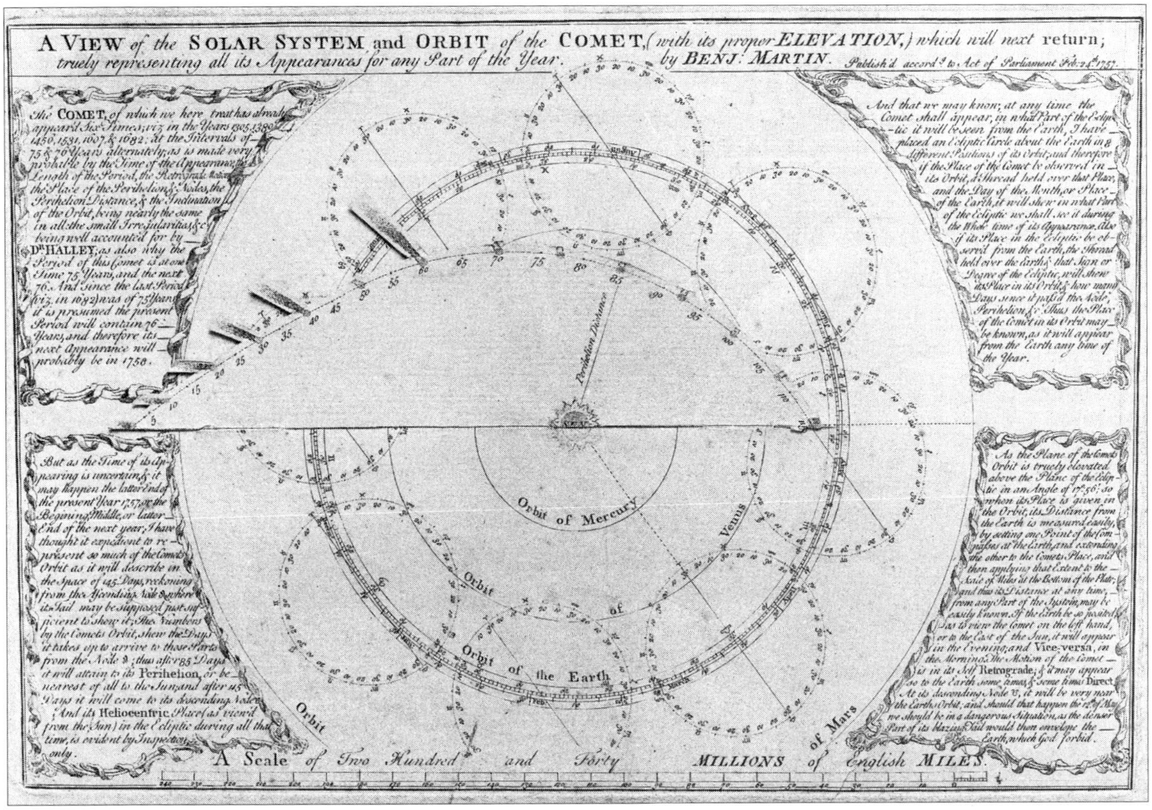

Figure 13. Benjamin Martin, *View of the Solar System and Orbit of Halley's Comet*, 1757, engraving, 248 × 373 mm, Museum of the History of Science, Oxford University.

(the plane in which the earth – and approximately the other planets – orbit the sun) for viewing the comet's orbit. Eight ecliptic circles around the Earth's orbit allow the user to hold a thread between one of those circles and the comet's multiple positions on the foldout to calculate where it is in its orbit. Martin warns, "At its descending Node, it will be very near the Earths Orbit; and should that happen the 12th of May we should be in a dangerous Situation, as the denser Part of its blazing Tail would then envelope the Earth, which God forbid."

During the years when Martin was active, the comet of 1769 (C/1769 P1) appeared. It was discovered by Charles Messier in Paris as part of his comet searches. When Captain James Cook saw it on 30 August in the South Pacific, he estimated that its tail extended 42°. By 11 September when it was seen by Pingré, its tail measured 90°. The comet, whose period is around 2,000 years, was carefully observed by astronomers at the Royal Observatory at Greenwich and was tracked through 3 December.[10] We have found no works of art by English artists that represent it, although it is recorded in the observing notebooks of Maskelyne, which were long on tables and short on illustrations.[11] Strangely, there were no reports about it in the *Philosophical Transactions*, whose astronomical articles in the preceding decade had been mostly about comets.

The popularization of science engendered a prosperous exhibition business to show off new technologies to an ever-eager public hungry for astronomical topics, including comets and meteors.[12] As lectures on science – both formal ones, such as those at the Royal Society and the Royal Institution, and informal ones by Ferguson and Martin – grew increasingly popular, a demand for devices to demonstrate scientific principles was created. Observatories were set up by professional and amateur astronomers alike, and their telescopes became larger. (Larger telescopes collect more light and thus allow the observer to see fainter objects.) Consequently, a new commercial market arose to respond to these demands, in which engravings and illustrations in the popular press and novel graphic media disseminated astronomical knowledge to a larger, seemingly insatiable audience into the new century. As a result, a general appreciation of astronomy, visits to science road-show spectaculars, and astronomical recreation were commonplace in Georgian and early Victorian society. High among Georgian astronomical delights illustrating the pleasures and rewards of science were discussions of the periodicity of comets.[13]

Cometaria (Cometariums)

Comets were such topical phenomena that devices called "cometaria" or "cometariums" were constructed to demonstrate their motion for exhibition purposes [Figs. 14-17]. Cometaria, along with orreries and planetaria, were among the instruments that Martin made. His machine represents a refinement of earlier types, though all were based on the discovery by Kepler (1609) that planets travel in elliptical orbits, later applied to comets by Newton and Halley. When Halley's Comet reappeared in 1758-59, it aroused keen interest, and the sales and use of cometaria boomed.

Cometaria demonstrate on a flat plane how the speed of a comet varies in its orbit around the sun according to Kepler's second law. (The law states that a line joining the sun and the comet sweeps out equal areas in equal times.) Actually, cometaria only approximate Kepler's laws, showing the more rapid motion of a comet when near the sun (at perihelion) than when far from the sun (aphelion). They also illustrate Kepler's first law, that planets travel in ellipses (Kepler thought that it did not apply to comets, which he believed went in straight lines). Martin explains the theoretical and technical operation of a cometarium in his tract *The Young Gentleman and Lady's Philosophy* (1756) through his erudite hero, Cleonicus, talking to his bookish sister: "Observe, when I turn the winch, the brazen Comet moves, and with a very unequal Pace in its elliptical orbit, about the *focal Sun at S.* – That when it is nearest the Sun (in what is called the *Perihelion* at B) it moves very quick – and, on the contrary, at its greatest Distance (or Aphelion A) it moves extremely slow."[14]

Martin's cometarium now at Harvard University consists of a mahogany box, with a lock and a key, which contains an elliptical-pulley mechanism driven by two gear wheels [Fig. 14].[15] A three-part brass plate – with a circle inside an oval inside a rectangle – is fastened to the lid. It conceals a mechanism operated by a hand-crank to turn two brass pointers above the lid. The longer one bears the brass ball of the comet's head that moves with the pointer along an elliptical orbit. The other measures time on a scale marked off in yearly units from 1 to 75 ½ (the circle), a scale dependent explicitly on Halley's calculations for Halley's Comet. Even though the eccentricities of comets like Halley's were too high for correct modeling, this cometarium was intended to demonstrate the movements of the comet that had recently returned and kindled such interest.[16]

Martin's instrument was neither the first cometarium nor the earliest type. Earlier ones used by scientific lecturers were based on two identical elliptical pulleys. They were first described, and probably invented, by J.T. Desaguliers (1683-1744) in his *A Course of Experimental Philosophy* (1734). Elliptical-pulley cometaria of this type, like the one owned by Stephen Demainbray [Fig. 15], one of his disciples and the tutor to the Prince of Wales, suffer from the disadvantage that when the cord comes off, it is difficult to get it working again.[17]

Figure 14. Benjamin Martin, *Cometarium*, c. 1766, mahogany and brass, 254 × 165 × 50 mm, Harvard University Collection of Historical Scientific Instruments, Cambridge, MA.

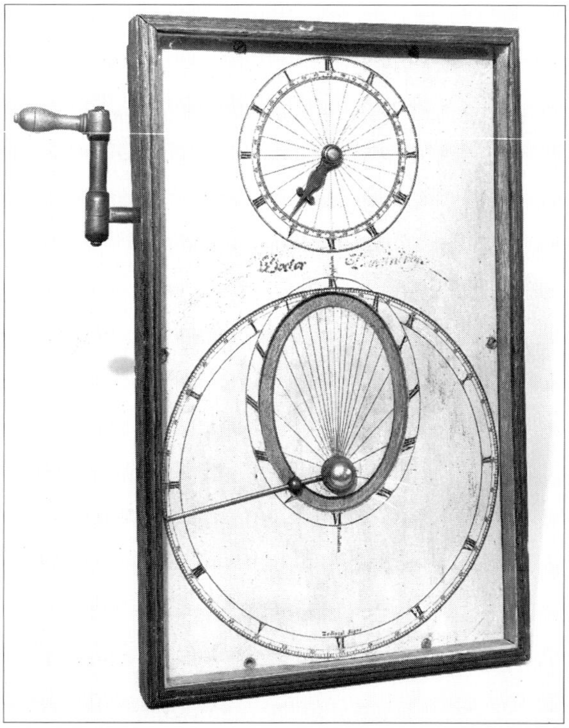

Figure 15. J.T. Desaguliers, *Cometarium*, c. 1755, wood, glass, and metal, 500 × 390 × 115 mm, King George III Collection, The Science Museum, London.

Figure 16. W. & S. Jones, *Cometarium*, early nineteenth century, wood, brass, and ivory, 300 × 230 × 150 mm, The Science Museum, London.

In the early nineteenth century, W. & S. Jones produced a cometarium [Fig. 16] constructed by William Jones that is similar to that of Martin. It uses an eccentrically-mounted circular gear quite different from the elliptical-pulley-cord device and was more expensive.[18] While its gear is relatively easy to make, it requires a complicated driving mechanism with sliding parts and means for keeping the gear constantly in mesh with its driving pinion. The fully-geared Jones version was intended to overcome the problem of the elliptical-pulley models from the previous century, but by the time they were introduced cometaria were going out of fashion. It features an ivory ball with a tail of bristles as its comet to make the device seem less theoretical. The comet moves in an elliptical groove around the brass ball of the sun that is both at the center of a large circle representing the heavens and at one focus of the ellipse of the comet's orbit. A brass pointer radiating from the sun shows, at its intersection with the large circle, how the comet appears to move against the background of the sky. The speed of the comet varies as is needed to counterbalance the effect of the force of gravity, which varies as an inverse square of the comet's distance from the sun. Figure 17 reveals the gear mechanism inside the box.

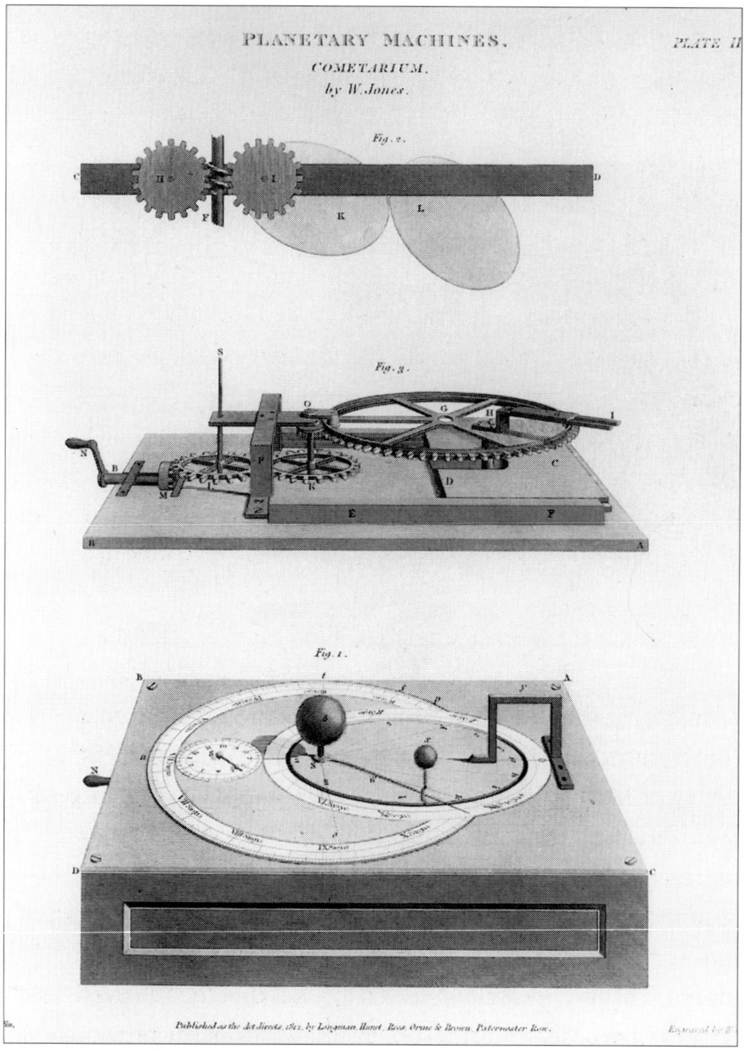

Figure 17. J. Farey and W. Lowry, *Cometarium by William Jones*, illustration from Abraham Rees, *The Cyclopaedia*, plate II from volume 9, 1819, engraving, 268 × 205 mm, Private Collection, New York.

Other cometaria include a vertical demonstration model by John Miller, used for teaching at the University of Edinburgh when astronomy formed a part of the ordinary arts curriculum,[19] and a later device used at the University of St. Andrews by Robert Fidler in association with the Reverend William Pearson.[20] Fidler's employs the same principle as the elliptical-pulley models but because it needed to be more reliable has a pair of elliptical gears instead of pulleys. These gears eliminate problems with the cord, although the manufacture of the gears was difficult. Despite the improvements, the craze for cometaria eventually died out but not the British interest in comets and meteors.

Eighteenth-Century Comets and Meteors, Hogarth and Fuseli

With the birth of journalism, celestial occurrences were reported with great frequency and zeal. Astronomical topics, including comets and meteors, were featured in periodicals of the day. Accounts of their appearances begin as early as 1737 in *The Gentleman's Magazine*, where it is clear that a learned person's knowledge extended to an interest in celestial phenomena. These articles also indicate that the nature of comets and meteors was still in flux. They are rarely illustrated and the emphasis is not as scientific as in the *Philosophical Transactions*, although there are occasional contributions by scientists.

Through an examination of articles and representations of various comets and meteors, we can better understand the excitement caused by these celestial fireworks and chart the discoveries made about them. One already notices the great change in the attitude toward comets in the post-Halley era in a large, anonymous print, the descendent of broadsides, that records the appearance of the comet of 1744 (C/1743 X1) near St. Martin's, London [Fig. 18].[21] In contrast to the more sensational nature of Gadbury's earlier announcement of comet apparitions [Fig. 3], the 1744 print emits great dispassion. The artist's sense of wonder has been transformed into a fascination with the scientific aspects of the comet's appearance and observance. This engraving combines a scenic view of the apparition with a more diagrammatic rendition of the comet's anatomy as well as matter-of-fact inscriptions noting that the comet appeared on 26 January 1744. Its large image shows the comet as seen with naked eye, while the insert at the left records the comet as it appeared in a telescope with its center of condensation illuminated on its side closest to the sun, its coma, and its long tail. In the large scene, observers atop a building gaze through telescopes at the apparition. The artist has increased the drama and the immediacy of the view by assuming an unusual vantage point from atop an adjacent building. In addition, the artist has heightened the stellar effects of the nocturnal scene by dotting the sky with hundreds of stars that seem kinetic.

The comet of 1744, which was discovered on 29 November 1743, is depicted here before perihelion, when it was judged to be brighter than any star, including even Sirius, the brightest.

Figure 18. Anonymous, *Comet seen above St. Martin's Church on January 26, 174¾*, 1744, engraving, 277 × 175 mm, The Science Museum, London.

It became a singularly spectacular comet a few days after perihelion in March 1744, when it exhibited a peacock-like tail, which at one time manifested eleven or twelve fanning rays. Although it remained brilliant for only a short period of time, this naked-eye comet attracted a great deal of attention from astronomers and amateurs throughout Europe, with a goodly number observing it

Figure 19. William Hogarth, Detail of *Marriage à la Mode*, plate I, 1745, engraving, 353 × 440 mm (whole image), Sir John Soane's Museum, London.

in England. Today it is believed that the multiple structures of its tail were generated from three active sources on the nucleus surface, which were successively exposed to solar radiation as the comet rotated.[22]

The comet of 1744 may have inspired the comet included by William Hogarth (1697-1764) in his scathingly critical series *Marriage à la Mode*, engraved in 1745 by Gérard Jean-Baptiste Scotin in reverse from Hogarth's paintings in the National Gallery, London.[23] In this series, which attacks the nouveau riche, the degenerate aristocracy, and the contemporary English mania for the French style, Hogarth's biting humor and psychological insight is so finely honed that nothing is accidental. Each detail constitutes a pointed allusion with an emblematic significance, as in the writings of Fielding and Samuel Johnson. Hogarth inserted a comet in the first print of the series, the "Marriage Settlement," which depicts the signing of the contract between Lord Squanderfield (on behalf of his sniveling, syphilitic son) and a newly rich merchant (on behalf of his weak, flirtatious daughter). It is filled with ironic, disparaging remarks about the Earl that ridicule his pretensions via a depiction of his family tree, a regal canopy, and a comet.

Hogarth placed a comet in the ridiculously pretentious portrait of Lord Squanderfield [Fig. 19], as though he were aware of the earlier British occasional practice of including comets, either heraldic or historical, in portraits [Figs. 4 color, 5]. He portrayed Lord Squanderfield full of meaningless flutter, parodying French portraits, assuming the identity of a sprightly Jupiter to satirize the French aristocracy's habit of depicting themselves in the guise of the classical gods. He holds Jupiter's lightning bolts, while pointing with a pinkie finger to the insignia hanging from his belt: the orders of the Golden Fleece and the French *saint esprit*, which testify to the Earl's vanity. (Prior to 1743, no Englishman had been awarded these prestigious orders of chivalry; their presence is evidence of the painter's outrageous flattery of his subject and the Earl's bad taste.) Like the architecturally unsound facade of the adjacent ill-fated neo-Palladian building, this outwardly impressive portrait is based on false assumptions and a number of contradictions. Most noteworthy for this study is that the sitter straddles a cannon firing in the opposite direction to a soaring comet; in turn, a zephyr puffs in opposition to this comet.[24] It is this ironic comet that drives home the sharp edge of Hogarth's satire, because a comet can also presage doom, herein that of Lord Squanderfield and his decadent progeny. Hogarth was cognizant of the superstitions attached to comets and in this minor detail exploits them wonderfully. Cowley has observed that Lord Squanderfield's portrait is an apt comment on the central situation, and that it is also one of the few pleasantly absurd and mock-heroic details in the entire series. But, like the view through the window with which it is juxtaposed, it has a serious point to make: that ridiculous causes can lead to catastrophic results.[25] Hogarth's versatile comet is, therefore, both an omen of the

Figure 20. Johann Heinrich Fuseli, *Fortune Suspended over the Earth* (or *The Mighty Mother Sails Through the Air*, from Theophilus Swift, *The Temple of Folly*), 1787, etching, 233 × 188 mm, The British Museum, London.

approaching ruin of the frivolous couple and a satirical comment on Squanderfield's pomposity. Moreover, the artist must have been aware of the more objective attitude towards comets that was surfacing and was, therefore, also satirizing the superstitions attached to them.[26]

Because of this objectivity, comets and meteors could be regarded less "seriously" and used not only in a humorous vein but also in a fantastic manner as metaphysical symbols, as in an irreverent print by Heinrich Johann Fuseli (1741-1825) depicting the tipsy figure of Fortune suspended over the Earth [Fig. 20].[27] Fuseli treated this traditionally ambivalent personification in a slightly menacing and satirical manner. She is dressed as a jester to render her not only more comic but also more foolish. A disembodied comet or meteor, whose head resembles a snowball,

Figure 21. W.P., *Allegory of Prosperity under Humphy Parsons Lord Mayor of London*, 1730-31, engraving, 388 × 295 mm, The Pierpont Morgan Library, Peel Collection, New York.

careens in to the right of the cosmic cocktail she mixes in her goblet. This comet or meteor adds another pernicious and unstable chord to Fuseli's representation of disharmony.

Traditionally Fuseli's unsigned etching dated 1787 was thought to represent *Fortune Over the Earth*, although Weinglass identifies it as *The Mighty Mother Sails Through the Air*, an illustration to Theophilus Swift's *The Temple of Folly*.[28] Weinglass believes that the artist derived Fortune's cane and the spray of the thistle-like plant eryngium that she holds from Albrecht Dürer's two prints of Fortune, combining them to form her wand. The blossoms or berries of the eryngo have become the round objects – whether seeds or worlds – being shaken into Folly's goblet. Weinglass suggests that the globe on which Folly stands is a balloon rather than the "sphere of uncertainty" in Dürer's engravings. He also believes that the print reflects the 1780's fascination with successful balloon ascents in France and England.[29] We suggest that Fuseli may have added the unpredictable comet or meteor, not present in the preparatory drawing, because of the current interest in comets and meteors. This celestial body's presence contributes another element of instability, while it simultaneously functions as a prescient symbol of doom, a sardonic comment on Fortune/Folly. Whizzing on a trajectory that will probably just miss Fortune/Folly's head, it lends a cosmic spin to this topsy-turvy allegory.

Because of the new attitudes toward comets in eighteenth-century England, the word "comet" became magical, leading to the exploitation of positive comet symbolism. For example, a comet apparition could presage prosperity and bounty, and was frequently associated with a ruler. This theme appears in an allegory of 1730-31 touting via symbols the just and prosperous times of Lord Mayor Humphy Parsons. One of the most prominent is a comet that literally shines through Parsons's visage [Fig. 21].[30] These positive cometary connotations were attached to all manner of objects. Even utilitarian items benefited from their upbeat qualities which frequently implied progress or swift directional motion. For example, a certain type of flour mill, constructed at Lyon on a boat in a river to utilize the current, was termed a "Comet Mill;" it is preserved in a model from 1753.[31]

Until the latter half of the eighteenth century, certain stock conventions were employed in Britain for rendering comets and meteors, just as they were for images of other celestial events. Gradually the substantial progress made in understanding these different but related phenomena allowed individuals to describe them more accurately and to depict them according to their distinctive characteristics. A significant case in point occurs with the very bright meteor of 18 August 1783, which we will soon discuss.

Prior to that time, even the manuscript drawings and plates for articles in the Royal Society's *Philosophical Transactions* sometimes featured amazingly bizarre celestial apparitions, especially those for meteors. An early illustration for Lord Paisley's account

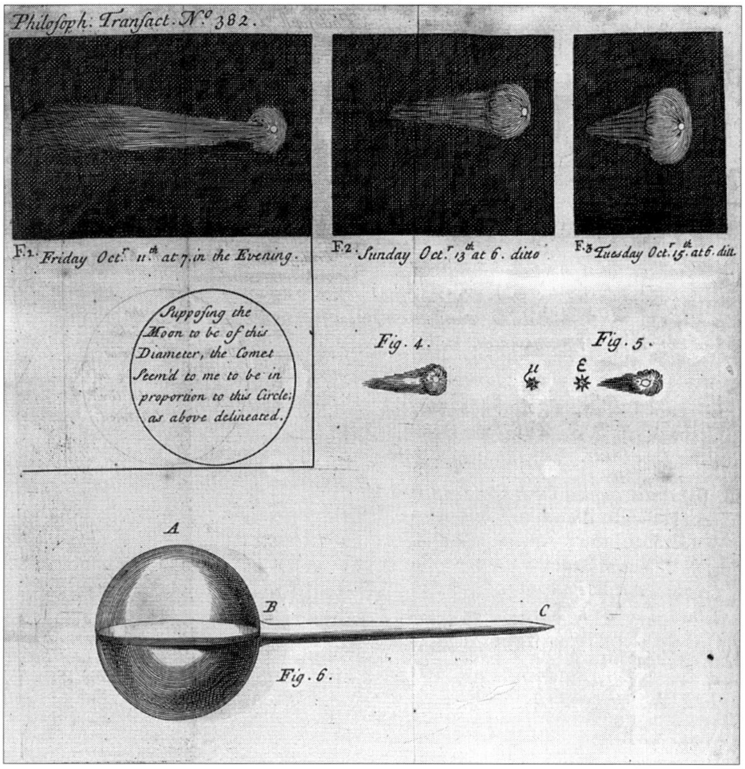

Figure 22. After Lord Paisley, *Figures of a Comet*, unpaginated plate opposite p. 39 from volume 33 of the *Philosophical Transactions*, 1733, measuring 221 × 160 mm, Archives of the Royal Society, London.

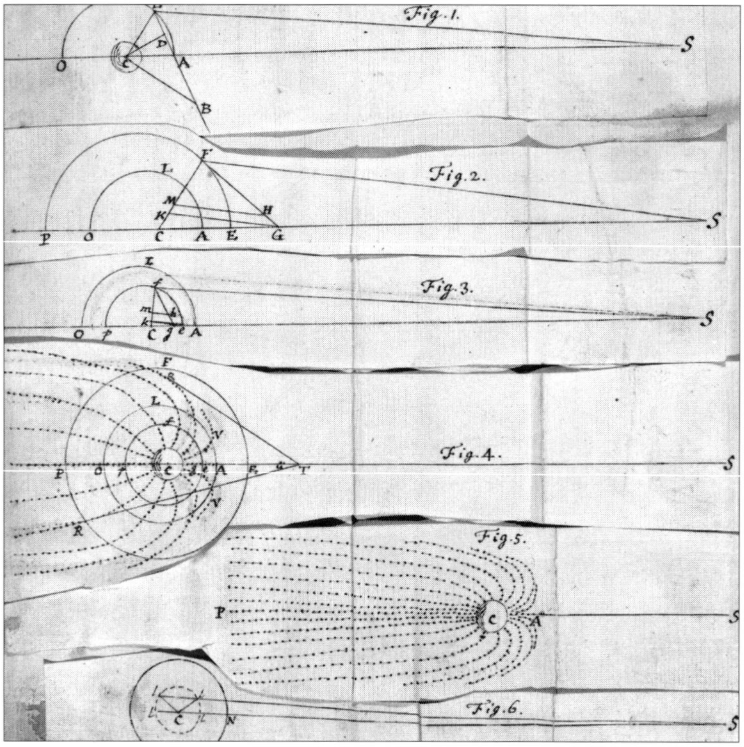

Figure 23. John Winthrop (transmitted by Benjamin Franklin), *Model for a Comet*, 1767, pen and brown ink, 201 × 205 mm, Archives of the Royal Society, London.

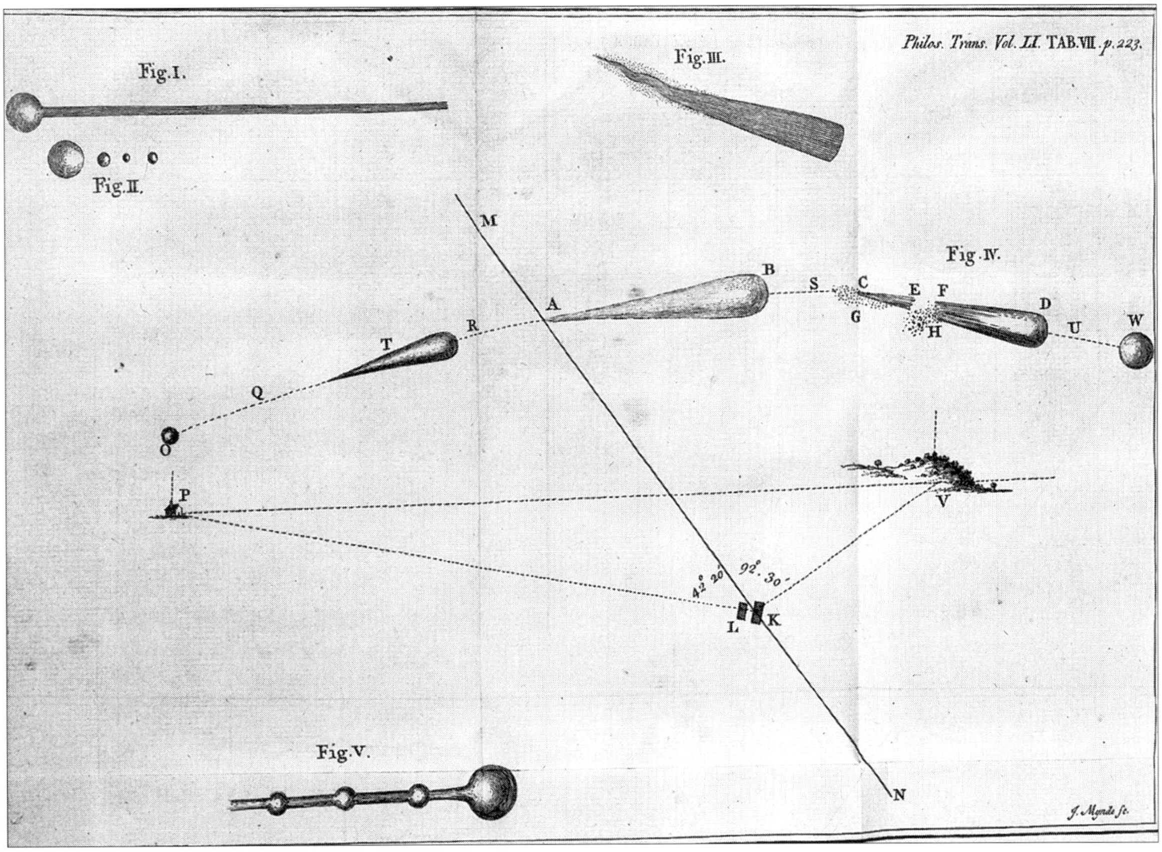

Figure 24. After John Pringle, *The Meteor of 26 November 1758*, plate VII from volume 51 of the *Philosophical Transactions*, 1759, measuring 224 × 310 mm, Archives of the Royal Society, London.

of a comet in 1733 seems quite analytically correct by comparison [Fig. 22].[32] In addition, John Winthrop's report in 1767 from "New-England" in the American colonies (communicated by Benjamin Franklin) includes a surprisingly accurate illustration of a geometrical model for a comet [Fig. 23] with its tail extending away from the sun, a fact known from observations.[33] Meteors assume various shapes, and their changing nature is illustrated in a plate showing the progression of a meteor in 1758 that accompanies an article by John Pringle, a physician and President of the Royal Society from 1772 until 1776 [Fig. 24].[34] Zach. Cradock drew the meteor he reported on 27 May 1744 in a trumpet-like form [Fig. 25],[35] as did Henry Baker representing a meteor of 28 July 1750 [Fig. 26].[36] The latter has the illusion that the bright head is larger than the trail left behind.

These events were what we today would still call "meteors," while other references to "meteors" cite entirely different phe-

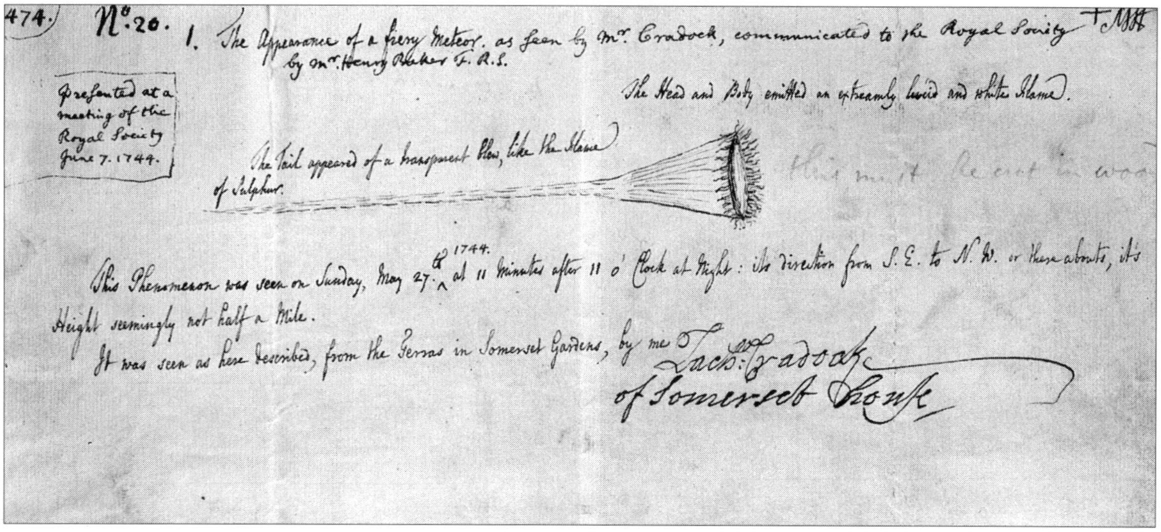

474. No. 20. 1. The Appearance of a fiery Meteor, as seen by Mr. Cradock, communicated to the Royal Society by Mr. Henry Baker F.R.S.

Presented at a meeting of the Royal Society June 7. 1744.

The Head and Body emitted an extreamly lucid and white Flame.

The Tail appeared of a transparent Blew, like the Flame of Sulphur.

this must be cut in wood

This Phenomenon was seen on Sunday, May 27th 1744 at 11 minutes after 11 o' Clock at Night: its direction from S.E. to N.W. or thereabouts, it's Height seemingly not half a Mile.

It was seen as here described, from the Terras in Somerset Gardens, by me Zach: Cradock of Somerset House

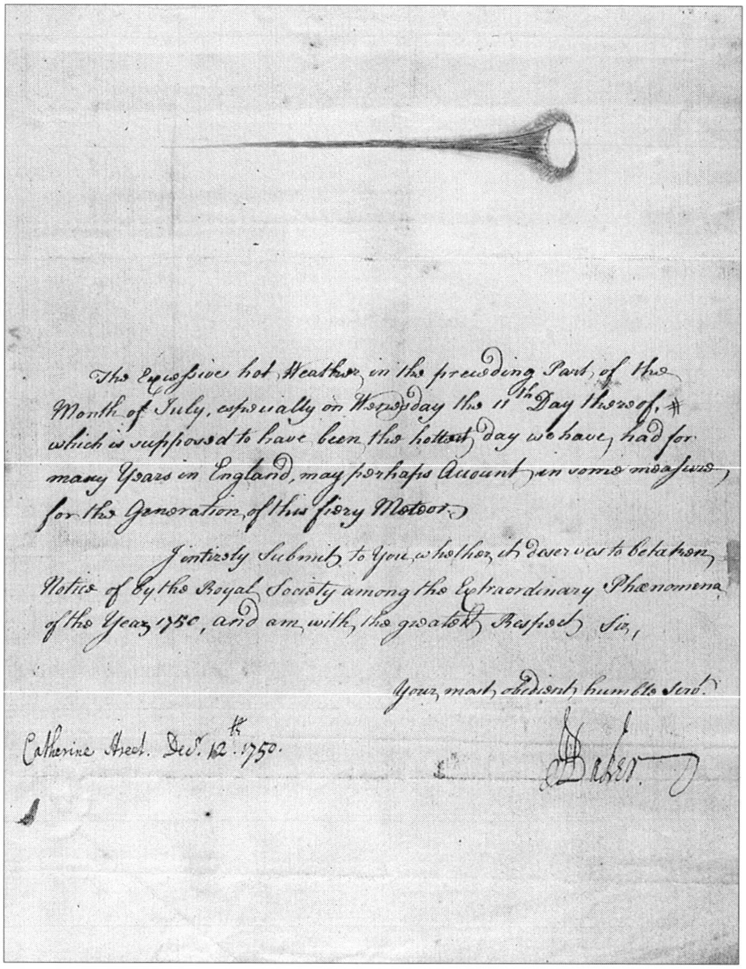

The Excessive hot Weather in the preceding Part of the Month of July, especially on Wednesday the 11th Day thereof, which is supposed to have been the hottest day we have had for many Years in England, may perhaps Account in some measure for the Generation of this fiery Meteor.

I intirely Submit to You, whether it deserves to be taken Notice of by the Royal Society among the Extraordinary Phænomena of the Year 1750, and am with the greatest Respect Sir,

Your most obedient humble Servt.

Catherine Street. Decr. 12th 1750.

H Baker.

Figure 25. Zach. Craddock, *A Fiery Meteor*, 1744, pen and brown ink, 179 × 310 mm, Archives of the Royal Society, London.

Figure 26. Henry Baker, *A Fiery Meteor*, 1750, pen and brown ink, 240 × 190 mm, Archives of the Royal Society, London.

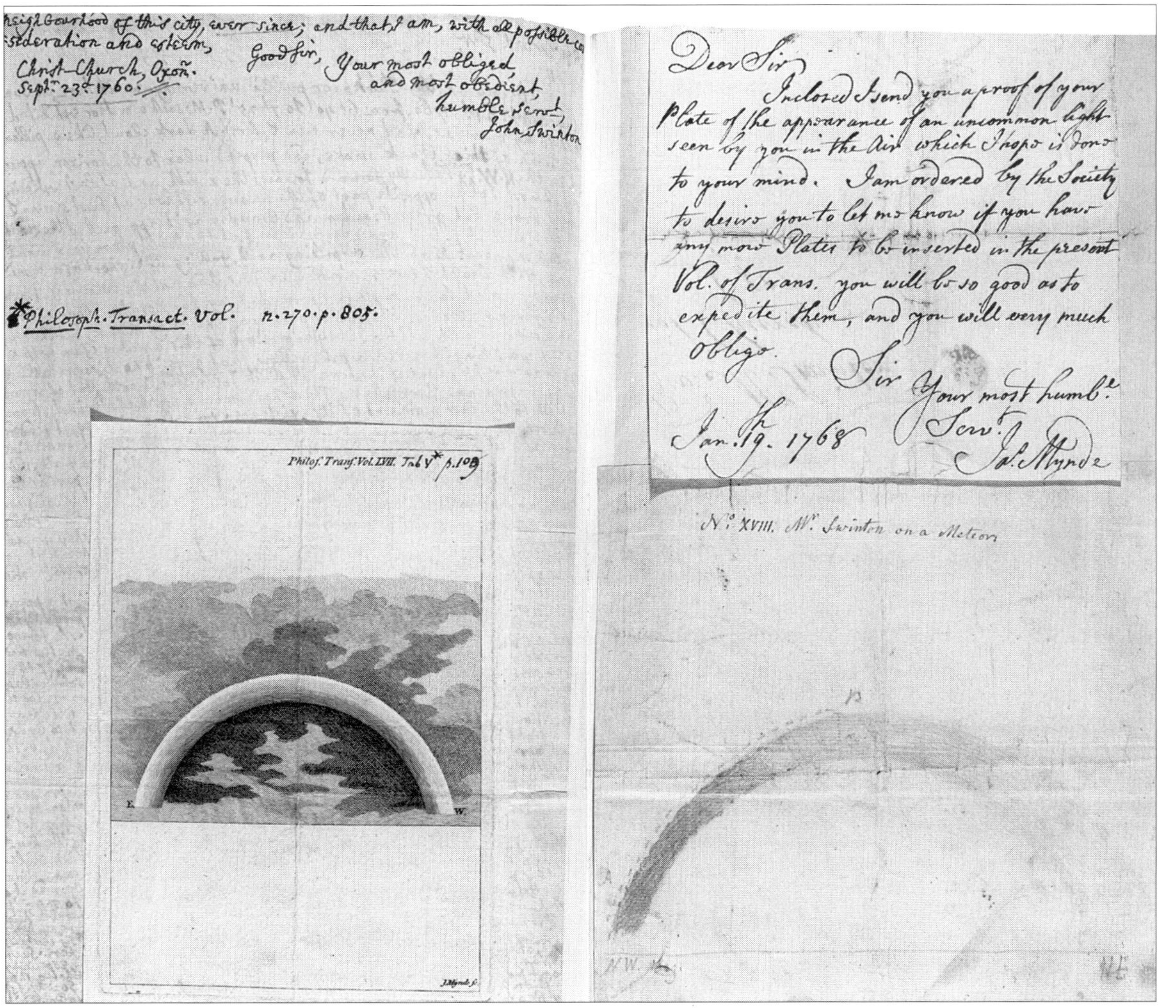

Figure 27. John Swinton, *Two Versions of "Meteors,"* 1760 and 1766, at the lower right in graphite, 179 × 191 mm, preparatory for plate III of volume 52; at the left engraved plate V from volume 57 of the *Philosophical Transactions*, Archives of the Royal Society, London.

nomena. The report of a "meteor" on 21 September 1760 by the Reverend John Swinton in the *Philosophical Transactions* allies the phenomenon with a waterspout [Fig. 27]. He describes it as a black column of smoke perpendicular to the horizon.[37] In fact, Swinton associated many celestial phenomena with the word "meteor," which he employed in the old Aristotelian sense still then in use, signifying any phenomenon that occurs in the air. Another such "meteor" is represented as an illustration to Swinton's report of a sky pillar on 5 March 1764 (it was not what we now call a "meteor" because it lasted four hours), which by association he links to the aurora borealis and the aurora australis. This phenomenon resembles a jet plane tail with ten

Figure 28. John Swinton, "*Meteor" of 5 March 1764*, plate 18 from volume 54 of the *Philosophical Transactions*, measuring 148 × 124 mm, Archives of the Royal Society, London.

horizontal, bristling icicle-like forms projecting from it [Fig. 28].[38] Yet another "meteor" was reported on 23 April 1764 as lasting around 32 minutes. Again, Swinton links this non-meteoric phenomenon with the aurora borealis and an *Iris* (rainbow).[39] Then, in 1767 Swinton published a nearly identical illustration with subtle variations to represent the "meteor" of 12 October 1766 [Fig. 27]. Swinton classifies it as a broad luminous arch, which he relates appeared shortly after an aurora borealis.[40] While many people remained confused about these natural occurrences, articles in the *Philosophical Transactions* reveal that a concerted attempt was being made to differentiate among them.

Because most eighteenth-century natural philosophers doubted that they had fallen to earth, meteorites were another burning issue. This denial of an unusual but nonetheless genuine natural phenomenon has subsequently been a great embarrassment to scientists, particularly since it persisted for about a century. One explanation for the lack of judgment, especially ironic since it occurred during the Enlightenment, is that scientists (notably members of the Académie Royale des Sciences in Paris) were overzealous in denying eyewitness reports in a desire to eradicate popular superstitions. This skepticism about meteorites relates to prevalent theories about meteors and fireballs, and the lack of knowledge about their relationship, size, and chemical composition. In fact, prior to the publication of Chladni's treatise in 1794, there were only a few fleeting conjectures that shooting stars and fireballs might be extraterrestrial matter.[41]

Nonetheless, the period was rich in reported meteors and meteorites. During the eighteenth century, there were nearly 200 reports of fireballs, two-thirds of which appeared after 1750. Then, in the first decade of the nineteenth century alone, after meteorite falls were finally accepted and an awareness of them had developed, the number of reports amounted to 65. One of the most famous incidents in the late eighteenth century was the meteorite weighing 56 stone that fell to earth on 13 December 1795 at Wold Cottage in the East Riding of Yorkshire. It was subsequently exhibited in London, where the President of the Royal Society, Sir Joseph Banks, observed it (he collected fragments of meteorites including this one).[42] Later, Banks asked Edward Charles Howard to examine some meteorites. In the 1802 report of his experiments using four chondrites, Howard concluded that the stones were unlike terrestrial rocks, that they fell from heaven, and that "native iron," which also contained nickel, provided the vital link to suggest their extraterrestrial origin.[43]

Perhaps the controversy over meteors finds reflection, or at least an awareness, in Benjamin West's (1738-1820) *Aeneas and Creusa* of c. 1771 [Fig. 29], where a tiny but freely painted meteor plunges through the turbulent sky framed by classical architecture.[44] West, who was born in the United States and is claimed by both British and American schools of painting, rendered this canvas in England and was certainly touched by the Sublime. His

Figure 29. Benjamin West, *Aeneas and Creusa*, c. 1771, oil on canvas, 185.5 × 139.5 cm, The New-York Historical Society.

literary source was not Homer's *Iliad* but Virgil's *Aeneid* (Book II, 940-945), perhaps John Dryden's translation, with its description of the meteor:

> There shot a streaming Lamp along the Sky.
> Which on the winged Lightning seem'd to fly'
> From o'er the Roof the blaze began to move,
> And trailing, vanish'd in th' Idaen Grove.
> It swept a path in Heav'n, and shone a Guide,
> Then in a steaming stench of Sulphur dy'd.[45]

The Meteor of 18 August 1783

One of the landmarks in meteor observation occurred with the meteor of 18 August 1783. It was still confused with a comet in some records, where it was identified with the ancient type of comet *Draco volans*, or the "flying dragon." It was what we now call a "fireball," a very bright meteor. Since both written and visual accounts note that it exploded near the end of its path, it can be identified more specifically as a bolide (a particularly bright meteor whose appearance is accompanied by an explosive sound or sonic boom).

This pivotal apparition was reported by an unusually large number of individuals from both the scientific and artistic communities. No fewer than six articles concerning it were published in volume 74 of the Royal Society's *Philosophical Transactions*. By comparison with descriptions of previous meteors, the fireball of 1783 was reported with an usual amount of care.[46] What is more, two of the articles on the meteor of 18 August 1783, those of Tiberius Cavallo and Nathaniel Pigott, contain fascinating three-stage, foldout engravings of it. Both are based on original drawings submitted with the authors' manuscripts [Figs. 30, 31].[47] A comparison of the original drawings with the final prints reveals that the engravers added few embellishments. In the drawings, the meteor is depicted progressing from the left to the right. This progression is reversed in the print accompanying Pigott's article, whereas the print with Cavallo's was correctly reversed by the engraver before being transferred to the plate. Other representations of the same phenomenon record its direction from left to right.

Figure 30. Tiberius Cavallo, *Observation drawing of the Meteor of 18 August 1783 from Windsor* (detail), pen and brown ink, 229 × 186 mm, Archives of the Royal Society, London.

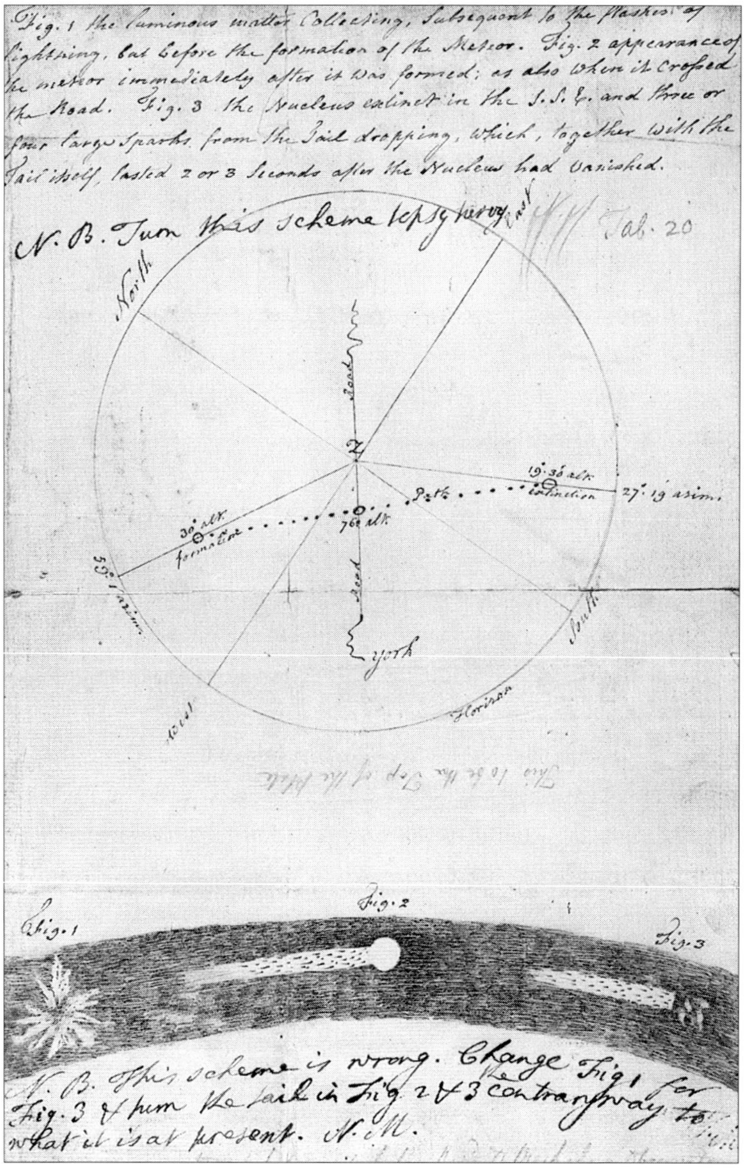

Figure 31. Nathaniel Pigott, *Observation drawing of the Meteor of 18 August 1783*, pen and brown ink, 525 × 175 mm, Archives of the Royal Society, London.

Four of the aforementioned papers on the August 18th meteor – those of Cavallo, Aubert, Cooper, and Edgeworth – were read at the Royal Society on 15 January 1784, in what constituted a mini meteor symposium. Subsequently, Blagden's was read on 19 February 1784, followed by Pigott's on 24 June 1784. Each paper empirically reports the weather and viewing conditions, and the objectivity of the scientific investigatory spirit is only clouded incidentally by traditional assumptions about meteors. So many people were outside to observe its transit across the sky because the weather was hot and sultry for England. Cooper, the Archdeacon of York and a Fellow of the Royal Society, informs us that the temperature ranged between 74° and 82° and relates that he observed the meteor on horseback:

> Towards nine o'clock at night it was so dark, that I could scarcely discern the hedges, road, or even the horses heads. As we proceeded, I observed to my attendants, that there was something singularly striking in the appearance of the night, not merely from its stillness and darkness, but from the sulphureous vapours which seemed to surround us on every side. In the midst of this gloom, and on an instant, a brilliant tremulous light appeared to the N.W. by N. At the first it seemed stationary; but in a small space of time it burst from its position, and took its course to the S.E. by E. It passed directly over our heads with a buzzing noise, seemingly at the height of sixty yards. Its tail, as far as the eye could form any judgement, was about eight or ten yards in length. At last, this wonderful meteor divided into several glowing parts or balls of fire, the chief part still remaining in its full splendour. Soon after this I heard two great explosions, each equal to the report of a cannon carrying a nine-pound ball. During its awful progress, the whole of the atmosphere, as far as I could discern, was perfectly illuminated with the most beautifully vivid light I ever remember to have seen. The horses on which we rode shrunk with fear; and some people whom we met upon the road declared their consternation in the most expressive terms.[48]

Alexander Aubert, also a Fellow of the Royal Society, reports in his paper on this spectacular meteor, together with another on 4 October 1783. He attempts to synthesize his observations because he feels that people are so little acquainted with meteors. To make his report more reliable, he revisits the places of his observations

Figure 32. Henry Robinson, *Representation of a Meteor seen on 18 August 1783*, mezzotint, 188 × 252 mm, The British Museum, London.

with proper instruments in order to reconstruct them more precisely. He notes important information at the close of his report: "it left behind several globules of various shapes; the first which detached itself being very small, and the others gradually larger and larger, until the last was nearly as large as the remaining preceding body; soon afterwards they all extinguished gradually, like the bright stars of a sky-rocket...." Aubert adds that after the meteor disappeared, the time was 9:14 (he believes his watch was 30 seconds slower than mean time). He calculates the entire time of visibility to around ten or twelve seconds, which is reasonable for a fireball given videotaped observations of one in 1994.[49] Like Cooper, he was on horseback, returning to his observatory in Kent, when the meteor of 18 August, the second in his report, which was of shorter duration, appeared. For both he notes color tinges of red and blue, although he states that he does not recollect any noise with either.

Figure 33. Anonymous (after Henry Robinson?), *Exploding Meteor of 18 August 1783*, c. 1783, pen and black ink with wash and watercolor, 255 × 313 mm (paper), Royal Astronomical Society Library, London.

Edgeworth, a third Fellow of the Royal Society, reports the time of the 18 August meteor as 9:30 p.m., although he agrees approximately with Aubert about its duration and color, stating that it appeared for between ten and fifteen seconds and exhibited "vivid colours" of red and blue.[50] Two other illustrations produced for circulation [Figs. 32, 33] support Edgeworth's contention that the breakup of this fireball exhibited these colors.[51] The watercolor is monochromatic except for the blue in its watercolor border (lined with yellow) and for the red in the fireball and its offspring.

Pigott seems to have been the most timely in his reportage of this event, because his paper, dated 18 October 1783, notes that on 19 August, the day after the apparition, he had first written to Maskelyne about it. (He read his paper on 24 June 1784.) Pigott remembers the apparition as occurring around 10:00 p.m., when, like the others, he was on horseback. He notes that:

> It suggested the idea of a highly brilliant comet, emitting a train or tail, but of a different colour from the ball itself, this last being of a most brilliant bluish white, and the tail of a dusky red, the length of which appeared to extend over fifteen or more degrees of the heavens.... It has been said, that the ball divided itself into three or four parts before its extinction. To me it appeared to vanish or gently die away: what confirms me in the opinion, that it did not divide, is, that the three or four scattering parts above-mentioned were not of the bright colour of the ball itself, but of the dusky red which the tail invariably showed.[52]

He estimates its duration as around twenty seconds, perhaps a few seconds less, and he too heard noises like the shot of a cannon. Later, he conscientiously returned to the spot near East Riding, where he had originally observed it, to make calculations.

One of the most interesting of the reports on the August 18th fireball is that of Tiberius Cavallo (1749-1809), another Fellow of the Royal Society. Cavallo, a natural philosopher and physician who was involved with electrical studies and experiments, relates that he watched the fireball from the Windsor Castle north terrace in the company of Dr. James Lind, Dr. Lockman, Thomas Sandby, and a few other persons, two of whom are known to have been women from the works of art depicting this viewing [Figs. 34-38].[53] Cavallo was also interested in art for, together with the daughters of Lind, he made silhouettes, including a portrait of Thomas Sandby, today in the Royal Library, Windsor.[54] Cavallo remarks that they had a perfect view of the fireball and that everyone of the company contributed something to his account:

> Some flashes of lambent light, much like the *aurora borealis*, were first observed on the northern part of the heavens, which were soon perceived to proceed from a roundish luminous body, nearly as big as the semidiameter of the moon, and almost stationary in the abovementioned point of the heavens.... It was then about 25 minutes after nine o'clock in the evening [Mr. Sandby's watch was seventeen minutes past nine nearest; it did not mark seconds]. This ball, at the beginning, appeared of a faint bluish light...but it gradually increased its light, and soon began to move, at first ascending above the horizon in an oblique direction towards the east...and

> moving in a direction nearly parallel to the horizon, reached as far as the S.E. by S. where it finally disappeared.... A short time after the beginning of its motion, the luminous body passed behind the above-mentioned small cloud, so that during this passage we observed only the light that was cast in the heavens from behind the cloud...but as soon as the meteor emerged from behind the cloud, its light was prodigious.... At this moment the body of the meteor appeared of an oblong form, like that represented in the figure; but it presently acquired a tail, and soon after it parted into several small bodies, each having a tail, and all moving in the same direction, at a small distance from each other, and very little behind the principal body, the size of which was gradually reduced after the division (see D in the figure). In this form the whole meteor moved....[55]

Both Cavallo's observational sketch [Fig. 30] and the plate engraved after it in his article correspond to the above description. Moreover, the letters *A*, *B* (*C* not drawn but by implication behind the cloud), and *D* correlate with his written observations. Cavallo's configuration *D*, the third stage, is remarkably similar to the forms rendered in the Robinson print, and its related watercolor [Figs. 32, 33], both of which also capture the slightly eerie quality of the meteoric illumination.

Furthermore, all this information corresponds to an aquatint by Paul Sandby (1730/31-1809) and to four fascinating, related watercolors variously attributed to him and his brother Thomas Sandby (1721/23-1798). Three of these watercolors contain a similar haunting depiction of the three stages of the 18 August 1783 meteor observed from the terrace of Windsor Castle. The third stage in all the Sandby works closely resembles the one illustrated by Pigott [Fig. 31]. The rare, early aquatint [Fig. 34], which Paul published in October of that year, was rendered after Thomas's observations and possibly after Thomas's studies, according to its inscription: *To Sir Joseph Bankes, President of the Royal Society, London. This Plate from Motives of Respect and Esteem, is inscribed by/his most obed.t humble Servts/Tho.s & Paul Sandby.* This inscription testifies to the collaborative nature of the print and the prominence which Paul accorded Thomas. The longer inscription in smaller letters below echoes the observations and drawings of Cavallo: *The* Meteor *of Aug. 18, 1783, as it appeared from the N.E. Corner of the Terrace, at Windsor Castle. 18. Min. after 9 in the Evening it's appeared Diameter was nearly*

Figure 34. Paul Sandby based on Thomas Sandby, *The Meteor of 18 August 1783 from the East Angle of the North Terrace of Windsor Castle*, 1783, aquatint, 272 × 493 mm (image), The British Museum, London.

equal to the Semidiameter of the Moon, but its light much more vivid, its greatest Altitude was 25 Deg./A. It's appearance soon after it emerged from a Cloud, in the N.W. by W. where it was first discovered. B. *It's further Progress when it grew more oblong.* C. *When it divided & formed a long train of small luminous bodies, each having/a tail: in this form it continued till it disappeared from the interposition of a Cloud in E. by S.* Its three stages are correspondingly faintly lettered inside the meteor: *A*, *B*, and *C*.[56]

Sandby's aquatint captures the strange light cast by the meteor and also conveys its motion in three stages, like montaged stills taken by a movie camera. When combined these stages create a progressive movement similar to contemporary reports. The work's wonderful light effects were achieved by the artist's technical innovations in this medium, whereby he brushed a liquid resin onto the plate rather like a watercolor wash to produce subtle qualities of tone and light. The print, one of a group of aquatints of Windsor Castle by Paul Sandby, is really a permanent historical

Figure 35. Thomas (?) Sandby, *The Meteor of 18 August 1783 from the East Angle of the North Terrace of Windsor Castle*, 1783, watercolor over graphite, 282 × 493 mm, The Royal Library, Windsor Castle.

record of a riveting contemporary event. The fact that the observations were made in the presence of natural philosophers, including Cavallo, and dedicated to Banks, President of the Royal Society as well as a patron of Paul Sandby, underlines the commerce between artists and scientists at the time.[57]

A watercolor that has been variously attributed to both Paul and Thomas Sandby and with nearly the same scene as the aquatint is in the Royal Library, Windsor Castle [Fig. 35].[58] It contains some slight but telling differences; for example, all the figures are positioned further to the right near the wall. In addition, they are arranged in a line rather than in a group of five with the sixth to the right. Moreover, the cloud formations are not as fully realized as in the aquatint, where they give a domical effect to the cloud break.

The British Museum owns two other watercolors related to the aquatint, which again have been attributed to both Sandbys. The first one has a depiction of Windsor Castle terrace with the meteor sputtering above; it is executed in broadly applied blue, gray, light green, and brown watercolor [Fig. 36].[59] Here, the light issues from the right instead of from the left, as in the print, and there is no shadow on the bench at the left. While it lacks the figures, this

Figure 36. Thomas (?) Sandby, *The Meteor of 18 August 1783 from the East Angle of the North Terrace of Windsor Castle*, 1783, watercolor over traces of graphite, 285 × 461 mm, The British Museum, London.

sheet features the three stages of the meteor, the landscape, and the clouds, all of which appear close to the aquatint. In fact, the delineation of the third stage of the meteor seems slightly closer to the way it is treated in the aquatint than the Windsor watercolor. The second British Museum watercolor contains a depiction of the terrace with the observers divided into a group of five with a sixth figure removed and with the light from the left casting shadows, as in the print. However, its sky and landscape, otherwise deftly executed in broad washes applied with the brush in browns and greys, are unique in featuring blank expanses of paper instead of the celestial display and landscape [Fig. 37].[60] This example is eerily surreal, because the leftmost figure points upwards to a void.

The two watercolors in the British Museum were obviously intended to study individual elements of the Sandbys' final composition and were combined in the final print. The figures and the terrace with its shadows were adapted from the latter [Fig. 37], while the sky and the landscape were based on the former [Fig. 36].

A more quickly executed drawing – with figures and landscape – differs in some respects from both the aquatint and the other watercolors [Fig. 38 color].[61] It lacks the separation of the one

Figure 37. Thomas (?) Sandby, *Studies of figures for aquatint "The Meteor of 18 August 1783 from the East Angle of the North Terrace of Windsor Castle,"* 1783 watercolor over traces of graphite, 321 × 510 mm, The British Museum, London.

figure from the group of six (like the Windsor sheet) as well as the detailed treatment of the sky and the subtle modeling and coloration of the meteor, especially in its third stage (although there are some globular forms delineated in graphite). This work is signed or inscribed at the lower right: *P.l Sandby*. Its verso carries the inscription: *East end of the Terrace/With the Meteor*, together with two sketches in graphite at the upper right. The top one consists of the head and linearly striated tail of a meteor-like form, while the lower one looks very much like the outline of a tadpole with a wiggling tail. Both of these sketches represent the artist's attempt to render the meteor. On its recto, the three stages of the apparition lack much interior definition, as though the artist left the paper blank to suggest that he was perhaps consulting another version. This treatment certainly underlines the fact that the sheet was executed quickly and perhaps left unfinished. The lack of detail is especially apparent in the third stage of the meteor, which in the other watercolors has pink tints, just as in the Royal Astronomical Society watercolor [Fig. 33] and reports by observers. Further, the Yale example does not have the staccato accents of the trailing projectiles that had broken off from the main body. Because the figures are not divided five to one, as in the print and the British Museum study, the Yale watercolor would seem to have been executed prior to the two in the British Museum, in which the more or less final configurations were reached.

Since we know that Thomas witnessed the meteor's fall, it would seem that he was responsible for reporting the event to Paul, who credits his brother as the originator of the print's composition in its inscription. Nevertheless, the question of the attributions of these works remains slightly problematic. In light of what we know about the genesis of the aquatint, this work should be viewed as a collaboration, not only between the two brothers but also among the individuals involved with science and art who witnessed the apparition.

Yet another individual, Charles Blagden, reported synoptically in the *Philosophical Transactions* on this sensational meteor and included a reference to Sandby's "drawing," but to which of the watercolors we do not know. Blagden, who was Secretary of the Royal Society, groups his report, sent from Edinburgh, with observations he made of the fireball that appeared on 4 October 1784. He attempts to synthesize his ideas in a lengthy paper that is for the time seemingly exhaustive and objective. It combines his own observations with those of others. He remarks that: "These meteors were of the kind known to the ancients by the names of... Bolide...from peculiar differences in their shape...and sometimes, I believe, under the general term of Comets."[62] His ambitious summary of this fireball agrees with the other written reports as well as with the visual data. He pinpoints the event to 9:16 mean time of the meridian in London and plots the meteor's course from Edinburgh onto the Continent from reported observations. But most importantly for this study, Blagden cites the "beautiful Drawing" by "Mr. Sandby," which he notes was since engraved (and then he also refers to plate IV, Cavallo's print – Fig. 30).[63]

Blagden's description informs us how contemporaries read the Sandby works. He comments that the August fireball was described by most spectators as three different forms, "and is thus represented by Mr. Sandby." He continues:

> but the two first of those do not imply any real variation in its shape, depending only on a difference in the point of view.

> Accordingly, in the first part of its course over Scotland, it was seen to have a tail, when the object is seen nearly in front; and, secondly, that the light of most part of the tail is so inferior a kind, as to be difficultly perceived at a great distance, especially when the eye is dazzled by the over-powering brilliancy of the body. The length and shape of the tail, however, were perpetually varying; nor did the body continue always of the same magnitude and figure, but was sometimes round, at other times elliptical, with a blunt or pointed protuberance behind....[64]

Subsequently, Blagden discusses the fireball's bright light, color, accompanying noises, like the hissing sound, in addition to sparks, height, size, analogies with electricity, and the impossibility of determining its velocity.

Blagden mentions that he believes Herschel (William, a member of the Royal Society who did not publish a paper on this meteor but observed it at Windsor) kept it in view longer than anyone else – 40 or 45 seconds. In fact, Herschel watched the bolide from his observatory one and a half miles to the northeast of Windsor Castle at Datchet, where he was established as the King's Astronomer. Since it was his practice to observe every hour of each clear night, the meteor did not escape Herschel's attention as it traversed an arc of 167°.[65] From Blagden's repeated references to Halley's earlier writings, it is apparent that the definitive nature of meteors was yet to be established and that they were still associated with comets as well as with the aurora borealis.

The importance of the observations and visual representations of the meteor of 18 August 1783 is apparent with Alexander Stewart Herschel's 1864 article on the progress of meteoric astronomy. In it, he not only cites "Sanby's" [sic Sandby's] and Robinson's prints but also reproduces them as figures 1 and 2.[66]

The meteor of 18 August 1783 must have created a great awareness of celestial phenomena. Subsequently, a bright meteor of shorter duration appeared on 4 October of 1783, followed by a comet five weeks later. Then, Edward Pigott wrote to Maskelyne on 11 December 1783 to report a comet he discovered on 19 November and observed until 6 December.[67] It is comet D/1783 W1, periodic (and defunct) Comet Pigott.[68] He read his paper about it to the Royal Society on 24 June 1784, the same day that Nathaniel Pigott read his on the August 18th meteor. Evident in

Edward Pigott's report is the luxuriously calm nature of cometary observation in contrast to the hurried observations of the earlier fleeting meteor.

The configuration of the third stage of the meteor of 18 August 1783 – after it had formed multiple projectiles – corresponds to a celestial apparition painted on a beautiful canvas, whose attribution has been questioned [Fig. 39 color].[69] Moreover, the specifically rendered opening of the clouds resembles the precise meteorology depicted in the Sandby watercolors and print. Traditionally the painting has been identified as depicting the apparition of Halley's Comet in 1759.[70] The event takes place on the River Thames with Westminster Abbey, the spire of St. Margaret's Church, Westminster Hall, and the trees of Cotton and Speakers Gardens in the left background. Old Westminster Bridge is in the right middle ground. The unusually dark nocturnal scene, a quality accentuated by the discoloring of its varnishes, features a horizontally disposed celestial body with a large head trailed by little staccato bits, exactly the configuration depicted by the Sandbys as the third stage of the August 18th meteor. In our inspection of the canvas, we discerned beneath the varnish reddish pink, yellow, and shaded areas in the tail's smaller globules as well as in the left part of the head. This information nearly clinches the identification of the apparition as the sensational meteor of 18 August 1783. It also suggests that the artist may have known at least one of the Sandby works. Given the dazzling passages of fireballs across the sky in seconds, this explanation would also explain the amazement registered by the majority of the observers in the foreground. This identification, given its date, would cancel automatically the traditional attribution of the canvas to Samuel Scott (c. 1702-1772), known as "the Canaletto of England."[71]

This conclusion supports part of an earlier theory postulated by Schmidt, who had doubted the traditional identification of the apparition in the canvas. Instead of representing Halley's Comet in 1759, he had suggested that it records a brilliant fireball, such as those of 11 December 1742 (he probably meant the meteor of 11 December 1741 – see Appendix I) or 26 November 1758.[72]

We also tackled the question of the painting's date in reference to the internal evidence presented by the costumes and buildings.

The costumes worn by the figures seem to point to a date in the latter part of the century, although the artist did not, in fact, focus on clothing details.[73] The buildings also help to date the canvas because its artist depicted both Hawksmoor towers of Westminster Abbey completed; thus it must postdate 1744. Moreover, Old Westminster Bridge – designed by the Swiss artist Charles Labelye, who was a pupil of Demainbray, the disciple of Desaguliers – was begun in 1738-39 and opened on 18 November 1750. Its inclusion in its completed form provides a *terminus post quem* for the painting. It thus rules out Schmidt's earlier meteor of 1742 [sic 1741], but does not help in deciding whether the artist represented Halley's Comet of 1759 or the meteor of 18 August 1783. Even though Scott may not have executed this canvas, he painted many similar views that include Old Westminster Bridge as well as Westminster Abbey. One of these features the barge of the Lord Mayor of London which seems to be the same vessel depicted in figure 39.[74] Another painting attributed to Scott with a nearly identical view and with the same diagonal railing in the foreground helps us to establish the location of the right foreground as Lambeth, although it does not help us attribute this canvas with the spectacular fireball.[75]

In summary, both visual evidence and written reports suggest that the artist of figure 39 recorded the meteor of 18 August 1783. One argument that no longer seems valid was that the celestial phenomenon seems more likely to be a meteor rather than a comet, since comets' tails nearly always point away from the sun and its tail has a horizontal orientation. In nocturnal scenes such as this, the sun is below the horizon, so that it might seem that the tail should point generally upwards if it were a comet. However, the dust tail of the recent Comet Hale-Bopp also appeared horizontal when it first became visible in the evening twilight. This alignment turns out not to be a problem, because it is the gas tail, when visible, that always points away from the sun. The gas tail can be at some angle to the dust tail. For Hale-Bopp, in particular, since the comet was at a different compass angle from that of the sun right after sunset, the gas tail was at the same angle to the horizontal, and the dust tail, further over yet, was horizontal. In the course of the night the angle of those tails to the horizon rotated as the sky turned around the poles. By morning, Hale-Bopp's dust tail was nearly vertical. Returning to the eighteenth-century canvas, it is also significant that there is a bright reflection of the light on the water, and comets are usually too faint for such an extensive reflection, although exceptions exist [Fig. 112]. A fireball, however, can be 10,000 times brighter than a comet, making such a reflection more likely. Thus, the unique characteristics of the celestial phenomenon of this canvas argue that it was intended to depict the fireball of 18 August 1783. The only other explanation possible would be that the artist based his depiction of whatever event he intended on his observation of the August 18th meteor or another representation of it. But the nature of the late eighteenth-century attitudes renders this hypothesis unlikely. In conclusion, we believe, therefore, that this canvas depicts the third stage of the brilliant fireball of 18 August 1783, one of the most riveting astronomical events of the eighteenth century.

Other Links Between Art and Astronomy

While we know of no comets and meteors by the artist John Russell (1745-1806), he made significant contributions to British art and science as both a fashionable portraitist and as an astronomical draftsman. His early mapping of the moon is considered by some to be the best ever accomplished. Although Russell was trained as an artist, he made a contribution to astronomy through the very skills that made him a good portraitist. For his observations he used a six-foot reflector with a six-inch mirror made by William Herschel, together with a Dollond instrument given him by the sculptor John Bacon. His selenography included pastels and prints of the moon's surface as well as the largest and most accurate lunar relief globe at that time. In addition, Russell painted a portrait of William Herschel, showing the astronomer holding a scroll inscribed with a diagram of the universe, replete with the planet he discovered – Uranus – as well as a portrait of Banks. Russell also painted other members of the Royal Society – for example, Maskelyne – as well as Edward Topham, publicist of the Wold Cottage meteorite, and his work was encouraged by the scientific community. Thus, Russell is yet another example of a British Enlightenment artist-scientist working with astronomical material.[76]

British Artistic Institutions

The singular nature of British art of our period and its numerous images of comets and meteors is the direct result of artistic freedom. This freedom resulted from the unique nature of the English Royal Academy and the loose system that evolved for exhibiting works of art. While the roots of the system lie in the eighteenth century, its effects continue through the time span considered in this study.

The unusual nature of the English Royal Academy resulted partially from a fiercely independent system, one to be expected from an insular culture, which encouraged artistic self-sufficiency. It provided the proper matrix for the maturation of the type of individuals who produced comet/meteor imagery. Free of the heavy demands of state patronage and control, British artists enjoyed the curses and blessings of this freedom. The Royal Academy sprang from a culture where no guild system was entrenched, already predisposing British artists to a liberty not enjoyed by artists in other European countries. Despite its name, the Royal Academy was a private institution, as opposed to the Italian academies, born in the frenzy of court *Maniera*, and the French Academy, where the traditional system of training and patronage reached its apogee. While the continental academies were the products of absolutism and controlled by a royal patronage, the origins of the Royal Academy were only loosely connected with a governmental initiative.[77] Although it was individual artists who first organized academies, the desire for a collective national academy was present long before its actual establishment on 10 December 1768, as the Royal Academy of Arts in London, for the Purpose of Cultivating and Improving the Arts of Painting, Sculpture, and Architecture.[78]

Although the Academy brought benefits to artists, there were also many disputes due to its rigid hierarchy of genres culminating with that of history painting.[79] During the first five decades of the nineteenth century, attempts were made to establish alternative societies, where artists specializing in genres accorded a lower status by the Academy could exhibit. These societies represented many sections of the profession, complemented the Academy, and courted public enthusiasm.[80] It was here where works without historical subjects containing comets and meteors could be exhibited, such as a watercolor by Samuel Palmer (see chapter V, n. 23). The British affinity for the medium of watercolor found expression in a society for watercolor painters formed in London in 1804, which is still active today. Its foundation marked the increasing differentiation of watercolor painting as a specific British area of expertise and the growing specialization of its practitioners. Except for Joseph Mallord William Turner, most of the distinguished British watercolor painters subsequently exhibited there. This body was known by several names, but it is commonly known as "The Old Water-Colour Society" to distinguish it from "The New Society of Painters in Water-Colours," founded in 1832.[81]

Several other societies encouraged the development of highly individual artists. One venue for works of art was the British Institution for Promoting the Fine Arts in the United Kingdom, founded 4 June 1805 (opened 18 June 1806).[82] Another was the Society of British Artists (1823) whose exhibitions, like those of the Academy and the British Institution, became part of the social season. It also de-emphasized history painting to favor landscape and genre pictures, subjects that appealed to a more socially diverse art public at a time of changing taste as well as of a broadening of the social basis of the art market.[83] These and other organizations became forums for the display and sale of works of art with unusual subject matter, like those including comets and meteors.

The Comets and Meteors of William Blake

William Blake (1757-1827), poet and painter as well as a member of the Society of Painters in Water-colours, was captivated by comet/meteor symbolism. In his visual oeuvre, we find at least twenty works with comets, meteors, or shooting stars. Blake, himself a mystic of great complexity and singular thought, was an artist with a keen interest in literature and art of the past, especially the looming, philosophical works of Michelangelo. He was not only an assiduous reader of English poetry from childhood, but also a literary artist who uniquely combined poetry and painting and who viewed the faculty of the imagination as a divine gift serving to equate God and man. In some ways, Blake, who was a transitional figure on the cusp between centuries, stands outside history. In other ways, he belongs to a trend reacting against the rationalism of the Enlightenment that began in the eighteenth century and developed with increasing rapidity in the nineteenth century, as shown in his highly imaginative images of comets and meteors.

At the time when comets and meteors were being demystified by the scientific community, these mesmerizing vagabonds of the solar system were transformed into popular symbols that haunted the imagination, a universe more real to many Romantics than physical reality. Artists and writers wishing to escape sordid reality frequently represented comets that did not refer to specific apparitions. Often these nonspecific comets were part of an existing literary and/or artistic tradition and thus symptomatic of the age's historicism. Blake, an innovative and subjective illustrator, often grafted his own highly original ideas onto those of the authors he illustrated. His unorthodox yet eclectic beliefs were personal and radical. In addition to his drawings, prints, and paintings, his ideas also find expression in his "illuminated books," as he termed his illustrated writings.[84]

To William Blake's beloved youngest brother Robert (1767-1787) – an artist who entered the Royal Academy Schools in 1782 and died around 20 years of age – has been attributed a drawing with a generic representation of a shooting star that may have been influenced by the spectacular meteor of 18 August 1783. The sheet is dated to c. 1785 and assuredly contains the partial image of a comet or meteor at the right [Fig. 40].[85] While a long tail with jagged edges is still visible, it seems that the sheet has been cropped at the right. A group of figures at the left observe the event with great fear; behind them a wall of trees opens where the comet/meteor streaks across the backdrop of the starry sky.

It has been suggested that the mysterious, amorphous white light in William Blake's early relief and white-line etching [Fig. 41] – roughly based on Robert's design – represents the bolide of 18 August 1783 and its luminous train.[86] It is unlikely, however, that William Blake's print, known in only one impression, was overtly intended to represent the meteor because his print has been dated on stylistic grounds to around 1792, nearly a decade after its appearance. This is not to say categorically that the print was not generally inspired either by a recollection of a celestial event or by the work attributed to Robert. In his print, William

has included neither a landscape setting nor a starry sky nor the telltale tail of a meteor, but only a glowing illumination to help evoke a universal terror of the unknown.

Figure 40. Robert Blake, *The Approach of Doom* (or *The Meteor of 18 August 1783*), pen and black ink with brown and gray wash over traces of graphite and black chalk, 335 × 475 mm, The British Museum, London.

Certainly William Blake was aware of the scientific ferment about comets and meteors. It was fitting that in the very year (1794) when Chladni published his treatise that makes a convincing case for the cosmic origin of meteors,[87] Blake published his famous *Songs of Experience*, whose memorable poem "The Tyger" has in its fifth stanza the following provocative lines:

> When the stars threw down their spears
> And water'd heaven with their tears....

Blake's images associating stars with spears and tears are related directly to some notable meteors that occurred in the poet's lifetime and, specifically, to a legend concerning the Perseid meteor showers. This famous shower, which still appears annually around 12 August, was known at least since A.D. 36, when it was

Figure 41. William Blake, *The Approach of Doom*, c. 1792, relief and white-line etching, 297 × 209 mm, The British Museum, London.

recorded by the Chinese. The shower seems to have exhibited a lively activity in the late eighteenth century, probably at least in part due to the increase in western astronomical expertise and to the fact that meteor showers can be more noticeable because of various atmospheric conditions and more active because of the earth-crossing clusters of comet dust in space. In fact, Blake may have seen one of the spectacular displays in the years just prior to the publication of "The Tyger."

The meteor shower was given the name "Perseids" in 1866, when Giovanni Schiaparelli named them after the constellation Perseus, from which they seem to issue. Around the same time, Schiaparelli theorized that the showers were associated with a periodic comet, P/Swift-Tuttle (109P/1862 O1, formerly 1862 III) and published his findings in 1873.[88] In 1794 they were known by their traditional name, the "tears of St. Lawrence," because they appeared near that saint's feast day. They were also known as "the

Figure 42. William Blake, *Edward III as the Scourge*, illustration to Gray's "The Bard. A Pindaric Ode," 1797, watercolor with pen and black ink over graphite, 420 × 326 mm, Yale Center for British Art, Paul Mellon Collection, New Haven.

Figure 43. William Blake, *Fame*, Title page of "Ode for Music," from *The Poems of Thomas Gray*, 1797-8, watercolor with pen and black ink over graphite, 420 × 327 mm, Yale Center for British Art, Paul Mellon Collection, New Haven.

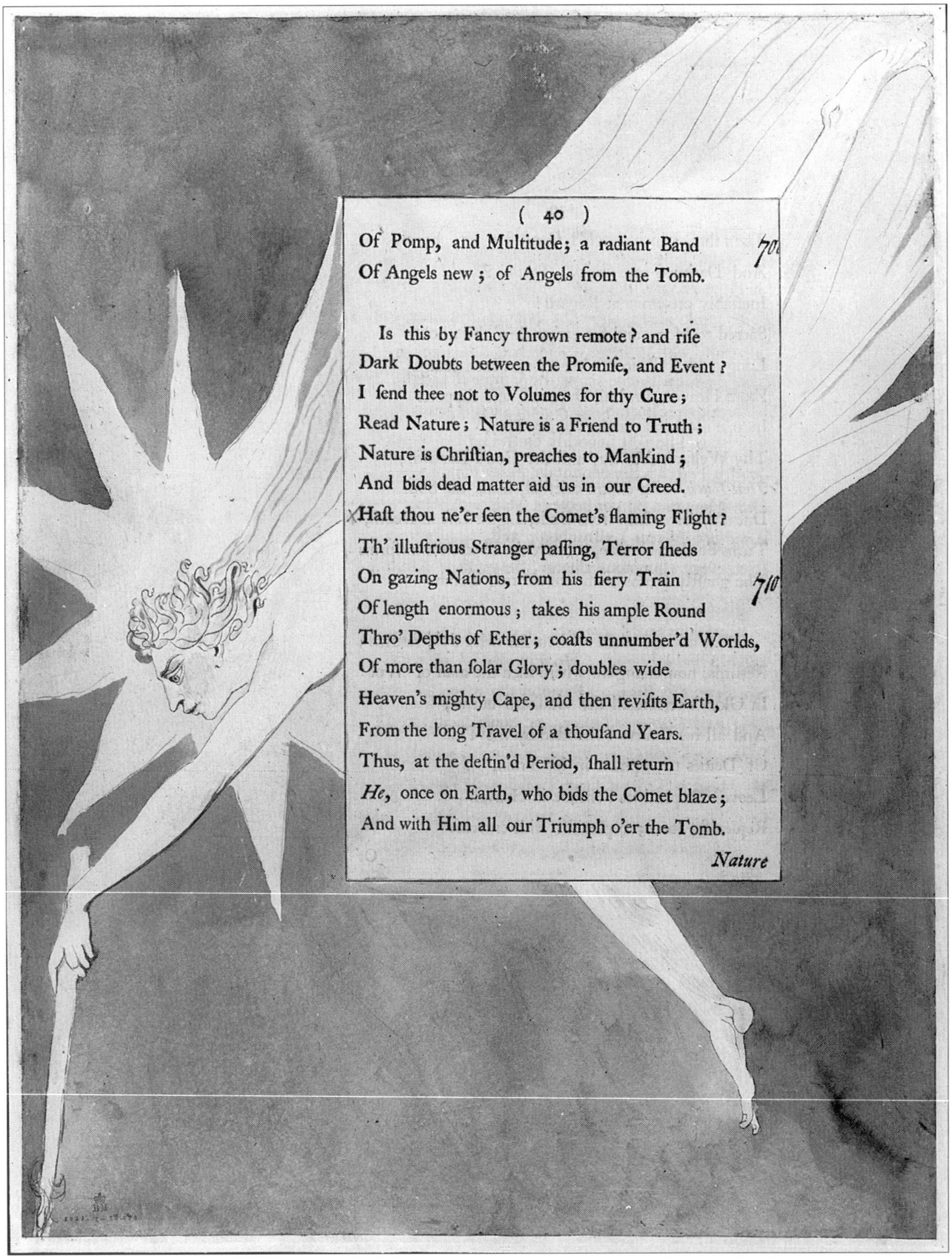

Figure 44. William Blake, *Illustration for Young's "Night Thoughts," 149, Night IV, page 40*, 1795-97, watercolor with pen and black ink over graphite, 412 × 320 mm, The British Museum, London.

Figure 45. William Blake, *Illustration for Young's "Night Thoughts," 313, Night VII, page 41*, 1795-97, watercolor with pen and black ink over graphite, 412 × 318 mm, The British Museum, London.

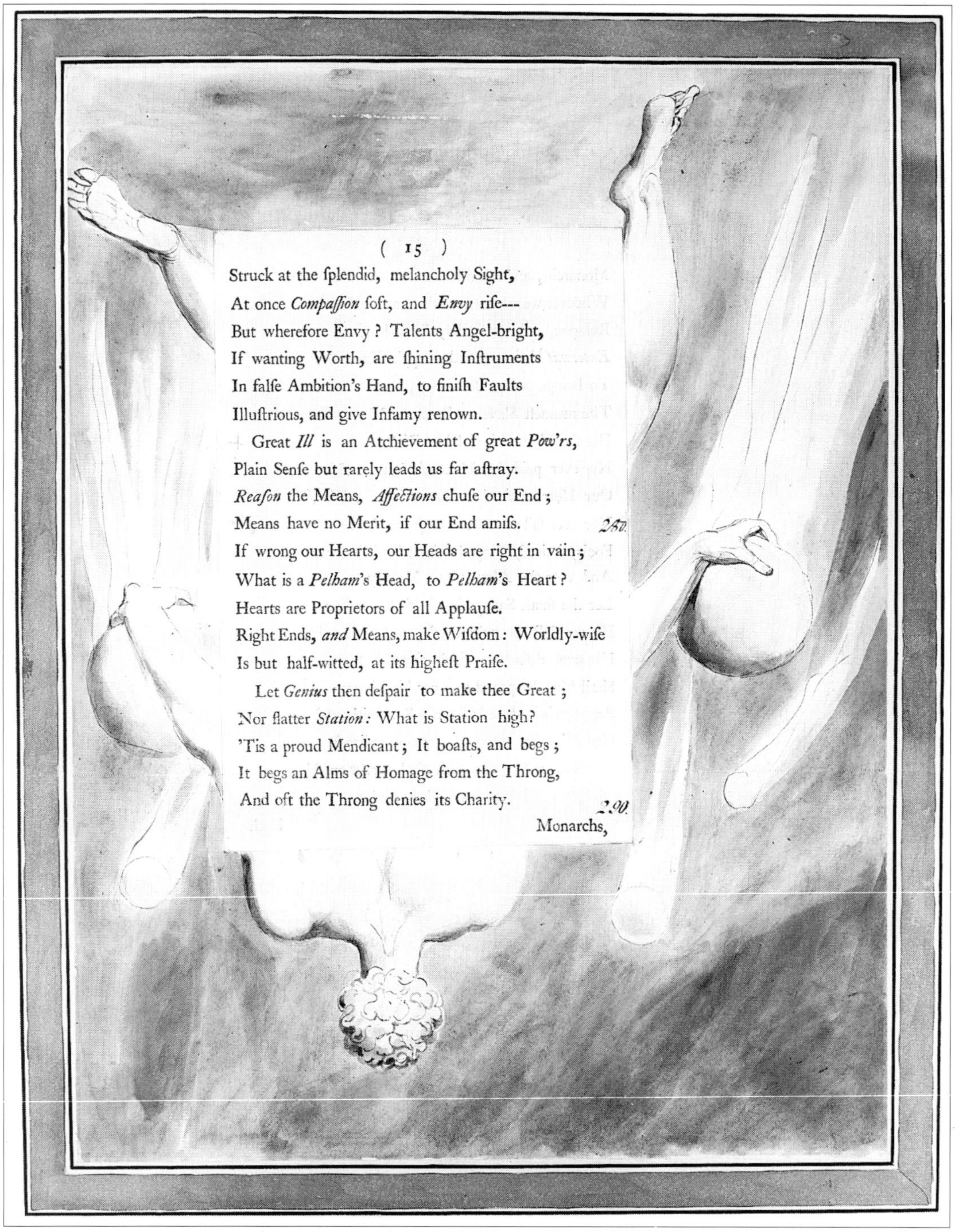

Figure 46. William Blake, *Illustration for Young's "Night Thoughts," 236, Night VI, page 15*, 1795-97, watercolor with pen and black ink over graphite, 415 × 320 mm, The British Museum, London.

Figure 47. William Blake, *Illustration for Young's "Night Thoughts," 320, Night VII, page 48*, 1795-97, watercolor with pen and black ink over graphite, 408 × 317 mm, The British Museum, London.

Figure 48. William Blake, *Illustration for Young's "Night Thoughts," 443, Night IX, page 25*, watercolor with pen and black ink over graphite, 418 × 328 mm, The British Museum, London.

Figure 49. William Blake, *Illustration for Young's "Night Thoughts," 463, Night IX, page 45*, 1795-97, watercolor with pen and black ink over graphite, 415 × 325 mm, The British Museum, London.

Figure 50. William Blake, *Illustration for Young's "Night Thoughts," 504, Night IX, page 86*, 1795-97, watercolor with pen and black ink over graphite, 410 × 318 mm, The British Museum, London.

fiery tears," because St. Lawrence was martyred by being roasted alive on a grill. One can thus see where Blake derived one of his poetic images in "The Tyger." The other he drew from the tradition going back to Classical Antiquity, wherein comets and meteors were described as swords or spears. Scientists were just realizing that meteors are not terrestrial phenomena, as first suggested by the Greeks (as are lightning, St. Elmo's fire, and will-o'-the wisp), but rather that they are cosmic objects entering the earth's atmosphere.[89] And Blake responded to these topical phenomena of his time in his unique, imaginative manner.

His fascination with evocative meteor images is revealed in other works. Around 1797, he illustrated the poems of Thomas Gray, providing 116 watercolors for this unfinished project which are now in the Yale Center for British Art.[90] Blake's seventh illustration to Gray's poem "The Bard. A Pindaric Ode," which depicts Edward III, avenger of his father Edward II, as the scourge [Fig. 42] utilizes Gray's description:

> Robed in the sable garb of woe,
> With haggard eyes the Poet stood;
> (Loose his beard and hoary hair
> Stream'd, like meteors, to the troubled air).

In his illustration, Blake transferred the meteor image from the hair of the bard in the poem to Edward's whip, whose three thongs terminate in star-like spikes resembling meteors, which have long been associated with a display of wrath.[91] Blake's watercolor title page of Gray's "Ode for Music" [Fig. 43] features a winged figure of Fame, whose trumpets also suggest a personification of music.[92] The floating figure is seemingly crowned by a star to imply her "higher," celestial qualities, while the rectangle bearing the text, functions as a banner. Behind her, the night sky is sprinkled with stars, including a comet or a shooting star at the lower left. It was probably intended as a metaphorical comment on ascendant, unpredictable Fame – swift, brilliant, and perhaps fleeting. Below, Blake drew the tops of buildings to simulate the University of Cambridge, where Blake had never been. Keynes believes that several of the other structures, like the pyramid, are clearly "mathematical forms," inspired by the thought of Newton, who had formerly been in residence in Cambridge.[93]

Immediately before the Gray project, Blake had been commissioned in 1795 by the London bookseller and publisher Richard Edwards to illustrate Edward Young's "Night Thoughts." For two years he worked on the 537, vaguely "neoclassical," watercolor illustrations, many of which feature comets or meteors. He had intended to engrave 200 plates for the ill-fated project but only 43 of the designs were engraved.[94]

In one of Blake's most powerful images, the artist has personified the comet of the poem as a plunging androgynous being holding a sharp object [Fig. 44].[95] His image visualizes the following passage lettered in the central rectangle:

> Hast thou ne'er seen the Comet's flaming Flight?
> Th' illustrious Stranger passing, Terror sheds
> On gazing Nations, from his fiery Train
> Of length enormous; takes his ample Round
> Thro' Depths of Ether; coasts unnumber'd Worlds,
> Of more than solar Glory; doubles wide
> Heaven's mighty Cape, and then revists Earth,
> From the long Travel of a thousand Years.
> Thus, at the destin'd Period, shall return
> *He*, once on Earth, who bids the Comet blaze;
> And with Him all our Triumph o'er the Tomb.

The passage reveals Young's awareness of periodic comets and the basically Christian tone of his thought. Blake's diagonal, threatening comet with its eight-pointed star-like head suggests that he may have been aware of the stylized comet in James Sayers's political satire of 1789, where the comet's direction is reversed [Fig. 69].

In "Night Thoughts," Young refers to comets five times. However, in the apocalyptic "Night Thoughts" 158 and 313, although the poet does not mention a comet or meteor, Blake the illustrator has drawn them. In the first, meteor showers symbolize death, destruction, and mortality.[96] In the second, Blake drew two meteors accompanying a tormented figure, symbolizing the life of the soul, who is self-destructing like a burning meteor plummeting toward annihilation past two shooting stars, whose more graceful trajectories contrast with his desperate, lethal fall [Fig. 45].[97] In this illustration, Blake relied on the tradition that meteors, meteor showers, and comets are linked with the Apocalypse. These ideas are powerfully related in the book of Revelation and visualized in illustrations of it.[98] Moreover, in the third verse of the Epistle of Jude, again an apocalyptic context, there is a reference to wandering stars, a euphemism in this case for comets, which seems to inform Blake's image: "wandering stars for whom is reserved the blackness of darkness for ever." Some of the ideas of the latter illustration seem to be repeated less explicitly in the eschatological "Night Thoughts" 236 [Fig. 46].[99]

Blake's remaining comet and meteor images in "Night Thoughts," have a decidedly different feeling. "Night Thoughts" number 320, wherein a babe is depicted in a star-like form with a zigzag tail [Fig. 47], engenders thoughts of creation and immortality.[100] A marked change of mood occurs with number 443, which corresponds with Young's idea that comets can be positive omens: "*Comets* good omens are, when duly scann'd...." Blake's illustration for this passage shows sage-like figures contemplating the huge more realistic comet, whose head nevertheless consists of seven points of a stylized star-like form [Fig. 48].[101] This particular illustration may reflect the earlier works of the brothers Blake featuring a similar cluster of figures [Figs. 40, 41]. Blake drew several shooting stars in number 463 [Fig. 49].[102] Like many of the private symbols in the artist's oeuvre, they are not found in Young's text. However, Blake's final dominant comet in "Night Thoughts" 504 is discussed by Young [Fig. 50].[103]

> With the bold *Comet*, take my bolder Flight,
> Amid those *Sovereign* Glories of the Skies;
> Of independent, native Lustre proud,
> The Souls of Systems! and the Lords of Life,
> Thro' their wide Empires! – what behold I *now*?
> A Wilderness of Wonders burning round;
> Where *larger* Suns inhabit *higher* Spheres,
> Perhaps the *Villas* of descending Gods!

From Blake's early works, it is clear that he was not actively observing astronomical phenomena, but rather responding to a symbolic tradition in literary and visual arenas. He probably was stimulated by the general cultural and intellectual interest in comets and meteors and the great rise in amateur and professional observation. In fact, in "Night Thoughts" 509, Blake drew a large telescope to illustrate the line "O for a Telescope, His Throne to make!"[104] Blake, like Young, thus linked telescopes and astronomy to God.

Telescopes and Blake's and Sandby's Contemporaries: William and Caroline Herschel

The eighteenth century saw not only the fruits of Halley's work on comets but also telescopes suitable to view comets. Galileo's refracting telescopes, with lenses about five centimeters in diameter, were hard to direct and suffered from chromatic aberration. Newton struggled with this problem and erroneously concluded that the effect could not be overcome and that refracting telescopes – those based on the refraction (bending) of light by lenses – would always suffer from chromatic aberration. Newton thus went on to invent the first reflecting telescope, an instrument that used only mirrors.

Newton's first reflecting telescope, or an early replica thereof, is at the Royal Society in London.[105] It works on the principle that the entering light bounces off a curved mirror at the bottom of the telescope, and a flat, secondary mirror placed diagonally near the top bounces the returning beam of light off to the side. The mirror's focal length is 15 centimeters (6 inches). Other schemes of reflecting telescopes, which used curved secondary mirrors to bounce the returning beam back through a hole in the center of the primary mirror, were invented by James Gregory in Scotland in 1663 and by Guillaume Cassegrain in France in 1672. But these designs were not widely used for decades. Only with the presentation of a telescope to the Royal Society by John Hadley in 1722 and the mass production of reflecting telescopes by James Short in Edinburgh, starting around 1740, did these telescopes become readily available.[106]

Sir William Herschel's own improved telescopes were an aid to him and his sister Caroline (the first female comet-hunter known). They, together with Charles Messier in France, known as "the comet-ferret" after Louis XV dubbed him that in recognition of the fifteen or more comets he discovered, were among the most assiduous observational astronomers. (Messier compiled a catalogue of nebulae in order that such objects would not be assumed to be comets.) And, observation to them meant a certain amount of comet-hunting. Extraordinary progress in the fields of optics and physics also aided in the sighting of at least sixty-two comets during the eighteenth century. During its last quarter, one comet was spotted each year. Consequently, the interest in celestial observation became so widespread that by the year 1800, observatories began to multiply all over Europe.

William Herschel (1738-1822), who was the most famous English scientist of the late eighteenth century, was an outstanding and prolific maker of telescopes. For his telescopes, which were widely distributed, he ground over 400 mirrors. Herschel's largest telescopes used mirrors with diameters of 48 centimeters (19 inches) and 122 centimeters (48 inches). Each was mounted on a wooden framework that could rotate, though motion up and down in the sky (altitude) was easier than the rotation (azimuth). This type of mounting is known as altazimuth, a type that is now coming back into vogue.[107] These telescopes were mounted at Herschel's home in Slough near Windsor Castle. Herschel used the 20-foot [focal length] telescope starting in 1783, the year of the spectacular meteor of 18 August, to sweep the sky in search of nebulae. His 40-foot telescope, which was completed in 1789 at Royal expense, was not as successful, in large part because it was so unwieldy.[108]

Herschel, who was born Friedrich Wilhelm Herschel in Hannover, Germany, was the first astronomer in a family that eventually included some of the most famous English astronomers of the eighteenth and nineteenth centuries. He came to England in 1757 as a musician and an amateur astronomer, and nine years later settled in Bath. On 13 March 1781, he spotted an unusual object in his telescope, which he thought might be a comet. He described the object to the Royal Society under the title "Account of a Comet." His letter begins:

> Sir – On Tuesday Evening the 13 March about half after 10 o'clock I directed my Telescope to one of the four small stars that form a quartile between the right feet of Auriga and the left foot of Castor, and discovered near it a star of an appearance very different from the rest. With a paler sort of light it had much more diameter than I thought it should have. At first I suspected some disorder in my telescope; but finding it in very good order I began to surmise that this star might be a Comet. However, as this opinion only struck me as a mere possibility I thought it sufficient to mark its place by the following estimations.[109]

By the meeting of the Royal Society at which his paper was read, Maskelyne suspected that the object might be a planet and by the end of 1781 further observations clinched that identification.

Other astronomers determined its orbit and showed that it was actually a major planet, the first to be discovered since antiquity. Herschel proposed the name *Georgium Sidus* (*Novum sidus Georgium*) to honor his patron, King George III, though the name Uranus ultimately gained worldwide favor.[110]

Herschel's original illustrated letter of his 19 March announcement of this "comet" contains several illustrations.[111] Essentially all the written material from the five-page letter, along with subsequent letters, was published in the *Philosophical Transactions*, accompanied by eight figures, although we have discovered that the manuscript contains additional unpublished figures. Only in it do we find (as the first, unnumbered figure) his sky map showing the position of the "comet" against the constellations.[112] At the top of the letter, he wrote that this drawing is based on Flamsteed's *Atlas Coelestis*.[113]

This discovery gained Herschel great fame and the favor and patronage of the King. In 1781, the Royal Society elected him a Fellow and gave him its Copley medal. When appointed the King's "personal astronomer," he was awarded a pension. (His sister Caroline, his assistant, was also given a stipend.) One of his duties was to show the sky to the Royal family. In order to be closer to the King, Herschel moved, as required, from Bath to a location near Windsor, eventually living in Slough, where he built the telescopes described above. The second telescope, with its 48-inch diameter mirror, was a reflecting telescope of a type now known as "Herschelian." In this type, the main mirror is slightly tilted in the telescope's tube so that the reflection could be viewed directly, without a secondary mirror. He eventually used the telescope to catalogue 2,500 nebulae published in 1785.[114] Herschel was particularly interested in whether all the nebulae were actually made up of stars, so he built bigger telescopes to try to resolve individual stars of the nebulae. By 1790, though, observations brought him to admit that at least some of the nebulae did not break up entirely into stars. This startling discovery showed him that there was a universe beyond our own galaxy, something that was not proven until the twentieth century.

Another Herschel letter in the Archives of the Royal Society contains a drawing with the position of the comet of 1795 (2P/1795 V1) against a constellation background from 7 through

10 November, including Cygnus and Lyra [Fig. 10]. It appears in an unpublished manuscript from 1795.[115] In that year, Herschel thought that comets could replenish the sun and play a role in the habitability of planets, an idea regaining popularity as we approach the third millennium.[116]

For many years, William Herschel worked with his sister Caroline, his amanuensis. He left the discovery of comets to her, although he also observed and published accounts of them from 1787, including his major reports of the comets of 1807 and 1811 (see Appendix I).[117] His idea, discussed in the context of the latter comet, that comets have atmospheres of nebulous material about their heads, has been proven to be essentially correct.

When he was 50, William Herschel married the widow of a neighbor. It took some time before Caroline adjusted to her displacement from the center of his affections and to her move to private lodgings, but eventually, Caroline and William continued their work together. In 1792, when William was 54, he and his wife Mary, had their only child, John Frederick William Herschel, who succeeded his father as one of the major astronomers of England (see chapter III).[118]

Caroline Herschel (1750-1848) was William's younger sister. She actively participated in astronomical observing and discovered three nebulae, including one of the satellite galaxies of the Great Galaxy of Andromeda, as we now know it. When the Herschels moved to the vicinity of Windsor Castle, Caroline began to look for comets with a telescope her brother had built especially for her. She was also occupied with writing down her brother's observations. During the day, besides managing the household and entertaining visitors, she carried out the necessary extensive routine calculations, prepared catalogues and papers for publication, and even ground and polished mirrors.

Caroline is especially known for the eight comets she discovered on her own, which are reported in her observational books.[119] Five of them are reported in articles in the *Philosophical Transactions* (see Appendix I).[120] In her observations of the comet of 1807 (C/1807 R1) and the comet of 1811 (C/1811 F1 [Fig. 51], she drew them in a round telescopic field, a frequent practice of both Caroline and William.[121] She also revised Flamsteed's star catalogue published by the Royal Society. After

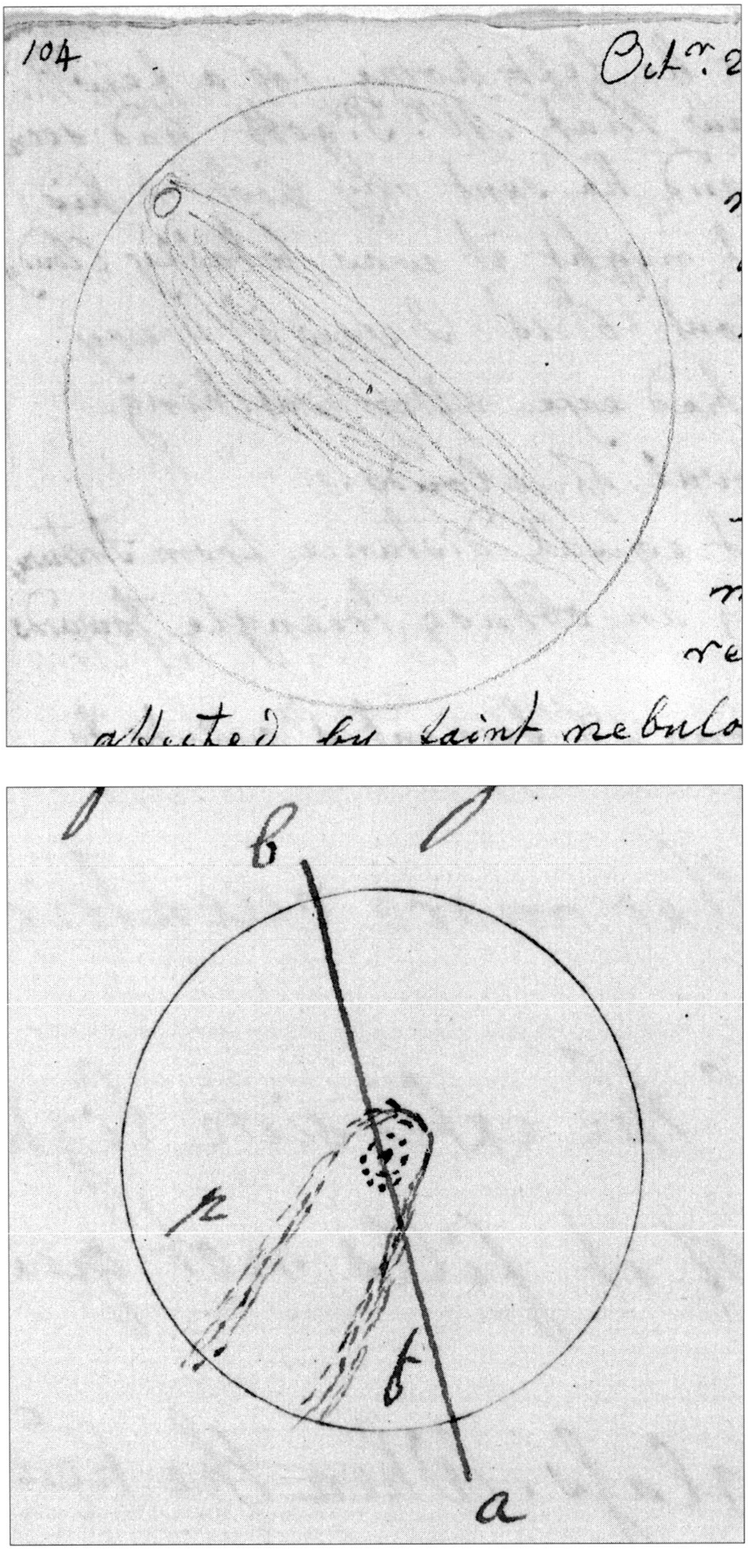

Figure 51. Caroline Herschel, *Observations on the Comets of 1807 (24 October) and 1811*, C1/1.3, details of pages 104-105, 1807 and 1811, graphite and pen and brown ink, 210 × 168 mm (each page), Herschel Archive, Royal Astronomical Society Library, London.

William's death, she received the gold medal of the Royal Astronomical Society, notably for her work on his star clusters.[122]

After Caroline Herschel's comet discovery in 1786, the Astronomer Royal, Maskelyne, wrote to her brother, "I hope that we shall by our united endeavours get this branch of astronomical business from the French, by seeing comets sooner and observing them later."[123] Maskelyne himself computed the orbits for the comets she observed. Though Caroline discovered several comets, the lead in discovering comets was to return in the next century to the French, with the work of Jean-Louis Pons.[124] The international rivalry between England and France would eventually reach a new height with the battle for the priority of the discovery of Neptune. For a time, the French even championed the restoration of Herschel's name for the planet Uranus so that they could use the name of the Frenchman Urbain Jean Joseph Leverrier for the planet Neptune in a parallel manner. It was not until over a century later that the name Uranus was universally adopted.[125]

With the discovery of Uranus, William Herschel had gained for Britain a place of honor in the lead of world astronomy. With his and Caroline's continued work on comets and on their mapping the sky with his ever-larger telescopes, the largest in the world, Herschel marked an important transition from a primary interest in the solar system to primary interest in the universe beyond. Indeed, today's star catalogues and a naming scheme for celestial objects are still based on Herschel's work.

Since the Herschels and Blake continued to work in the next century, chapter III will begin with a consideration of the comets and meteors that pepper the oeuvre of Blake after 1800. It will be followed by additional discussion of the Herschels' contributions in the early decades of the nineteenth century, which we call "the comet-crazed century."

Notes

1 Johnson, 1986, 16.

2 D. Spadafora, *The Idea of Progress in Eighteenth-Century Britain*, New Haven and London, 1990.

3 *Ibid.*, 28ff. Spadafora thinks that the Scots differed from the English because they did not believe in indefinite progress. North of the Tweed, the notion of progress began later but resulted in a more dramatic transformation from an almost medieval way of life. See also R. Mitchison, "Scotland 1750-1850," and T.C. Smout, "Scotland 1850-1950," in *The Cambridge Social History of Britain*, 1, ed. F.M.L. Thompson, Cambridge, 1990, 155-207, 209-280. Spadafora's study remains controversial; he takes sides in a methodological controversy in the history of ideas (the validity of writing the history of individual ideas, which has been accused of being textualist rather than contextualist). See also S. Shapin, *The Scientific Revolution*, Chicago and London, 1996.

4 Spadafora, 66ff. On the idea of progress, it is telling that the Irish artist James Barry's theme for his six large canvases, begun in 1777 for the Great Room of the Royal Society of the Arts, was the "Progress of Human Culture." See N.L. Pressly, *The Fuseli Circle in Rome. Early Romantic Art of the 1770s*, exh. cat., New Haven, 1979, 3. Although France may have been the focal point of the Enlightenment's criticism of the establishment, Spadafora views Britain as the eighteenth-century center of historical optimism. Historical pessimism, he believes, was stronger in France despite its revolution, whereas the English had a sense of living in a political environment that had successfully changed for the better without violence. He argues that a notion of progress developed independently in England and was strongest during the mid-Georgian period, whereas a fragile, almost pessimistic quality appeared in Victorian times. Certainly Romantic writers related to this idea – for example, Samuel Taylor Coleridge, Percy Bysshe Shelley, and Sir Walter Scott (for whom progress was the organizing principle of history and who was indebted to the Scottish Enlightenment and the Utilitarianism of John Stuart Mill). Spadafora thinks that the optimism of eighteenth-century Britain sparked a broader spirit of improvement and a belief in the value of hard work, contributing to the social and economic changes of the modern Britain. Not everyone was enthusiastic about these developments; for example, Jonathan Swift's *Gulliver's Travels* satirically disparaged institutions like the Royal Society via the Grand Academy of Lagado.

5 Inv. no. 1885 M2709 in the City Museums and Art Gallery, Birmingham, in jasper, white relief on blue, 92 × 71 mm. The four are reproduced in R. Reilly and G. Savage, *Wedgwood, the Portrait Medallions*, London, 1973, 257-259, a-e. The plaque is generally attributed to Hackwood, remodelled by John Flaxman. For the circle of Wedgwood, see B. Wedgwood and H. Wedgwood, *The Wedgwood Circle 1730-1897. Four Generations of a Family and Their Friends*, Westfield, NJ, 1980, 42ff.

6 See R.F. Schofield, *The Lunar Society of Birmingham*, Oxford, 1963; *idem*, "The Industrial Orientation of Science in the Lunar Society of Birmingham," *Science, Technology and Economic Growth in the Eighteenth Century*, ed. A.E. Musson, London, 1972, 136-147.

7 In the Derby Art Gallery, in oil on canvas measuring 147.3 × 203.3 cm; see J. Egerton, *Wright of Derby*, exh. cat., London, 1990, 54-55, no. 18, and the essay in the same volume by D. Fraser, "Joseph Wright of Derby and the Lunar Society," 15-23.

8 Inv. no. 1951-685/53 in the Science Museum, London, an engraving measuring 203 × 238 mm (paper). See H.C. Calvert, *Scientific Trade Cards in the Science Museum Collection*, London, 1971, 31, no. 258; H.C. King and J.R. Millburn, *Geared to the Stars. The Evolution of Planetariums, Orreries and Astronomical Clocks*, Toronto, Buffalo, and London, 1978, 195-197; M. Daumas, *Scientific Instruments of the Seventeenth and Eighteenth Centuries and their Makers*, trans. and ed. M. Holbrook, London, 1992, 160, 172, 237-239; A.Q. Morton and J.A. Wess, *Public and Private Science. The King George III Collection*, Oxford, 1993, 54, 84-85.

9 In the Museum of the History of Science, Oxford University, an engraving measuring 248 × 373 mm. It is inscribed in its upper border: *A View of the Solar System and Orbit of the Comet (with its proper Elevation,) which will next return;/truely representing all its Appearances for any Part of the Year. by Benj. Martin.* For interest in this comet, see C.B. Watt, "Predicting the mid-eighteenth-century return of Halley's Comet," in R. Taton and C. Wilson, *General History of Astronomy*, 2: *Planetary astronomy from the Renaissance to the rise of astrophysics, pt. B: The eighteenth century*, Cambridge and New York, 1995, 281.

10 Vsekhyvatskii, 130-131; Kronk, 1984, 19-20; *idem.*, forthcoming, 241-243.

11 Ms. RGO 4/61, fol. 12v, in the Royal Greenwich Observatory Archives, Cambridge University Library, in pen and brown ink measuring 225 × 90 mm.

12 Morton and Wess, 39-87, point out that lectures on natural philosophy became a regular feature of life in large cities in the early eighteenth century; during the latter half of the century they faced competition from newspapers and periodicals as well as from institutions. H. Cunningham, "Leisure and culture," in *The Cambridge Social History of Britain 1750-1950*, 2, ed. F.M.L. Thompson, Cambridge, 1990, 310-330, distinguishes the commercialism of leisure for the well-to-do in the eighteenth-century from its different character in the mid-nineteenth century, in part fed by the needs of an increasingly urban society. See also D.J. Olsen, *The Growth of Victorian London*, London, 1976.

13 These ideas were explored in an exhibition (October 1993-January 1994) and preserved in London, Science Museum, "Astronomical Recreation," exh. checklist, 1993. We are grateful to Wendy Sheridan for notice of it. For an investigation of related ideas, see B.M. Stafford, *Artful Science: Enlightenment Entertainment and the Eclipse of Visual Education*, Cambridge, MA, and London, 1994. The importance of comets is clear in the frontispiece of a book by the instrument maker George Adams, where one of the three women surrounding a telescope points to a comet (*Astronomical and Geographical Essays: Containing a Comprehensive View of the General Principles of Astronomy...; The Use of the Celestial and Terrestrial Globes...Armillary Sphere, Planetarium, Tellurian, and Lunarium; An Introduction to Practical Astronomy...*, London, 1789). We thank John R. Millburn for the reference.

14 B. Martin, *The Young Gentleman and Lady's Philosophy, in a Continued Survey of the Works of Nature and Art; By Way of Dialogue*, 1, London, 1759, 109. Plates X and XI show the exterior and interior of the cometarium. It is obvious that Martin wrote before the return of Halley's Comet in 1759. In a letter of 26 June 1995, John R. Millburn relates:

> Martin's account of his cometarium was originally published in his *General Magazine of Arts and Sciences*, a complex monthly

publication begun in 1755 in which each number consisted of a few pages from six separate works, subsequently bound individually in volume form under their own titles.... The first Part was called *The Young Gentleman and Lady's Philosophy, Vol. I*, with a title page dated 1759.

See J.R. Millburn, "Martin's Magazine: the General Magazine of Arts and Sciences, 1755-65," *The Library (Transactions of the Bibliographic Society)*, fifth series, 28:3, London, 1973, 221-239; also D.P. Wheatland, *The Apparatus of Science at Harvard 1765-1800*, Cambridge, MA, 1968, 60.

15 Inv. no. 6 in the Collection of Historical Scientific Instruments of Harvard University, Cambridge, MA, in mahogany and brass measuring 254 × 165 × 50 mm. Harvard purchased it at a cost of £3.13.6 in 1766 to restock its new philosophical chamber. A fire in 1764 had destroyed nearly all its instruments, including an earlier cometarium.

16 In a letter of 26 June 1995, John R. Millburn remarks: "his device only illustrates the motion of a comet in general terms.... Also, the Harvard model does not have the large circle of stars, centred on the 'Sun', shown in his Plate X [n. 14 above]. The purpose of this was to explain why the comet seen from the Earth (or Sun) appears to move at different speeds relative to the stars, according to its distance from the Sun; the device itself represents the comet's motion as seen by an observer in outer space."

17 Inv. no. 1927-1254 in the King George III Collection, the Science Museum, London, in wood, glass, and metal, measuring 500 × 390 × 115 mm. It bears an inscription on its face: *Doctor Demainbray.* See Morton and Wess, 160, no. D27, who date it c. 1755; also King and Millburn, 172-173. Desaguliers's cometarium evolved from his planetarium, which also features a device to show comet orbits (*ibid.*, 170; fig. 10.2). His cometarium consists of a rectangular wooden box with a glazed cover and three scales on its surface – two circles and an elliptical groove drilled at one focus to take the vertical stem of a sun-ball. Attached to the latter point and acting as a horizontal arm is a smooth brass rod on which slid a small ball representing the head of the comet. The smaller upper circle, divided into degrees and twelve equal signs, has a movable pointer which demonstrates uniform circular motion about the sun.

John R. Millburn, in his letter of 26 June 1995, writes that Martin was the first person to use the term "cometarium" in connection with a Desaguliers-type mechanism in his *Lectures on Natural and Experimental Philosophy* of 1743. Lecture XI is entitled "The Copernican or Solar System explained by the Orrery and Cometarium." [His explanation is similar to that cited in n. 14 above.] When James Ferguson lectured in London in the late 1740s, he initially called his elliptical-pulley device an "equi-areal machine," but soon followed Martin and used the term "cometarium" (illustrated in his *Astronomy Explained upon Sir Isaac Newton's Principles* of 1756). See also H.M. Barkla, "Ingenuity Misdirected," unpublished manuscript, n.d.

We thank John Millburn for his expansion on these ideas in a letter of 19 December 1994. He notes that the motion in the cometariums is such that only four times in each orbit is the comet in the correct position for a given time, although its velocity and acceleration are not accurate at this point. While the magnitude of these errors may appear significant to a mathematician, they are insignificant to someone attending a demonstration. Millburn has worked on the error-curve problem of cometaria and the true Keplerian position of comets with J.E. Nuttall, resulting in "The Elliptical-Pulley Cometarium and its Accuracy," unpublished manuscript, 1985. Both the manuscript of their research and that of another by H.M. Barkla ("Cometarium: made by R. Fidler, early 19th century," unpublished manuscript, 1994) are on deposit in the Royal Museum of Scotland and referenced in A. MacDonald and A.D. Morrison-Low, eds., *A Heavenly Library. Treasures from the Royal Observatory's Crawford Collection*, exh. cat., Edinburgh, 1994, 67, 70.

18 Inv. no. 1909-202 in the Science Museum, London, in wood, brass, and ivory measuring 300 × 230 × 150 mm. The W. & S. Jones cometarium was reproduced in A. Rees, ed., *Cyclopaedia: or Universal Dictionary of Arts, Sciences and Literature*, 9, London, 1819, pls. I, II. Millburn in a letter of 26 June 1995, added that the Jones example was more expensive than the elliptical-pulley models; the catalogue price was five guineas (£5 5s). He does not know whether the firm sold any, because the only example he knows of is the one in the Science Museum.

19 Inv. no. NMS NPM A78 in the National Museums of Scotland, Edinburgh, in wood, silvered brass, and brass with steel pinions measuring 445 × 245 × 205 mm. See MacDonald and Morrison-Low, 37, no. 4.5. It uses an elliptical system of pulleys like that of Desaguliers's cometarium.

20 The University of St. Andrews, Department of Physics and Astronomy, in wood and metal measuring 596 × 345 × 110 mm with lid removed (lid adds 50 mm). See *ibid.*, 37, no. 4.6. Its skillful cutting of the elliptical gear mechanism is more sophisticated than the Jones example.

21 Inv. no. 1982-1473 in the Science Museum, London, an engraving measuring 277 × 175 mm (plate). It is inscribed below the image: *A Representation of the COMET that Appear'd on Janry: ye26...174 ¾ in the Evening./Taken near St. Martin's-Church in the Strand./ A. the Comet as it Appear'd to the naked Eye. B as it Appear'd through a Telescope. C the Person making ye Observation./ Publish'd Febry. the 11th. 174 ¾. According to an Act of Parliament. – Price 3d.*

22 Vsekhsyvatskii, 126-127; Kronk, 1984, 16-17; *idem*, forthcoming, 220-222. Perihelion passage occurred 6-9 March. On 18 March the comet's tail measured 90°. Although the comet was discovered on 9 December by Dirk Klinkenberg in Harlem, four days later it was sighted by the Swiss astronomer J.P. Loys de Chéseaux. See Olson, 1985b, fig. 61, for a plate with the comet's stupendous tail from De Chéseaux's 1744 treatise *Traité de la comète....*

23 In Sir John Soane's Museum, London, an engraving measuring 353 × 440 mm. See R.L.S. Cowley, *Hogarth's Marriage A-La-Mode*, Ithaca, 1983, 27-55. The model for the six vignettes was John Dryden's comedy *Marriage à la mode* (1633). See also R. Paulson, *Hogarth's Graphic Works*, 1, New Haven and London, 1970, 267-270, no. 228; *idem.*, *Hogarth: His Life, Art, and Times*, 1, New Haven and London, 1971, 465-49; *idem.*, *Hogarth*, 3 vols., New Brunswick, NJ, 1993. Johnson, 1986, 20-21, points out that most of Hogarth's paintings were designed for translation into engravings, through the sale of which he could free himself from the traditional reliance on aristocratic patronage and address a genuinely popular audience. Like his friend Henry Fielding, Hogarth aspired to elevate the genre of comic realism to the stature of history painting.

24 Cowley, 27ff.

25 *Ibid.*, 44.

26 For contemporary attitudes, see P. Stewart, "Science and Superstition: Comets and the French Public in the 18th century," *American Journal of Physics*, 54, 1, 1986, 16-24.

27 Inv. no. 1863-5-9-6 in the British Museum, London, an etching measuring 233 × 188 mm. See G. Schiff, *Johann Heinrich Füssli, 1741-1825*, 1, Zurich and Munich, 1973, 508, no. 844. For its preparatory pen and ink drawing in the British Museum, *ibid.*, 508, no. 843.

28 D.H. Weinglass, *Prints and Engraved Illustrations By and After Henry Fuseli*. Aldershot, Hants, 1994, 84-85, no. 76. The imprint of the British Museum example sets publication on 1 June 1787 by J. Johnson. The eccentric Irish controversialist Theophilus Swift (1746-1815) was the son of Deane Swift, the cousin and biographer of Jonathan Swift.

29 *Ibid.*, 85.

30 Vol. 9, p. 46v, no. 169, in the Peel Collection, The Pierpont Morgan Library, New York, an engraving measuring 388 × 295 mm. It is inscribed *W.P. fecit*. See George, 2, 710, no. 1845. The comet-star rays shine upon Londonia (a variant on the Roman name for the city, Londinium), the personification of the city. These rays are inscribed: *Dono Coeli Felicitor*. The standing female opposite holds an escutcheon with the armorial bearings of Parsons.

31 Inv. no. 1929-125 in the Science Museum, London; Morton and Wess, 170-171, no. D39.

32 Lord Paisley, "With some curious Figures of the same Comet," *Philosophical Transactions*, 33, 1733, 50-51, supplementing an earlier article (appendix I). The illustration opp. p. 39 measures 221 × 160 mm.

33 No. L&P IV.375 in the Archives of the Royal Society, London, in pen and brown ink measuring 201 × 205 mm. See J. Winthrop (communicated by B. Franklin), "*Cogitata de Cometis*," *Philosophical Transactions*, 57, Part I, 1767, 132-137, pl. opp. p. 133. For early astronomy in Cambridge, MA, see S.S. Genuth, "Blazing Stars, Open Minds, and Loosened Purse Strings: Astronomical Research and Its Early Cambridge Audience," *Journal for the History of Astronomy*, 21, 1990, 9-20; *idem*, "From Heaven's Alarm to Public Appeal," in *Comets and the Rise of Astronomy at Harvard. Science at Harvard University. Historical Perspectives*, eds. C.A. Elliott and M.W. Rossier, Bethlehem, London, and Toronto, 1992, 28-54.

34 J. Pringle, "Several Accounts of the fiery Meteor which appeared on Sunday the 26th of November 1759, between Eight and Nine at Night," *Philosophical Transactions*, 51, Part I, 1759, 218-259, pl. VII opp. p. 223 measuring 224 × 310 mm.

35 No. L&P I.298 in the Archives of the Royal Society, London, in pen and brown ink measuring 179 × 310 mm. See Z. Cradock, "The Appearance of a fiery Meteor," *Philosophical Transactions*, 43, no. 473, 1744-45, 78, where it is reproduced.

36 No. L&P II.80 in the Archives of the Royal Society, London, in pen and brown ink measuring 240 × 190 mm. See H. Baker, "An Account of the same Meteor," *Philosophical Transactions*, 47, 1751-52, 3-4, pl. I, fig. 1 opp. p. 3.

37 J. Swinton, "An Account of a remarkable Meteor Seen At Oxford," *Philosophical Transactions*, 52, 1762, 99-102, pl. III opp. p. 99. The preparatory drawing is reproduced at the right; it is MM.I.51 in the Archives of the Royal Society, in graphite measuring 179 × 191 mm. It is inscribed centrally above the arch: *No. XVIII Mr. Swinton on a Meteor*, as well as: *A B C*, designations not adopted by the engraver. This study dates from the same time as the letter from Swinton to Thomas Birch in pen and brown ink (1760) at the upper left. The engraving below it on the left measures 192 × 139 mm and was mistakenly pasted in the sequence. It is related to another article by Swinton published in vol. 57 of the *Philosophical Transactions* cited in n. 40 below. Mary Sampson (verbal communication, 1994) believes that it is a proof and was probably sent by J. Mynde to Swinton to accompany Mynde's letter of 19 January 1768, reproduced at the upper right, referring to a proof he is sending Swinton.

38 J. Swinton, "An Account of a remarkable Meteor seen at Oxford, March 5, 1764," *Philosophical Transactions*, 54, 1764, 326-329, pl. XVIII opp. p. 326 measuring 148 × 124 mm.

39 J. Swinton, "An Account of a remarkable Meteor seen at Oxford, April 23, 1764," *Philosophical Transactions*, 54, 1764, 332-336, pl. XIX opp. p. 332 measuring 165 × 125 mm.

40 J. Swinton, "Description of a Meteor seen at Oxford, October 12, 1766," *Philosophical Transactions*, 57, 1767, 108-110, pl. V opp. p.109 measuring 192 × 139 mm. Swinton also observed at least three other phenomena he termed "meteors:" one on 26 February 1769; another on 9 September 1769 (the second report of which he published under the subheading "Letter II" in his article "An Account of Two Aurorae Boreales observed at Oxford," *Philosophical Transactions*, 59, 1769, 367-373); and a third on 24 October 1769 (linked to an aurora borealis in his report in the *Philosophical Transactions*, 60, 1770, 532-535).

41 Burke, 5ff.

42 *Ibid.*, 17-18, 45-46.

43 See E. Howard, "Experiments and Observations on certain stony and metallic Substances which at different Times are said to have fallen on the Earth; also on various Kinds of native Iron," *Philosophical Transactions*, 92, 1802, 168-212. See also Burke, 47-50; Pillinger and Pillinger.

44 Inv. no. 1865.2 in the New-York Historical Society, oil on canvas measuring 185.5 × 139.5 cm. See H. von Erffa and A. Staley, *The Paintings of Benjamin West*, New Haven and London, 1986, 256, no. 179, ill. West's depiction of the meteor cinches the subject (it had been identified as the parting of Hector and Andromache from the *Iliad*).

45 *Ibid.*

46 T. Cavallo, "Description of a Meteor, observed Aug. 18, 1783," *Philosophical Transactions*, 74, Part I, 1784, 108-111; A. Aubert, "An Account of the Meteors of the 18th of August and 4th of October, 1783," *Philosophical Transactions*, 74, Part I, 1784, 112-115; W. Cooper, "Observations on a remarkable Meteor seen on the 18th of August, 1783," *Philosophical Transactions*, 74, Part I, 1784, 116-117; R. L. Edgeworth, "An Account of the Meteor of the 18th of August, 1783," *Philosophical Transactions*, 74, Part I, 1784, 118; C. Blagden, "An Account of some late fiery Meteors; with Observations," *Philosophical Transactions*, 74, Part I, 1784, 201-232; N. Pigott, "An Account of an Observation of the Meteor of August 18, 1783, made on Hewit Common near York," *Philosophical Transactions*, 74, Part II, 1784, 457-459. Other reports occur in *The Gentleman's Magazine*, 53, 1783, 711-992. See also D. Howse, *Nevil Maskelyne: The Seaman's Astronomer*, Cambridge, 1989, 149-151, who notes that Maskelyne probably wanted to publish a definitive paper on the hotly debated subject of meteors. In November he sent out 100 copies of a three-page pamphlet ("A Plan for observing the Meteors called Fire-

balls," dated 6 November 1783, Greenwich, and published at his own expense), in which he instructed people to send him their observations. (Maskelyne's article, under the same title, appeared later in *The Gentleman's Magazine*, 53, 1783, 991-992.) To Maskelyne's chagrin, he was completely upstaged by Blagden, who on 19 February 1784 read his paper to the Royal Society.

Many of the earlier meteors were described as appearing like a flame, or a cone-shaped object, or a ball of fire that burst in an explosion with sparks like a sky-rocket, or a shooting star, or even a comet. Beginning in earnest with the meteor of 11 December 1741, these apparitions occur frequently (1741, 1744, 1745, 1750, 1758, 1760, 1763, 1764, 1766, and 1772). Some of the articles about them have illustrations, most of which are small in format. The only large illustration of an earlier meteor is contained as the foldout pl. VII in the very detailed, synoptic article by Pringle cited in n. 34 above. It is followed directly in the same volume by another related article by the same author: "Some Remarks upon the several Accounts of the fiery Meteor (which appeared on Sunday the 26th of November, 1758)," 259-274. Pringle's articles mark a watershed in the careful observation of meteors. They analytically attempt to probe the nature and genesis of meteors. Nevertheless, both descriptions and illustrations demonstrate the unique configurations of the 18 August 1783 meteor.

47 Nos. L&P VIII.50 and VIII.92, respectively, in the Archives of the Royal Society, London. Cavallo's in pen and brown ink measuring 229 × 186 mm, Pigott's in pen and brown ink measuring 525 × 175 mm (for pls. IV, XX, respectively).

48 Cooper, 116-117

49 Aubert, 113. For video of a fireball, see J.M. Pasachoff, *Saunders Videodisc on Astronomy, 1: The Solar System,* Philadelphia, 1994.

50 Edgeworth, 118.

51 The relationship between them – a print in the British Museum, London, by Henry Robinson (inv. no. 1917-12-8.4032, a mezzotint measuring 188 × 252 mm) and the anonymous watercolor and wash drawing in the Royal Astronomical Society Library, London (Ms. Add. 100 in pen and black ink with wash and watercolor measuring 255 × 313 mm [paper] and 164 × 239 mm [image]) – has not been resolved. The mezzotint is inscribed at the left: *H. Robinson delin.* In the center: *South.* On the next line (mostly in italics): "*An accurate Representation of the* Meteor *which was seen on Aug.t 18th 1783. – At first it appeared as one Ball of Fire, but in a few Seconds, broke into many small ones./It's Course was from N. W. to S.E. – this extraordinary Phaenomenon was of that Species of Meteor which the great Phisiologist D.r Woodward and others call the/* Draco volans or Flying Dragon. – *The above view was taken at Winthorpe near Newark upon Trent, by Henry Robinson, Schoolmaster. – and Published by him/as the Act directs, 14 Oct.r 1783. – This plate is inscribed to Roger Pocklington Esq.r by his much obliged humble Servant/Henry Robinson.*" It is inscribed on the four borders with the directions of the compass. Ms. Add. 100 contains a nearly identical image, inscribed: *An accurate Representation of the* Meteor *which was seen Aug.st18th 1783/At first it appeared as one Ball of Fire, but in a few Seconds, broke into many/small ones – It's Course was from N.W. to S.E. This extraordinary Phanomenon/was of that Species of Meteor which the great Phisiologist D.r Woodward/and others call the Draco volans or Flying Dragon.*

52 N. Pigott, 458.

53 Dr. James Lind (1736-1812), a physician to the Royal Household and a surgeon, accompanied Sir Joseph Banks on expeditions, while Dr. John Lockman was a canon of Windsor, where he later became a royal physician.

54 See A.P. Oppé, *The Drawings of Paul and Thomas Sandby in the Collection of His Majesty the King at Windsor Castle*, Oxford and London, 1947, 28, no. 53.

55 Cavallo, 109-110. See also T. Cavallo, *A complete treatise of electricity in theory and practice; with original experiments*, London, 1777.

56 Inv. no. 1872-7-13-481 in the British Museum, London (another impression is inv. no. 1991-307 in the Science Museum, London), an aquatint whose image measures 272 × 493 mm (paper, c. 332 × 495 mm). The plate mark has been trimmed in both, but an untrimmed impression is in a London private collection. The aquatint is inscribed below the image at the left *T. Sandby Delint* and at the right *P. Sandby sculp*. Another faint, partially cut inscription at the lower margin reads: *Publish'd according to Act of Parliament by P. Sandby St. Georges Rons [Row?] October 1793*. A first state without the inscriptions (but added in a nineteenth-century hand in pen and brown ink after those in the final state of the aquatint) is in the British Museum (inv. no. 1904-8-14-758). See Oppé, 1947, 28, no. 53, who states that the print was published by Paul after Thomas; J. Roberts, *Views of Windsor. Watercolours by Thomas and Paul Sandby*, exh. cat., London, 1995, 68-69. For Paul Sandby's important contribution to the early aquatint, see A. Griffiths, "Notes on Early Aquatint in England and France," *Print Quarterly*, 4:3, 1987, 268-70. For the artists, see W. Sandby, *Thomas and Paul Sandby Royal Academicians*, London, 1892; J. Faigan, *Paul Sandby Drawings*, exh. cat., Sydney, 1981; J. Ball, *Paul and Thomas Sandby, Royal Academicians*, Bath, 1985, who on p. 246 writes that the Sandbys' widening circle of scientific friends at Windsor revived Thomas's interest in astronomy, which had been implanted in Nottingham; New Haven, Yale Center for British Art, *The Art of Paul Sandby*, exh. cat., New Haven, 1985, 79ff, for Paul, and 131ff, for Thomas; L. Herrmann, *Paul and Thomas Sandby*, exh. cat., London, 1986; B. Robertson, "Paul Sandby and the early development of English watercolor," 2 vols., Ph.D. diss., Yale University, 1987. Howse, 1989, 151, notes that "Maskelyne had two impressions of Sandby's aquatint of the 1783 meteor coloured, one for Lady Clive, one for himself (NM to Dr. J. Lind [one of those on the terrace with Sandby], 11 Nov. 1783, R. Soc. MS 244/5)."

57 It was to Sir Joseph Banks that Cooper, Edgeworth, and Blagden all addressed their reports on the meteor of August 18th (n. 46). See also P. Hughes, "Paul Sandby's Tour of Wales with Joseph Banks," *The Burlington Magazine*, 117, 1975, 452-457.

58 Inv. no. 14577 in the Royal Library, Windsor Castle, in watercolor over graphite measuring 282 × 493 mm. The back of the mount is inscribed: *EIE/W.E.P.93. The representation of a fiery meteor which pass'd over London abt 9 Oclock in the evening of 18th August 1783/Taken from Windsor Terrace. P. Sandby.* Oppé, 1947, 28, no. 53, remarks that the Windsor sheet "may be by Thomas." See Roberts, 68-69, who is inclined to attribute all the sheets with figures to Paul, because of Thomas's known weakness with figures, and the only one without figures to Thomas. In a letter of 9 June 1995, the Hon. Jane Roberts wrote, "I remain somewhat uncertain about the responsibility of the two

Sandby brothers for these watercolours. As was often the case, there appears to have been an almost seamless collaboration."

59 Inv. no. 1904-8-19-34 in the British Museum, London, in watercolor over traces of graphite, measuring 285 × 461 mm and laid down on board. It is inscribed in graphite at the center of the lower border: *Windsor Terrace*; and at the right: *by Paul Sandby*. See *ibid.* (Roberts believes it may be by Thomas); Oppé, 1947, 28, no. 53, who attributes it and its pendant to Paul, and who cites William Sandby, the donor. These sheets were once placed on consecutive pages of an album. See also L. Parris, *Landscape in Britain c. 1750-1850*, exh. cat., London, 1973, 48, no. 74, who attributes it to ?Thomas.

60 Inv. no. 1904-8-19-427 (the last sequence of numbers is sometimes given as 35) in the British Museum, London, in watercolor over traces of graphite, measuring 321 × 510 mm and laid down on board. It is inscribed in graphite at the lower right: *Windsor Terrace*; further to the right: *by ThS.Sandby*. Oppé, 1947, 28, no. 53, notes that William Sandby attributed this sheet to Thomas, although Oppé assigns it and its pendant to Paul. See also Roberts, 68-69.

61 Inv. no. B1993.30.115 in the Yale Center for British Art, Paul Mellon Collection, New Haven, in watercolor with touches of pen and ink over graphite measuring 318 × 484 mm. See Christie's, London, sale cat., 12/VII/88, no. 53; Spink, sale cat., London, 1989, no. 10; Roberts, 68-69.

62 Blagden, 201.

63 *Ibid.*, 205.

64 *Ibid.*

65 *Ibid.*, 217. Ball, 246, notes that William Herschel was one of the scientific acquaintances of the Sandby brothers. He adds that Banks induced the King to confer on Herschel the appointment of court astronomer. One of the terms was that Herschel would move closer to Windsor. Ball believes that the Herschels and the Sandbys were introduced by their common interests in music rather than their joint interests in astronomy.

66 A.S. Herschel, "Progress in Meteoric Astronomy during the Year 1863-64," *Monthly Notices of the Royal Astronomical Society*, 25, 1864, 158-162. Alexander S. Herschel, an important astronomer and physicist, was the grandson of William Herschel and the son of John F.W. Herschel.

67 E. Pigott, "Observations of the Comet of 1783," *Philosophical Transactions*, 74, Part II, 1784, 460-462.

68 See Vsekhsvyatskii, 134; Kronk, forthcoming, 257-258.

69 In a private collection, England, in oil on canvas measuring 81.9 × 111.8 cm. Since the apparition has the identical cracklure as the rest of the canvas, it was not added at a later time. Its attribution to Samuel Scott was made before it was purchased in the 1950s from the Tooth Gallery in London. See R.J.M. Olson and J.M. Pasachoff, "Recording the 'wonderful meteor' of 18 August 1783: The Sandbys, 'Samuel Scott,' and heavenly bodies," *Apollo*, 146:430, 1997, 12-19.

70 Olson, 1985b, 20-21.

71 For Scott, a marine painter and a painter of Thames views, see R. Kingzett, "A Catalogue of the Works of Samuel Scott," *The Walpole Society*, 48, 1980-82, 1-134, as well as E.B. Chancellor, "London in the Works of Canaletto and Scott," *Connoisseur*, 37, 1913, 127-132; London, Guildhall Art Gallery, *Samuel Scott Bicentenary*, exh. cat., London and Sheffield, 1972. Kingzett wrote in a letter of 23 January 1987 that he believes the painting is not by Scott himself. In a telephone conversation (1988), Kingzett added that only two nocturnal pictures in Scott's oeuvre are known and they are battle pictures. Scott painted similar views of the Thames, for example, *A View of Somerset House and the Thames*, with the same approximate dimensions (in the collection of the Shell International Petroleum Company, reproduced in the Witt Library, the Courtauld Institute of Art, London). Nevertheless, the formulaic composition of figure 39 is typical of the time and is found in other works, for example, several oil paintings also of the Gardens of Somerset House attributed to Paul Sandby, reproduced in the Witt Library.

72 R.E. Schmidt, in a talk abstract published as "Samuel Scott's Comet Halley of 1759," *Abstract: Bulletin of the American Astronomical Society*, 17, 1985, 844. In a letter of 20 November 1986, Schmidt elaborated that the object in the painting was decidedly uncometary in nature, e.g. its sputtering tail. It is so bright that it illuminates clouds, reflects on the water, and casts shadows of the bystanders. Because the figures look startled, he too believes that the object is not merely hanging in the sky like a comet.

73 Phyllis Tortora has stated in a letter of 23 November 1990, that based on the clothing depicted, a date of 1783 is more likely than 1759: "I am basing this not on any one item of costume, but on the inclusion of a number of items, any of which might perhaps have been seen at the earlier date, but which considered *in toto* seem to me to tend to support the later date." See C.W. Cunnington and P. Cunnington, *Handbook of English Costume in the Eighteenth Century*, London, 1957, 205, 187, 199, 347; E. Ewing, *History of Children's Costume*, New York, 1977, 46, 151.

74 See Kingzett, 55-66; E. Einberg, *Manners and Morals: Hogarth and British Painting 1700-1760*, exh. cat., London, 1987, 188-189, nos. 174-175.

75 See the London, Burlington Fine Arts Club, *Catalogue of the Collection of Early Drawings and Pictures of London*, exh. cat., London, 1920, 53, no. 92, pl. XXXIX, a daylight scene, listed in the collection of Mr. P. Norman as attributed to Samuel Scott (?). This oblique composition with the Thames was popular from the 1750s and was painted by Canaletto and others, for example, Paul Sandby (see L. Stainton, *British Landscape Watercolors 1600-1860*, exh. cat., London, 1985, 23, no. 21a; Birmingham, Museum and Art Gallery, *Canaletto and England*, exh. cat., Birmingham, 1994, no. 66).

76 See G.C. Williamson, *John Russell, R.A.*, London, 1894; E.J. Stone, "Note on a Crayon Drawing of the Moon by John Russell R.A. at the Radcliffe Observatory, Oxford," *Monthly Notices of the Royal Astronomical Society*, 55, 1895, 88-95; W.F. Ryan, "John Russell, R.A., and Early Lunar Mapping," *The Smithsonian Journal of History*, 1, 1966, 27-48.

77 N. Pevsner, *Academies of Art Past and Present*, London, 1940, 124-126; S.C. Hutchison, *The Royal Academy 1768-1986*, London, 1986 (second edition).

78 Pevsner, 184-185, also discusses the later history of the Royal Academy as well as the Normal Schools, the influence of Prince Albert and the Great Exhibition of 1851, the founding of the Victoria and Albert Museum, etc; Hutchison, 7-32. See also F. Rinder, *The Royal Scottish Academy 1826-1916*, Glasgow, 1917.

79 See P. Funnell, "The London Arts World and Its Institutions," and A. Wilton, "Painting in London in the Early Nineteenth Century," in *London – World City – 1800-1840*, ed. C. Fox, exh. cat., New Haven and London, 1992, 155-166, 167-186.

80 Funnell, 157.

81 See J.L. Roget, *History of the Old Water-Colour Society, now the Royal Society of Painters in Water-Colour*, 2 vols., London, 1891; I.A. Williams, *Early English Watercolours*, Bath, 1952, 211-216; S. Wilcox, *British Watercolours: Drawings of the 18th and 19th Centuries from the Yale Center for British Art*, exh. cat., New York, 1985.

82 See A. Graves, "The British Institution," *The Art Journal*, no vol. no., 1910, 336-341, 377-381; *idem.*, *The British Institution 1806-1867. A Complete Dictionary of Contributors and Their Work from the Foundation of the Institution*, London, 1908.

83 Funnell, 163-167, notes that in the 1830s the art world became a matter for debate on the floor of the House of Commons. It was also accorded scrutiny by a Select Committee on Arts and Manufactures in 1835 and 1836. Ways of reform were looked at and the Academy was scrutinized.

84 See K. Raine, *Blake and Tradition*, 2 vols., Princeton, 1968; D. Bindman, *William Blake. His Art and Times*, exh. cat., London, 1982; E.P. Thompson, *William Blake and the Moral Law*, New York, 1994, as well as M. Butlin's books on Blake, especially *The Paintings and Drawings of William Blake*, 2 vols., New Haven and London, 1981. For his book illustrations see: D. Bindman, *The Complete Graphic Works of William Blake*, New York, 1978; *idem.*, ed., *Blake's The Illustrated Books*, 5 vols., Princeton, 1991-93; R.N. Essick, *William Blake's Commercial Book Illustrations. A Catalogue and Study of the Plates Engraved by Blake after Designs by Other Artists*, Oxford, 1991; J. Mee, *Dangerous Enthusiasm. William Blake and the Culture of Radicalism in the 1790's*, Oxford, 1992.

85 Inv. no. 1894-6-12-16 in the British Museum, London, in pen and black ink with brown and gray wash over traces of graphite and black chalk measuring 335 × 475 mm. See M. Butlin, *William Blake*, exh. cat., London, 1978, 41, no. 36; D.W. Olson and M.S. Olson, "William Blake and August's Fiery Meteors," *Sky and Telescope*, 78, 1989, 194.

86 Inv. no. 1894-6-12-17 in the British Museum, London, a relief and white-line etching measuring 297 × 209 mm (image). See Butlin, 1978, 41, no. 35; R.N. Essick, *The Separate Plates of William Blake. A Catalogue*, Princeton, 1983, 12-13.

87 Chaldni, 1794, who later published *idem*, *Uber Feure-Meteore und über die mit denselben herangefallen Massen*, Vienna, 1819.

88 J.V. Schiaparelli, "Sur la relation qui existe entre les comètes et les étoiles filantes," *Astronomische Nachrichten*, 68, 1873, 331-334. See E. King, *Remarks Concerning Stones Said to Have Fallen from the Clouds, both in These and in Antient Times*, London, 1796.

89 Olson and Olson, 192.

90 See G. Keynes, *William Blake's Water-Colour Designs for the Poems of Thomas Gray*, Jura, 1971. Blake used as his text the 1790 edition of Gray published by John Murray in London.

91 Inv. no. B1992.8.11 (59) in the Yale Center for British Art, Paul Mellon Collection, New Haven, in watercolor with pen and black ink over graphite measuring 420 × 326 mm. See *ibid.*, 1-2; P. Noon, *The Human Form Divine*, exh. cat. New Haven and London, 1997, 80. Among the five watercolors Blake exhibited at the Royal Academy in 1785, one was entitled "The Bard, from Gray" (no. 607). This drawing is not known to be extant, but the image of the Bard reappears in Blake's work; see M. Butlin, *Tate Gallery Collection 5: William Blake 1757-1827*, London and Seattle, 1990, 150-152, no. 60.

92 Inv. no. 1992.8.11 (93) in the Yale Center for British Art, Paul Mellon Collection, New Haven, in watercolor with pen and black ink over graphite measuring 420 × 327 mm.

93 Keynes, 1971, 133, no. 93. Gray's "Ode for Music" was composed in 1768 for the installation of the Duke of Grafton as Chancellor of the University of Cambridge and was performed in the Senate house at Cambridge on 1 July 1769.

94 See Bindman, 1978, nos. 337-380, for the 43 engravings (none include comets and meteors); D.V. Erdman et al., eds., *William Blake's Designs for Edward Young's Night Thoughts*, 1, Oxford, 1980, 3-93.

95 No. 149, Night IV, p. 40 in the British Museum, London, in watercolor with pen and black ink over graphite measuring 412 × 320 mm.

96 No. 158, Night V, title page in the British Museum, London, in watercolor with pen and black ink over graphite measuring 414 × 325 mm.

97 No. 313, Night VII, p. 41 in the British Museum, London, in watercolor with pen and black ink over graphite measuring 412 × 318 mm.

> *Undrawn no more?* Behind the Cloud of *Death*,
> Once, I beheld a Sun; a Sun which gilt
> That sable Cloud, and turn'd it all to Gold;
> How the *Grave's* alter'd? Fathomless, as Hell!
> A *real* Hell to Those, who dreamt of Heav'n.
> ANNIHILATION! How it yawns before me?
> Next Moment I may drop from *Thought*, from *Sense*,
> The Privilege of Angels, and of Worms,
> An Outcast from Existence! And this Spirit,
> This all-pervading, this all-conscious Soul,
> This Particle of Energy divine,
> Which travels Nature, flies from Star to Star,
> And visits Gods, and emulates their Pow'rs,
> For ever is extinguisht. Horror! Death!
> Death of *that* Death I *fearless*, once, survey'd.
> When Horror *Universal* shall descend,
> And Heav'n's dark Concave urn all Human Race,
> On that enormous, unresounding Tomb,
> How just this Verse? this monumental Sigh!
> *Beneath the Lumber of demolisht Worlds,*
> *Deep in the Rubbish of the gen'ral Wreck,*
> *Swept Ignominious to the common Mass.*

98 See Olson, 1985b, 39-40, for woodcuts by Albrecht Dürer and Matthias Gerung which illustrate the following passages. Revelation 6:12-13: "And I beheld when he had opened the sixth seal, and, lo, there was a great earthquake; and the sun became black as sackcloth of hair, and the moon became blood. And the stars of heaven fell unto the earth, even as a fig tree casteth her untimely figs when she is shaken in a mighty wind." Revelation 8:10-11: "And the third angel sounded and there fell a great star from heaven burning as it were a lamp.... And the name of the star is called Wormwood." Revelation 9:1-3: "And the fifth angel sounded, and I saw a star fall from heaven onto the earth: and to him was given the key of the bottomless pit and there arose a smoke out of the pit...locusts...scorpions...." For connections of the Sublime to the Apocalypse, see M.D. Paley, *The Apocalyptic Sublime*, New Haven and London, 1986.

99 No. 236, Night VI, p. 15 in the British Museum, London, in watercolor with pen and black ink over graphite measuring 415 × 320 mm.

100 No. 320, Night VII, p. 48 in the British Museum, London, in watercolor with pen and black ink over graphite measuring 408 × 317 mm.

101 No. 443, Night IX, p. 25 in the British Museum, London, in watercolor with pen and black ink over graphite measuring 418 × 328 mm.

102 No. 463, Night IX, p. 45 in the British Museum, London, in watercolor with pen and black over graphite measuring 415 × 325 mm.

103 No. 504, Night IX, p. 86 in the British Museum, London, in watercolor with pen and black ink over graphite measuring 410 × 318 mm.

104 No. 509, Night IX, p. 91 in the British Museum, London, in watercolor with pen and black ink over graphite measuring 415 × 325 mm.

105 R.L. Bishop, "Newton's Telescope Revealed," *Sky and Telescope*, 59, 1980, 207. Newton donated it in 1762.

106 L. Bell, *The Telescope*, New York, 1922 (1981 ed. with intro. by J.M. Pasachoff), 25-27.

107 D.B. Hermann, *The history of astronomy from Herschel to Hertzsprung*, trans. K. Krisciunas, Cambridge, 1984, 159, which on p. 160 has a photograph of the telescope in the garden of the Royal Greenwich Observatory. See also W. Herschel, *The Scientific Papers of Sir William Herschel*, London, 1912, pl. XVI, which was first published opp. p. 408 as pl. XXIV in the *Philosophical Transactions*, 85, 1795. His telescopes and discoveries with them are discussed in M.A. Hoskin, "The Astronomy of the Universe," in *Cambridge Illustrated History of Astronomy*, ed. M. Hoskin, New York and Cambridge, 1996, 231-247.

108 For telescopes see G. Abetti, *The History of Astronomy*, trans. B.B. Abetti, New York, 1952, 139-143; J.A. Bennett, "'On the power of penetrating into space': The telescopes of William Herschel," *Journal for the History of Astronomy*, 7, 1976, 75-108; Westfall, 392; L. Bell; Hermann, 158-167; M.A. Hoskin, "Rosse, Robinson, and the Resolution of the Nebulae," *Journal for the History of Astronomy*, 21, 1990, 331-344.

109 Quoted from no. L&P VII.192 in the Archives of the Royal Society, London, published in W. Herschel, "Account of a Comet [1781]," *Philosophical Transactions*, 71, Part II, 1781, 492-500. See also A.F. O'D. Alexander, *The Planet Uranus: A History of Observation, Theory and Discovery*, London, 1965, 27.

110 The discussion of the name started in 1781. See *ibid.*, 1-30, as well as W. Herschel, "A Letter [on the name of the Georgian Planet]," *Philosophical Transactions*, 73, 1783, 100-101.

111 No. L&P VII.192 in the Archives of the Royal Society, London, in pen and brown ink on a page measuring 330 × 210 mm (195 × 120 mm boxed diagram). It is a five-page letter followed by a paragraph of transmission by one W. Watson and one page with four figures. L&P VII.192 includes two further epistles (n.d.) with observations.

112 The first three figures from the manuscript page (the first is unnumbered) were not engraved for the article in *Philosophical Transactions*. The first plate (pl. XXIV opp. p. 500) in the published version is a variation of fig. 4 in Herschel's letter. The rethinking between this letter and the article probably resulted in part from the oral delivery, because the differences seem more due to language than to changes in substance.

113 J. Flamsteed, *Atlas Coelestis Britannica*, London, 1729, pl. IV.

114 Already in 1760, Messier in France had begun to make his list of approximately 100 nebulous objects to keep observers from confusing them with comets.

115 No. L&P X.135, p. 2 in the Archives of the Royal Society, London, in pen and brown ink on a page measuring 225 × 185 mm. After Encke figured out that several appearances of this comet were the same periodic comet, it was named after him (2P/Encke); see Vsekhsvyatskii, 139; Kronk, forthcoming, 268-269.

116 S. Schaffer, "'The Great Laboratories of the Universe': William Herschel on matter theory and planetary life," *Journal for the History of Astronomy*, 11, 1980, 96-100.

117 For Herschel's comet observations from 1781 until 1815, in the Herschel Archives of the Royal Astronomical Society Library, London, see J.A. Bennett, "Catalogue of the Archives and Manuscripts of the Royal Astronomical Society," *Memoirs of the Royal Astronomical Society*, 85, 1978, 65-67 (inv. nos. W3/1.12; W3/1.14; W3/36.1; W3/36.2; W3/37.1; W3/37.2; W3/38.1; W3/38.2; W3/39.1-4).

118 For William Herschel, see: C.D.P. Davies, "Herschel's 18 3/4 in. Speculum (the '20 ft')," *Monthly Notices of the Royal Astronomical Society*, 84, 1923-24, 23-26; M.A. Hoskin, *William Herschel, Pioneer of Sidereal Astronomy*, London, 1953; A. Armitage, *William Herschel*, London, 1962; M.A. Hoskin, "William Herschel," *Dictionary of Scientific Biography*, ed. C.C. Gillispie, New York, 1974, 328-336.

119 Inv. no. C1/3 of the Herschel Archives of the Royal Astronomical Society Library, London, contains the notes and observational drawings of the eight comets she discovered at Slough. They appeared in 1786, 1788, 1790, 1790, 1791-92, 1794, 1795, 1797. Kronk, forthcoming, lists them as: C/1786 P1, formerly 1786 II (p. 262); 35P/1788 Y1, formerly 1788 II (p. 264); C/1790 A1, 1790 I (p. 265); C/1790 H1, formerly 1790 III (p. 266); C/1791 X1, formerly 1792 I (p. 266); C/1793 S2, 1793 I (p. 268, an independent discovery of a comet first discovered by Messier 10 days earlier); 2P/1795 V1, formerly 1795 (p. 268); C/1797 P1, 1797 (p. 270). See Bennett, 1978, 55. We thank David Dewhirst and Alan Kucia for their assistance with the Herschel Archive.

120 Two of which are under her brother William Herschel's name ("Observations on a Comet," *Philosophical Transactions*, 79, 1789, 151-153; "Account of a comet," *Philosophical Transactions*, 82, 1792, 23-24).

121 Inv. no. C1/1.3, pp. 104-105, in the Herschel Archives of the Royal Astronomical Society Library, London; p. 104 in graphite and p. 105 in graphite and pen and brown ink, both measuring 210 × 168 mm. C1/1.3, *Book of Observations*, covers the time from 25 August 1797 to 31 January 1824. See Bennett, 1978, 54-55. The wrapper for C1 is inscribed: *This is what I call/ the Bills & Rec.ds /of my Comets*. For these comets, see chap. III, nn. 34, 42 (1811, 1807, respectively).

122 M.A. Hoskin, "Caroline Herschel," *Dictionary of Scientific Biography*, ed. C.C. Gillispie, New York, 1974, 322-323.

123 Howse, 1989, 155.

124 E. Roemer, "Jean Louis Pons: Discoverer of Comets," *Astronomical Society of the Pacific, Leaflet*, no. 371, May 1960, 1-8.

125 O. Gingerich, "The Naming of Uranus and Neptune," *Astronomical Society of the Pacific, Leaflet*, no. 352, October 1958, 7; R.W. Smith, "The Cambridge Network in Action: The Discovery of Neptune," *Isis*, 80, 1989, 395-422.

III

The Comet-Crazed Century Opens

The nineteenth century in Great Britain witnessed an unusually large number of spectacular comet apparitions, an unprecedented record that has not been matched since.* Perhaps this is not surprising given the accelerated interest in science and things scientific, although it may be a statistical fluke. Most of the institutions associated with modern science and newly defined disciplines, such as geology, biology, electromagnetism, and stellar astronomy, were established in the early part of the century. In addition, a large number of scientific objects, devices, and methods were offered to a fascinated public. And the very term "scientist" was coined in 1833.[1] This situation created a cultural climate in which comets and meteors remained objects of intense fascination. Moreover, journalism and satirical illustrations, which had their roots in the eighteenth century, became an even more influential force in which these phenomena were used to lampoon the foibles of contemporary politicians and culture. The comet and meteor images of the first three decades, which we will examine in this chapter, fall neatly into three categories: satirical and humorous, realistic and descriptive, and visionary.

The Comets and Meteors of Blake and His Circle and the Great Comet of 1811

The Romantic Blake, a link between the eighteenth and nineteenth centuries, was attracted to the writings of the seventeenth-century author John Milton, whose poetry is infused with comet/meteor symbolism. As a result, Blake executed twelve watercolor illustrations for Milton's poems *L'Allegro* and *Il Penseroso* on paper watermarked 1816.[2] These illustrations, together with Blake's commentary, both celebrate and critique the Puritan poet. In his watercolor of the Wandering Moon observed by Milton dressed in academic garb [Fig. 52], Blake rendered fanciful meteors to illustrate *Il Penseroso II*, 67-70:

> To behold the wandring moon
> Riding near her highest noon
> Like one that had been led astray
> Through the heaven's wide pathless way.

However, the terror that Blake ascribes to the passage in his notes is not in the poem:

> Milton, in his Character as a Student at Cambridge, Sees the Moon terrified as one led astray in the midst of her path thro heaven. The distant Steeple seen across a wide water indicates the sound of the Curfew Bell.[3]

Here, Blake has personified the Moon as a youthful maiden, related to the classical goddess of the moon, Artemis/Diana. On either side of her, he drew stars connected by tails resembling plant tendrils. Dunbar believes that they represent the constellations Orion and Ophiuchus and that the lines linking them are similar to those arbitrarily drawn to connect stars on astronomical charts and sky maps.[4] Neither configuration, however, approaches those of the two constellations. We believe that Blake's forms are both more whimsical and more in line with Milton's theme of disorder, because Blake intended them to be meteors. In fact, they resemble those Blake drew earlier as Edward III's meteor-like whip [Fig. 42].[5] Moreover, their identification with meteors or meteor showers – that is, with "errant" stars – harmonizes with the wandering quality Milton ascribes to the Moon.

Blake's illustrations to his own Prophetic book *Milton* provide commentary on this illustration and insights into how Blake regarded Milton. Despite his visionary gifts, the eventually blind poet was also, according to Blake, the proponent of two signal "errors": a belief in tyranny, both personal and domestic, and an attitude of self-righteous sexual morality. Blake had exposed these errors in his illustrations for *L'Allegro* and *Il Penseroso*, but in his final plate represented Milton as having overcome both. Blake offers a similar interpretation in his *Milton*, in which the Puritan

Figure 52. William Blake, *The Wandering Moon*, from Milton's *Il Penseroso II*, 67-70, 1816, pen and black ink and watercolor, 161 × 123 mm, The Pierpont Morgan Library, New York.

poet descends to earth a second time to correct the errors of his previous existence.

Certainly Blake was inspired not only by Milton's epic poetry but also by his prolific use of comet and meteor references. Blake depicts shooting stars on four pages of *Milton: A Poem in Two Books* (1810), a marriage of his personal poetry and visual art [Figs. 53-55].[6] Each time a meteor appears in *Milton*, it symbolizes poetic inspiration and embodies the soul of the great English poet, a bright star in the literary constellation. The initial one is on the first page of text, lettered at the top in a large, bold script *MILTON/ Book the First*. These letters are encased in the flowing tail of an ascendent shooting star with five points, whose upward

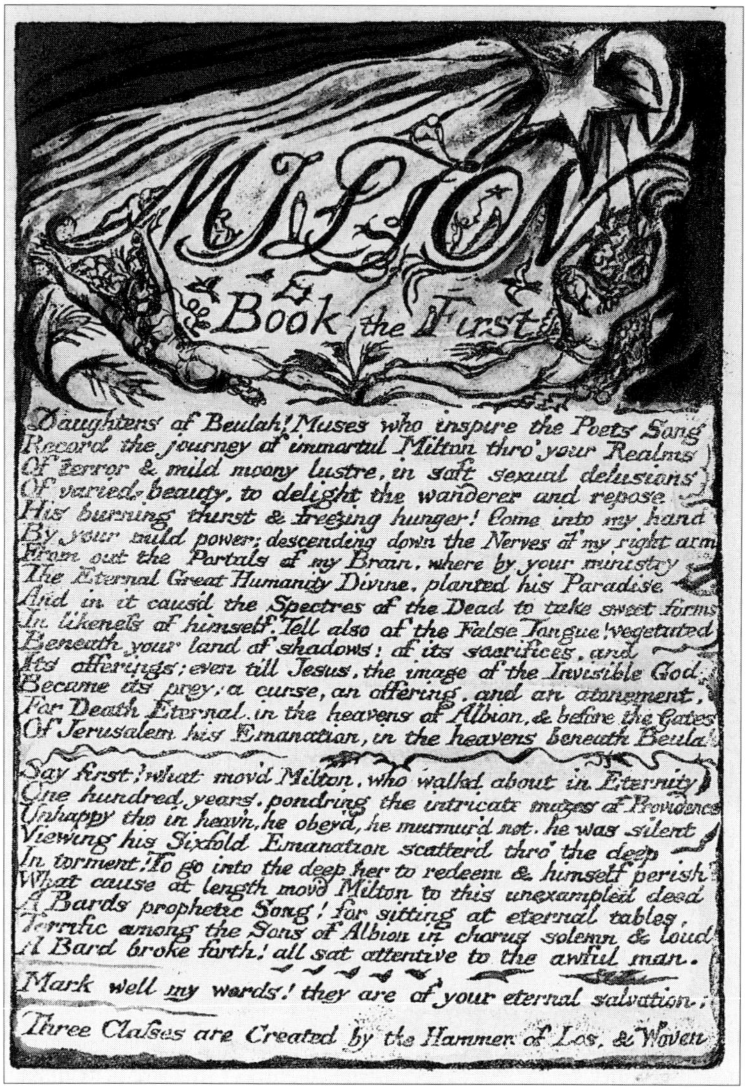

Figure 53. William Blake, *Illustration for "Milton: A Poem in Two Books,"* plate 3 (changed to 1), 1804-11, relief etching with watercolor and pen and ink, 230 × 149 mm (sheet) and 149 × 104 mm (image), The New York Public Library, Rare Books Division, Astor, Lenox and Tilden Foundations.

motion hints at Milton's redemption [Fig. 53]. Another on plate 14 features a small image of a six-pointed star with a fanning tail about to hit an ecstatic nude male (the artist) on the left foot, as described in the following lines:

> Then first I saw him in the Zenith as a falling star....
>
> Descending perpendicular, swift as the swallow or swift:
>
> And ere my left foot falling on the tarsus, entered there....

In the most dramatic illustration [Fig. 54], Blake rendered an enlarged version of the previous image in which Milton, represented as a five-pointed star with a flaming tail – hand-colored in

Figure 54. William Blake, *Illustration for "Milton: A Poem in Two Books,"* plate 29, 1804-11, relief etching with watercolor and pen and ink, 230 × 149 mm (sheet) and 154 × 107 mm (image), The New York Public Library, Rare Books Division, Astor, Lenox and Tilden Foundations.

Figure 55. *(Top right)* William Blake, *Illustration for "Milton: A Poem in Two Books,"* plate 33, 1804-11, relief etching with watercolor, pen and ink, and gold, 230 × 149 mm (sheet) and 169 × 114 mm (image), The New York Public Library, Rare Book Division, Astor, Lenox and Tilden Foundations.

delicate blues, pinks and electric greens in copy C – boldly calls out Blake's name, "*WILLIAM*," as it charges like an electrical current through the artist's spasmodically racked body. A mirror image on plate 33 [Fig. 55] shows William's deceased brother, Robert, about to be hit on the foot by another plummeting star. The letters *ROBERT* above the figure indicate that Milton is paging him. This image is much darker in tonality, perhaps because Robert is deceased. Further, in copy C, it is enlivened by gold pigments that lend it a vivid, rich, and other-worldly quality. In both of these, the stairs promise a symbolic ascent to which Milton's inspiration presumably leads.

Other related comet and meteor images appear in Blake's *Jerusalem*, the last of his major illuminated books (c. 1804-20).[7] The illustrations, integrally woven into the unorthodox religious text, reveal Blake's fascination with comet/meteor symbolism.[8] On plate 20 [Fig. 56], Blake knit at least five such celestial forms into the friezes of the dense composition, including two in the lower

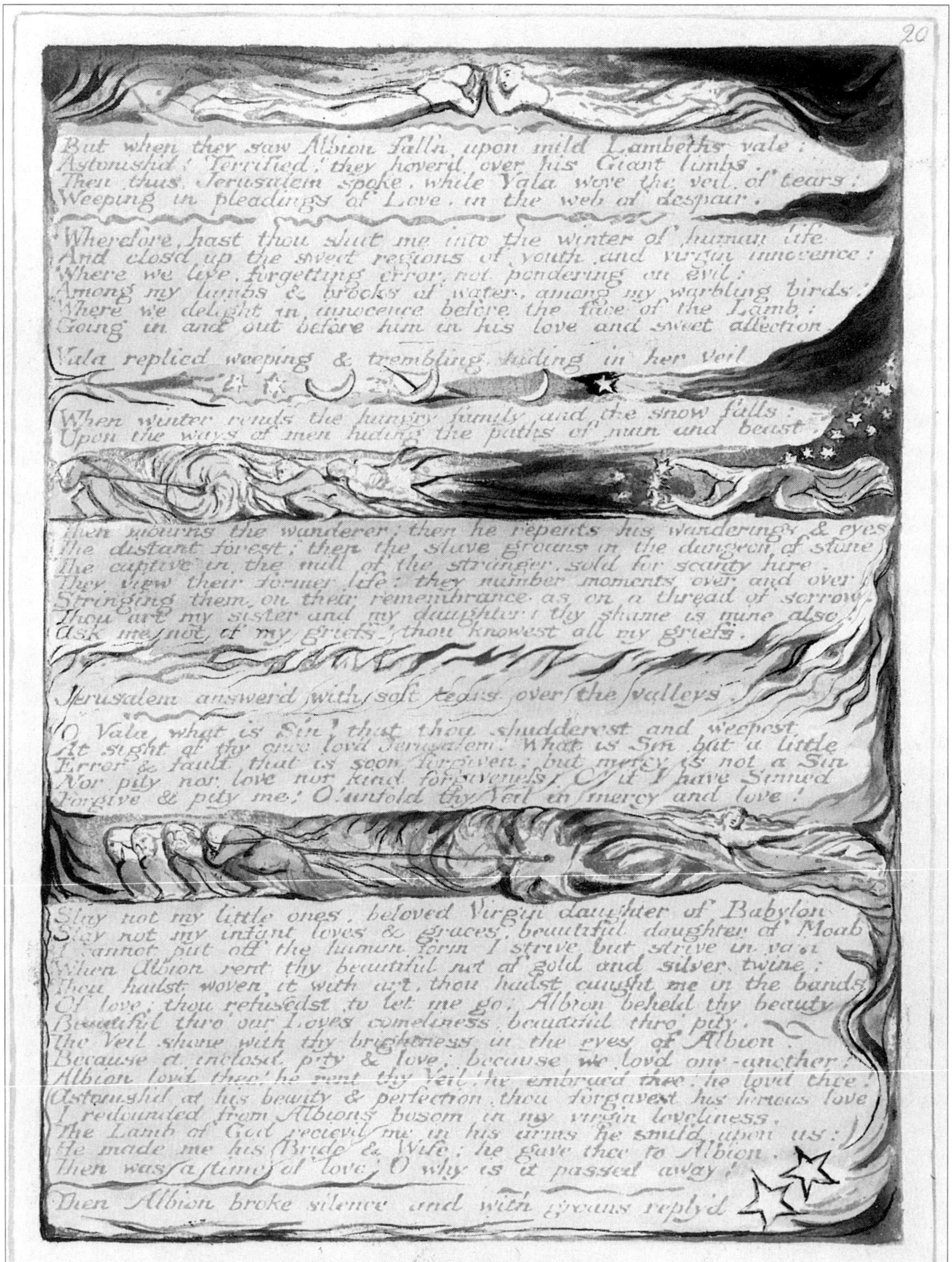

Figure 56. William Blake, *Illustration for "Jerusalem,"* plate 20, 1804-20, relief etching, printing color orange, with pen and ink and watercolor, 222 × 160 mm (plate), Yale Center for British Art, Paul Mellon Collection, New Haven.

right corner with watercolor highlighting their tails. Like many symbols in Blake's work, they held a personal meaning for him. On plate 85 in the lower frieze [Fig. 57], a comet dominates the sky. Unlike the comets and meteors in his earlier works, those in *Jerusalem* do not illustrate images in Blake's text but rather infuse a visual cosmic dimension to the artist's poetic speculations.

One of Blake's most powerful images in *Jerusalem*, which may or may not be a comet or shooting star, occurs on plate 97 [Fig. 58], where the male nude lunges forward at the brink of a chasm. To his right hang a crescent moon and a star, while his left hand touches a ball that resembles the head of a comet-like form plunging downward off a precipice. Paley identifies this ball as a globe of light and believes that it may refer to a tool used by engravers: they placed a globe of water between a candle and the copper plate to illuminate their work. He also thinks that this "globe of radiance" may represent the sun.[9] Because the head of this cosmic form is round rather than star-shaped, Paley's interpretation may be correct, although there is a star-shaped corona around it. Nevertheless, the form has affinities with comets and shooting stars, especially those in Blake's *Milton* [Figs. 53-55] and may be equivocal. The text only hints at the meaning this cataclysmic image had for him.

Blake also employed comets in his rarer paintings, such as *The Spiritual Form of Pitt Guiding Behemoth* [Fig. 59].[10] The painting was exhibited in Blake's only single-person exhibition as a companion to *The Spiritual Form of Nelson Guiding Leviathan.*[11] In this visionary, arcane political allegory of Pitt, Blake demonstrates that he was no stranger to the tradition in which rulers and their destinies were associated with comets. His apocalyptic scenario of war is based on his own "vision," where war appears as a perversion of energy (in itself a virtue) and as a prelude to the Last Judgment. The figure of William Pitt the younger, Prime Minister of England when these pictures were painted, dressed in quasi-biblical robes and wearing a halo, dominates the composition. He stands above Behemoth, controlling the monster by means of a rein held in his right hand, and is surrounded by other motifs of war and cosmic chaos. Many of the forms relate to Blake's other works as well as to Revelation and Michelangelo's Last Judgment fresco. Some people divine a secondary meaning in the painting and its pendant,

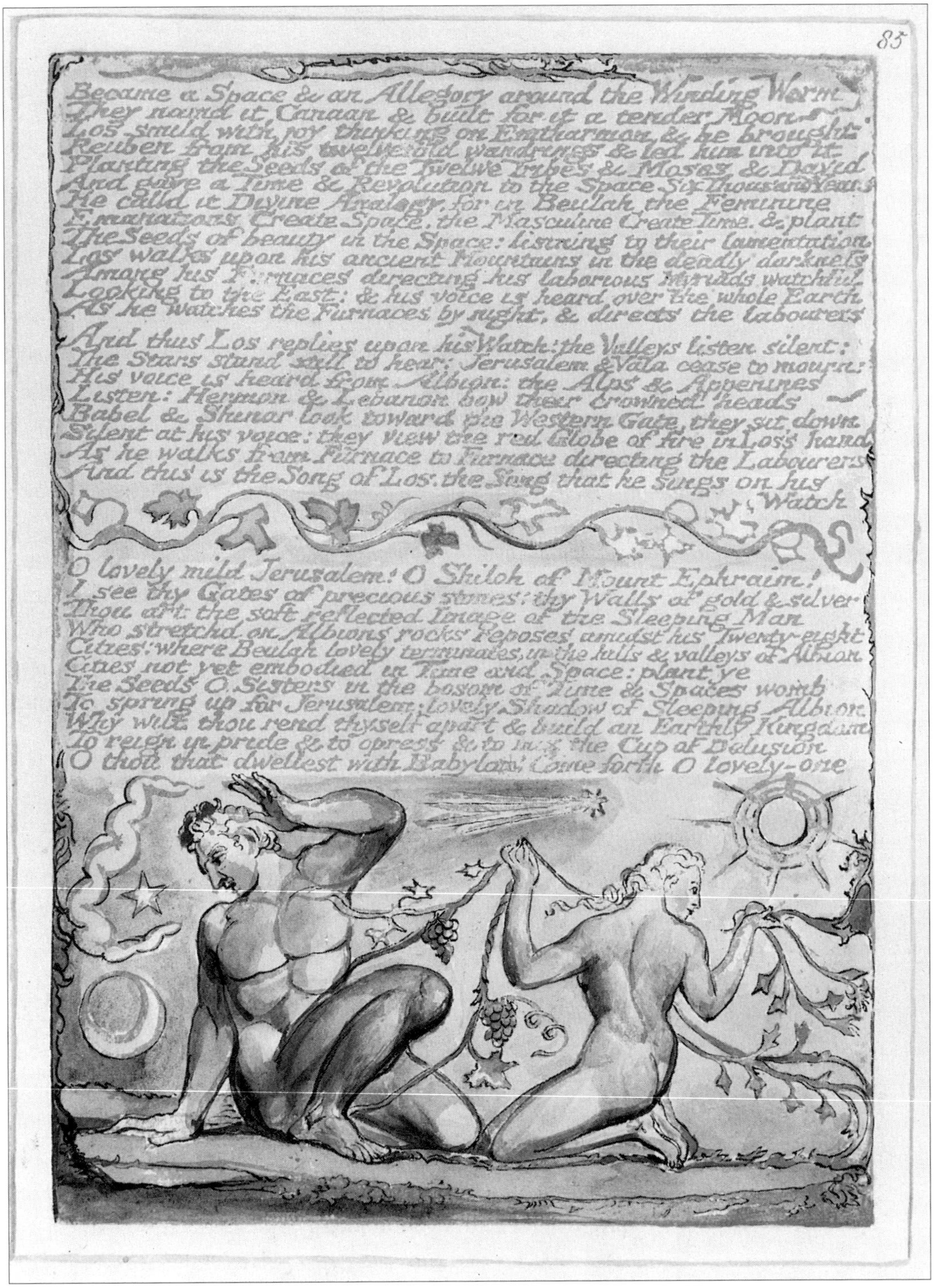

Figure 57. William Blake, *Illustration for "Jerusalem,"* plate 85, 1804-20, relief etching, printing color orange, with pen and ink and watercolor, 207 × 150 mm (plate), Yale Center for British Art, Paul Mellon Collection, New Haven.

Figure 58. William Blake, *Illustration for "Jerusalem,"* plate 97, 1804-20, relief etching, printing color orange, with pen and ink, watercolor, and gold, 207 × 149 mm (plate), Yale Center for British Art, Paul Mellon Collection, New Haven.

Figure 59. William Blake, *The Spiritual Form of Pitt Guiding Behemoth*, 1805 (?), tempera heightened with gold on canvas, 74 × 62.7 cm, The Tate Gallery, London.

believing that they express Blake's belief in the righteousness of the war then being waged against Napoleon.[12] Only Olson has commented on the comet with its six-pointed star and vaporous tail at the far left, which leads the viewer into the work.[13] It supplements the apocalyptic tenor of this allegory – echoing the one in Revelation (chapter II, n. 98) – and refers to the links between rulers and comets as well as between war and comets.

Nothing quite prepares one for Blake's strangest and most haunting image, *The Ghost of a Flea*, with his most memorable comet [Fig. 60 color].[14] It was painted c. 1819-20 after Blake had befriended John Linnell and the Varley brothers, all primarily landscapists with interests in celestial matters. John Varley (1778-1842) – who was involved with astrology and attended the séance at which this most famous of all Blake's apparitions appeared – spoke of its creation long afterwards to the biographer Allan Cunningham.

> I called on him one evening, and found Blake more than unusually excited. He told me he had seen a wonderful thing – the ghost of a flea!... He looked earnestly into a corner of the room, and then said, "here he is – reach me my things—I shall keep my eye on him. There he comes! his eager tongue whisking out of his mouth, a cup in his hands to hold blood and covered with a scaly skin of gold and green;" – as he described him so he drew him.[15]

Varley also related that while Blake drew a preparatory sketch, "The flea told him that all fleas were inhabited by the souls of men, as were by nature bloodthirsty to excess." Note how this relates indirectly to the huge comet that Blake represented in the panel, because comets have also been referred to as "bloodthirsty." How insidiously terrifying is Blake's vision even today. His image of human wickedness, a scaly specter, half-human and half-beast, dances onto the stage in an incongruous posture, his tongue flicking about like that of a lizard, his right hand holding a double-pronged stinger, and his bulging eye aglow with infernal light.

Several drawings allied to the painting illuminate its origin. One page of the Blake-Varley sketchbook (c. 1819) contains a drawing of the head of the flea, whose genesis John Varley describes in his *Treatise on Zodiacal Physiognomy* (1828), accompanied by an enlarged sketch of its open, predatory mouth.[16] Varley further relates that Blake also drew the whole figure of the flea, which Varley intended to engrave for his *Zodiacal Physiognomy*, but never did. This drawing was formerly page 94 in the Blake-Varley sketchbook and is currently in a private collection.[17]

It has been suggested that Blake's drawings and, therefore, his painting may have been influenced, consciously or unconsciously, by Robert Hooke's *Micrographia, or some Physiological Descriptions of Minute Bodies made by Magnifying Glasses, with Observations and Inquiries thereupon* (1665), in which there is an engraving of a greatly enlarged and detailed flea.[18] Klonsky notes that Hooke observed in his treatise the flea's "curiously polish'd suit of sable Armour, neatly jointed" and the tongue or sucker as well, which was perceived "to slip in and out," just as Blake painted it in his apparition. Klonsky believes but does not fully explain that the "cup in his hand to hold blood" (according to Varley) was really the aperture of Hooke's microscope, and reproduces an engraving of the microscope.[19] In light of the visual evidence he offers, such a fusion would be possible only via a kind of hallucinatory association, of which Blake was clearly capable.

While most people discuss the large specter dominating Blake's panel, which is the ghost of a flea, they, save Klonsky and Singer, ignore the tiny, barely discernible corpse of the flea, which lies on the floor of the proscenium space between the striding legs of the demonic ghost. This juxtaposition implies the flea is dead, hence the looming ghost, and indirectly alludes to the fact that the bubonic plague, which was carried by fleas, was traditionally associated with negative comet portents. While the tiny flea in Blake's painting could be allied to the one in Hooke's treatise, it is so minute and undetailed that it probably was not patterned after the highly magnified, detailed engraving. It seems more likely that Blake may have been exposed to Hooke's treatise and its engraving of the flea, which made such a strong impression on him that it caused his associative and hallucinatory faculties to conjure up his horrific visitation.

Blake's portrayals of the flea, together with other supposed "manifestations" of historical personages made at the request of John Varley, also involved the third member of the artistic triumvirate associated with these visions, John Linnell (1792-1882). Later, Linnell engraved some of the drawings in the Blake-Varley sketchbook – including two heads of the flea for the first and only part of Varley's projected, pseudo-scientific *Treatise on Zodiacal Physiognomy* (1828). Varley's note criticizing the first state of "The Ghost of the Flea" mentions the existence of tracings. It seems likely, therefore, that some drawings were replicated as aids to the engraving. Perhaps some were drawn with the aid of John Varley's brother Cornelius's (1781-1873) invention, the Graphic Telescope. Five examples of the Graphic Telescope, which was patented in 1811, are now in the Science Museum, London, including Varley's own apparatus.[20] The Graphic Telescope developed out of the *camera lucida*, which evolved from the *camera obscura*, and reproduced a magnified image on paper where it could be traced. This information demonstrates that Blake knew someone who had a reason to own Hooke's treatise and that there was interest in magnification in the Blake-Varley-Linnell circle.

Blake's huge comet in *The Ghost of a Flea* is directly linked to the flea and its more frightening ghost because it functions as a potent allusion to the blood and destruction caused by the flea's rapacious nature. Moreover, it is the most convincing comet of any in Blake's oeuvre. Despite the five-pointed star Blake rendered in its head, the form's phosphorescent glow, coma, and tail convey the vaporousness of actual comets, unlike his earlier stylized comets. Most likely, it reflects Blake's memory of the Great Comet of 1811-12 (C/1811 F1, formerly 1811 I), the only spectacular apparition of his lifetime, as suggested by Olson.[21] His experience of it was probably etched in his memory, as it was in that of everyone who saw it.

Certainly the impact of Blake's new friendship with Linnell, together with Linnell's ideas about nature, finds reflection in Blake's more naturalistic comet in *The Ghost of a Flea*. Blake met Linnell, a student of John Varley, in 1818. Before their meeting, Blake's comets had been stylized and heavily indebted to literary tradition. Moreover, we know that both Blake and Linnell shared the conviction that all natural phenomena harbor allegorical meanings and that Linnell was involved with the *Ghost of the Flea* because he engraved Blake's drawings for Varley's treatise.[22] In addition, it is significant that Linnell himself studied and drew the Comet of 1811 several times. Whereas Blake's comet communicates a feverish demonic energy, Linnell's comet drawings reveal a great reverence for nature and an objective, scientific, point of view.

Three of Linnell's drawings of the Great Comet of 1811 are gathered in an album [Figs. 61, 63, 64].[23] The leaves, a collection of independent sheets, were registered as "Three sketches and diagrams of the comet of September 17, 1811; and three drawings of the eclipse of the moon [*sic*], November 19, 1816" and were subsequently bound together. In these annotated studies of scientific phenomena, Linnell demonstrates links between science and art that are not apparent in his landscapes and portraits.[24]

Linnell has been described as "our greatest English naturalist."[25] His naturalism followed from his religious conversion in 1811, the year of the comet depicted in this series. Beginning as a nominal Anglican, Linnell eventually joined a Baptist community in Bloomsbury, although his beliefs were rooted firmly upon those of the Utilitarian philosopher and theologian William Paley.

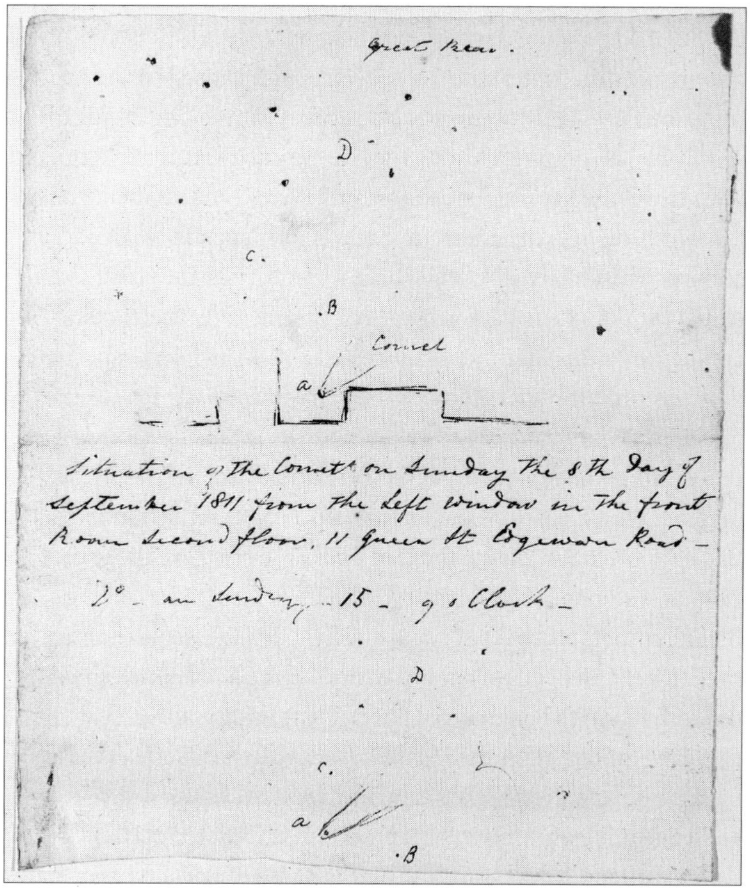

Figure 61. John Linnell, *Studies of the Comet of 1811*, 1811, pen and brown ink on white paper, 197 × 165 mm, The British Museum, London.

This dichotomy led him to understand the meticulous design of nature as direct proof of God's existence, so that truth to nature was not just an aesthetic criterion but also a moral obligation.[26] With this information, we can better understand his motivation to record the awesome spectacles of these two astronomical events. Linnell did not regard them superstitiously but rather felt that they revealed the divine presence in the universe.

Linnell's first sheet studying the Great Comet of 1811 [Fig. 61], the only one in ink, is the most precise and scientific.[27] In fact, it resembles diagrams drawn by astronomers of the time in manuscripts and in the observational records of the Astronomers Royal.

It is not known, unfortunately, whether Linnell associated with astronomers, although his friends had allied interests. And, it is significant that in his unpublished *Journals* Linnell records in the early days of September 1811 meetings with both John and

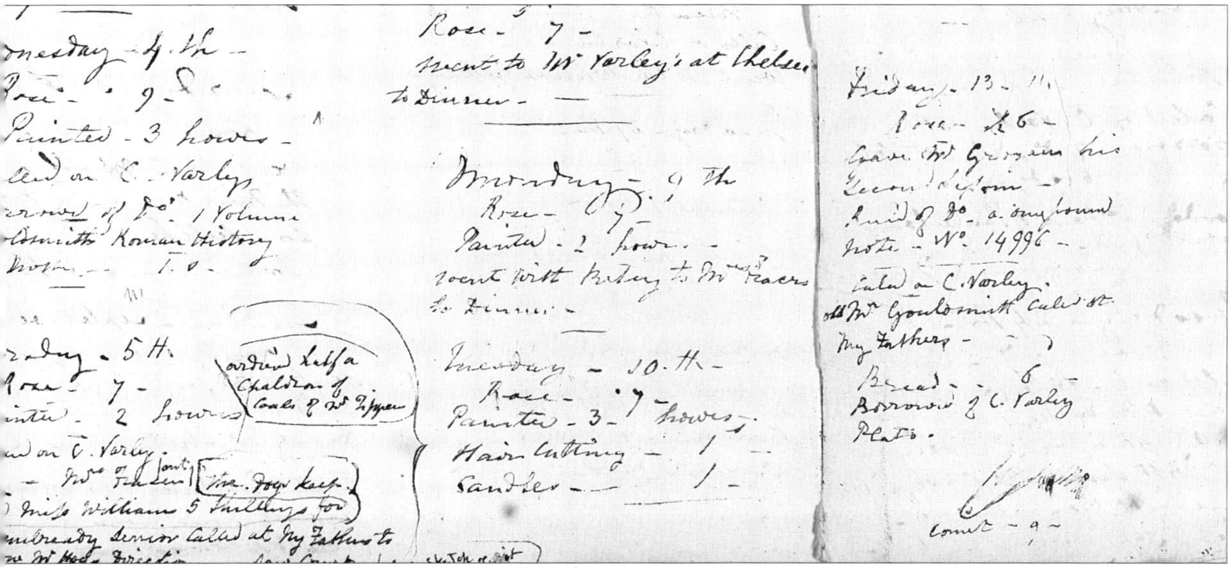

Figure 62. John Linnell, *The Comet of 1811*, detail showing the 5 and 13 September entries of the artist's *Journals*, folios 31v and 32r, 1811, pen and brownish gray ink on pages around 254 × 212 mm (irregular), The Fitzwilliam Museum, Cambridge.

Cornelius Varley, the latter of whom he was seeing on nearly a daily basis during the most intense phase of their friendship in this the very year Cornelius Varley patented his Graphic Telescope. From his three comet drawings, especially the first, it is apparent that Linnell possessed technical astronomical knowledge but where he acquired it we do not know.

Although Linnell's *Journals* are comprised mostly of brief and erratic domestic jottings and notations about paintings (many of them bread and butter portraits), one passage records his observation of the comet. He notes on Thursday 5 September at the end of a list: *Saw. Comet. (See sketch of it)*. The parenthetical phrase appears to have been appended later with a different pen nib. At the lower left of the opposite page under the listings of Friday 13 September is a sketch of the comet [Fig. 62]. It is inscribed beneath the pen and ink drawing: *Comet -9-*, meaning that Linnell observed it at 9:00, as noted on other studies. This small sketch is different than the artist's three drawings of the comet in the British Museum, but probably was a shorthand reference to them and perhaps others that are still unknown or lost.[28] It seems strange that in his *Journals* Linnell did not record his other observations of the comet, which from the British Museum drawings we know he drew on the 8th, 12th, 15th, and 17th of September. With the information from his journal, however, we now know that he observed it from the 5th to the 17th of September, three days longer than noted in the album.

Linnell's first comet drawing in the album sets the scene for the others and the *Journals* entry because it is the only one to plot the comet's celestial position with reference to labelled constellations, stars, and asterisms. At its center, dividing the two studies on the page, is the following inscription: *Situation of the Comet on Sunday the 8th day of/September 1811 from the Left window in the front/Room Second floor, 11 Queen St Edgeware Road-/ 2°. Sunday - 15 - 9 o Clock-.*[29] On this sheet, Linnell drew comet 1811 I twice on two successive Sunday nights, the 8th and 15th of September. Both times he pinpointed the apparition (labelled "a" in lower case) by relating it to some of the surrounding stars ("B," "C," and "D"). In the upper sketch of 8 September, the comet is positioned low on the horizon and shown with two rectangular shapes, which appear to be either the outlines of buildings or chimneys seen from the artist's vantage point. Near the top center of the paper, Linnell has indicated with his inscription the *Great Bear*. Below, labelled "D" at the center of its cup, is the asterism in Ursa Major (the Great Bear), known in the United States as the Big Dipper and in Great Britain as the Plough. Most interestingly, Linnell has also drawn the middle star of the three in the Big Dipper's handle correctly, as a double star, with the fainter star Alcor above and the brighter star Mizar below. (Whether he knew of the split of Alcor and Mizar from observation or from references we do not know. The split is barely visible to the naked eye but is readily so in binoculars or a telescope; we know that Linnell used a telescope [glass] as noted on another drawing he made of the comet discussed below.) Beneath the Big Dipper are two other stars of Ursa Major. The upper one on the left, inscribed "C," is the star *kappa* of the constellation. Below it to the right, inscribed "B," is *phi*. The second sketch of this first sheet – which falls below the central inscription – was made, according to the artist's notations, a week later and represents a quicker view of the comet's position on 15 September 1811. It lacks the celestial and earthly setting of the earlier one. However, it contains the same sidereal signposts, inscribed "B," "C," and "D." In a comparison with the upper sketch, it is apparent that the comet's position in the lower has

Figure 63. John Linnell, *Two Studies of the Comet of 1811*, white chalk on gray paper, 304 × 221 mm, The British Museum, London.

shifted in relation to the stars; the comet is now above *phi* and close to *kappa*. In both, Linnell has indicated the comet's anatomy: the head with its center of condensation (the nucleus is hidden and, in any case, invisibly small), indicated with a dot of ink, and the coma that sweeps back to form a broadly fanning dust tail. From the evidence in this sheet, it is probable that Linnell had gleaned knowledge of both comets and fixed stars from an astronomical source. It is noteworthy and unusual that this drawing is so analytical and that in it he is more interested in the comet's celestial position and accurate anatomy than in its stunning visual effects.

Linnell's second comet sheet in the album [Fig. 63] contains two diverse studies of comet 1811 I.[30] A quick inscription in black chalk, just below center identifies the time: *½ past 9 O Clock/ Tuesday Evening September 17 – 1811* (the "1811" has been gone over in brown ink). Below at the left is the second study of the comet's head with an inscription under it in the same black chalk: *Glass*.[31] The upper study once again shows the Big Dipper and the two fixed stars *kappa* and *phi* Ursae Majoris, drawn as white chalk dots without alphabetical labels. Linnell continued to record the comet's position accurately; the tail, drawn in a light pinkish chalk, now encompasses *kappa*. The lower study on the left near *Glass* is a closer view of the comet's head, and its inscription tells us that its magnification results from using a telescope. Viewing the comet thusly, Linnell corrected his rendering of its contours in the earlier study and deftly suggested the hazy, pinkish configuration of gases forming the coma around its center of condensation, which he noted as a dot of white chalk. His careful recording underlines Linnell's scientific methodology. By his use of predominantly white chalk, he created a more subtle study to emphasize the comet's mesmerizing luminescence, qualities that pen and ink could never capture.

Linnell's third album sheet studying the Great Comet of 1811 [Fig. 64] has a tripartite inscription.[32] The first section, in black chalk below the landscape and the comet that occupies two-thirds of the entire sheet, reads: *9 o Clock. Thursday Evening.* By analyzing the comet's position in relation to the fixed stars and asterisms, unlabeled as in the second sheet, we can conclude that this study falls between the two dates indicated on the first sheet (8 and 15 September). Thus it dates to Thursday, 12 September 1811, the day of perihelion. Directly below, Linnell drew an almost hallucinatory study of light as it falls on a small sphere that casts a shadow, which he has calculated with a line suggesting a triangular form. This sketch, rather like a doodle, indicates the artist's analysis of light, a thought tangential to his focus on the comet. Below this illuminated sphere (which bears comparison with Blake's sphere in Fig. 58) is the second installment, probably a repetition, of his inscription in black chalk and graphite: *9 o Clock. Evening - Thursday - September/ 13th* (the *13th* is written above the last letters of September when the artist ran out of space). No doubt he

Figure 64. John Linnell, *The Comet of 1811 and a Sketch of a Sphere*, black and white chalk on gray paper, 295 × 221 mm, The British Museum, London.

meant the 12th since the " – September/13th", written in graphite in a tighter hand in a crowded space, looks as if it might have been added as an afterthought, thus making the mistake more understandable. The third and last part of the inscription is in brown ink and a tighter penmanship, suggesting that it also was added at a later time. It is partially damaged or rubbed so that the middle is illegible: *Queen St. .....St. 1811.* Here, Linnell has once again, as in the first sheet, indicated rectangular forms on the groundline. But rather than the mere outlines of his first sketch [Fig. 61], the structures are hatched in black chalk conveying the solidity of the buildings or chimneys that they represent. For the first time in this series, Linnell has also indicated the murkiness of the night sky by a web of parallel chalk strokes, making this his most poetic and dramatically lit drawing of the group. It emphasizes and successfully captures the gossamer quality of the comet's tail, as well as its volatile, vaporous nature.

Linnell's three larger drawings of the Great Comet of 1811 record dispassionately his wonder at the magnificence of the created universe. He accomplishes this goal powerfully but unsentimentally by using a natural phenomenon instead of symbols. Thus, his comet studies are the perfect embodiment of his philosophical/theological and aesthetic beliefs and contrast with Blake's more imaginative ones.[33] The Great Comet of 1811 that Linnell studied was, significantly, one of the first whose structure was analyzed by astronomers during this Golden Age of British astronomy that began in the wake of Newton, Halley, and Herschel. Linnell's analysis of it suggests that he was knowledgeable about astronomical practices, although we can only speculate about whom he may have known in the scientific community.

The naked-eye comet of 1811-12 (C/1811 F1, formerly comet 1811 I) until C/1995 O1 (Hale-Bopp) held the record for being the longest naked-eye apparition in recorded history. It was visible telescopically from 25 March 1811 to 17 August 1812 and to the naked eye from 11 April 1811 to 20 January 1812. Its 512 days of telescopic visibility was surpassed only in 1889 when larger telescopes were available. Reputedly, its tail reached a visible span of 70° in December 1811 (20 to 30 million miles long), although Linnell studied it six weeks earlier when it was already spectacular enough to be an arresting sight. When Linnell drew the comet, it was about magnitude 4.5 (three times brighter than the faintest naked-eye stars), but it increased to nearly magnitude zero by 20 October (equal to the third brightest star in the sky). After Linnell's studies, its tail was reported to have split, and both straight and curved tails were measured by William Herschel (see below). This comet also reputedly displayed the largest coma ever recorded, whose maximum diameter is estimated to have been 1,250,000 miles, considerably larger than the sun yet with only a tiny fraction of the sun's mass. Around October, about two weeks after Linnell's drawings, the tail reached a maximum length of c. one hundred million miles and a breadth of fifty million miles.[34] No wonder that Linnell was enthralled and

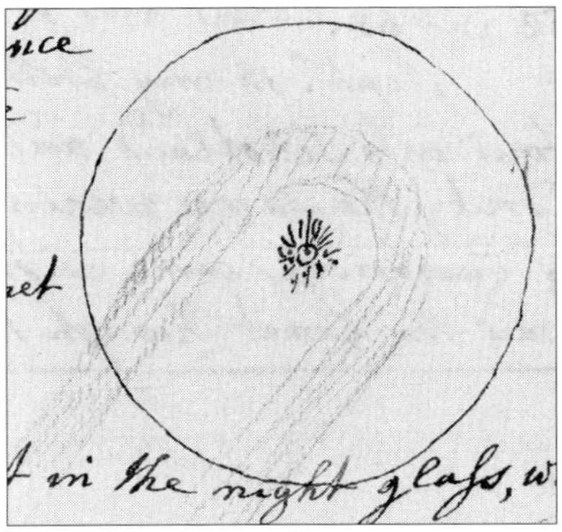

Figure 65. William Herschel, *Drawing of the Comet of 1811 (21 October)*, W3/1.12, detail of page 37, pen and brown ink, 334 × 210 mm, Herschel Archive, Royal Astronomical Society Library, London.

drawn to this sublimely awesome spectacle as tangible evidence of the divine, for his favorite theme was "man's impotence in the face of God."[35] It is truly puzzling that no articles about this comet were published in the *Philosophical Transactions of the Royal Society*.

However, Caroline and William Herschel's drawings of its apparition [Figs. 51, 65] demonstrate that it was one of the first comets whose structure was analyzed quite so carefully in astronomical drawings and text. Both Herschels drew it several times and some of their observations are recorded inside a circle to indicate the viewing field of their telescope. In the unpublished notes that accompany his drawing of this comet, William Herschel comments on the coma surrounding the head:

> I saw the head of the Comet in great perfection. By the head I mean that part which to the naked eye appeared like a star, or rather like a small planetary disk, or nucleus. This, in the telescope where its diameter was 65 × magnified, has only the appearance of an extremely round nebula.... The light of the tail was disengaged from the head and seemed to arise from a luminous envelope or halo surrounding the head at some little distance from it.... The length of the tail was about 9 or 10°. I pointed out its being curved to Mr. Smith....[36]

In the final analysis, Linnell's drawings of comet 1811 I are different in tone from other comet representations by artists of the period, even the most dispassionate and least stylized. They are closer in spirit to those by astronomers. Certainly Linnell's series, employing the artist's uncanny observational abilities, is the most visually correct and scientifically advanced study of any comet by an English artist at that date. Moreover, by comparing Linnell's third, highly naturalistic study of comet 1811 I with the comet represented by his friend Blake in his *The Ghost of the Flea* [Fig. 60], one can definitely gauge the impact of Linnell's naturalistic studies of the comet on the visionary Blake.

To appreciate the dispassion of Linnell's sketches, we compare them with the illustration of H.R. Cook (active 1800-1845), the English engraver, depicting the comet of 1811 after a drawing by the landscapist Abraham Pether (1756-1812) [Fig. 66].[37] Here, the comet is only one element in a larger topographical landscape with romantic overtones, whose small figures emphasize the

Figure 66. H.R. Cook after Abraham Pether, *The Comet of 1811, As seen at Day break the 15th October from Otterbourne Hill, near Winchester*, 1814, engraving, 100 × 170 mm, Collection of Gerald H. Morris, London.

Sublime *frisson* of its awesome spectacle, unlike the precise astronomical description in Linnell's studies devoid of figures and specific settings.

Cornelius Varley, Linnell's friend and inventor of the Graphic Telescope, may also have been interested in comets if the fascinating architectural drawing attributed to him is autograph [Fig. 67].[38] The four-storey facade – decorated for a festival set, a theater design, or an allegory – juxtaposes elements of the earth – rocks, vines, trees and vegetal growth in the lower storey – with those of the heavens – a large sun, Saturn with its rings and moons balancing Jupiter with its satellites, flanked by two other planets on each side. At the far left hangs the realistic comet, which resembles Linnell's drawings of comet 1811 I. The artist delineated its anatomy correctly – the envelopes of light around the head and a long, vaporous tail that points correctly away from the sun. The comet balances sundry stars and constellations sprinkled on the right of the facade. Because these forms are astronomical rather than astrological, as they have been described, they testify to the artist's and/or the patron's interest in celestial bodies.

Figure 67. Attributed to Cornelius Varley (?), *Palace Facade Design (Allegorical or Festival) with Earthly and Astronomical Elements*, pen and black ink and gray wash over graphite, 305 (311 uneven) × 557 mm, Yale Center for British Art, Paul Mellon Collection, New Haven.

Comets Prior to 1811

While the brilliant comet 1811 I was part of our discussion of Blake and his circle, we now need to backtrack to survey its predecessors and works of art allied to them. The only comet to appear in 1801 (C/1801 N1) is thought to have been bright since it was simultaneously discovered by several French observers.[39] In insular Great Britain, plagued or blessed by a maritime climate, the visibility of heavenly phenomena depended on the weather. This comet may have been seen in England, because Durs Egg (1770-1834), the celebrated London gunsmith who obtained the Royal Warrant of the Prince Regent, produced his *Flintlock Fowling Piece* with comet decorations in that year.[40] We can only speculate whether Egg placed the trajectory of a stylized comet – represented in silver inlay on the side of the wooden area of the barrel, forward of the trigger and hammer, as a flower-like form with a tail – to suggest the munition's direction or to function talismanically to insure that its ammunition would hit the targeted fowl. In addition, three more geometrical comets in gold are on three facets of the black metallic upper area of the barrel.[41] They are related by their placement and color to the signature *D. Egg* nearby at the breech and to the crescent moon and other stars on this area.

Between 1801 and the Great Comet of 1811-12, other comets appeared. Among them were comets in 1802 (C/1802 Q1), 1804 (C/1804 E1), and 1805 (2P/1805 U1, the third recorded apparition of Encke's Comet), the first two of which were not naked-eye.

In 1806 there were two: 3D/1805 V1 (formerly 1806 I, the second documented apparition of Comet Biela, which was observed with the naked eye only in France, although Maskelyne at Greenwich observed it with a meridian circle); and C/1806 V1 (formerly 1806 II, discovered by Pons and observed low in the sky by Caroline and William Herschel). When the very bright comet of 1807 (C/1807 R1) appeared, Caroline Herschel made sketches of it in her beautiful script [Fig. 51].[42]

What these sketches and works of art prove is that spectacular comet apparitions during the nineteenth century attracted the attention not only of astronomers but also of artists and the wider public. Every time there was a brilliant apparition, comets became even more topical and references to them crop up in the visual arts and other areas of the culture. One of most unusual is recorded by George Garrard (1760-1826), the animal painter and engraver, who painted a portrait of the celebrated bull named "Comet" and engraved it.[43] The bull's name must have been inspired by the general cometomania in the first decade of the century.

Humorous Comets and Meteors in Satirical Prints and Caricatures

Because of their topical nature, comets and meteors pepper the satirical prints and caricatures of the early nineteenth-century. Before turning to these predecessors of modern cartooning, we need to investigate their eighteenth-century roots. Prior to the birth of the satirical journal *Punch* (1841) or illustrated journalism in the 1830s, engraved social satires in Great Britain were sold separately by printmakers from the later years of George II's reign (1727-1760) through that of George III (1760-1810/20) and George IV (1820-1830, after a nearly ten year regency). This vogue for satirical prints – political as well as social – belonged to the Golden Age of English engraving, in which "straight" prints – military, topographical, sporting – were also produced. They were virtually the only pictorial rendering of the flow of events and fashions, and were displayed and sold in caricature shops. Understandably, they reflected the manias of the day.

When the word "caricature" came into use in the 1750s, print shops applied it indiscriminately to any print with comic or satirical intent, although Hogarth fought vehemently against an association with it. The word derives from the Italian verb *caricare*, meaning to load or to surcharge. The art of caricature involves exaggerating aspects of a person's appearance, costume, or situation, sometimes to the point of the ridiculous. It can range from inventive subtle statements to vulgar farce. By 1755 the word was included in Dr. Johnson's first dictionary as a verb meaning to ridicule and as a noun signifying a drawing intended as humor, satire, or comment.[44] The English were nationalistic about caricature, whose overriding theme was the progress of folly, considering it a third dimension of portraiture.[45]

The word "satire," like "caricature," is frequently used quite loosely, although it is a more all-encompassing word than caricature. It derives from *satura*, meaning medley.[46] Political satires preceded social satires and seem to have outnumbered them, although frequently there is an overlap between the two. The development of English social satire was indebted to Hogarth and the influence of Dutch prints, as well as to political or "emblematical" prints. And how appropriate that these prints began when satire was the language of the age – the epoch of Alexander Pope, Thomas Swift, and Henry Fielding – in a malicious world, where wit and ridicule were admired and feared to a degree that seems odd to us today. Because the eighteenth-century world was small in comparison to ours, people's references were fewer and their allusions were immediately apparent, keyed to contrasts of social status.

In her consideration of the satires in the British Museum, Mary Dorothy George has divided the era between the rise of these satirical prints c. 1720 and their decline in the 1830s into three subdivisions. Hogarth, who increased the prestige of his profession and agitated vehemently for the Copyright Act of 1735, dominated the first forty years. With his death in 1764 a sharp break occurs, followed by, but not directly, the Golden Age of English satire. The denouement begins with both Gillray and King George III becoming hopelessly insane c. 1810 and ends with the Regency and the rule of George IV.[47]

Independent satirical prints, which proliferated before periodicals, were related to broadsides, as is apparent in an undated etching by the Scottish-born caricaturist Isaac Cruikshank (1764-1810?), *The Flying Bishop Or Tale of a Pope* [Fig. 68]. Its poem,

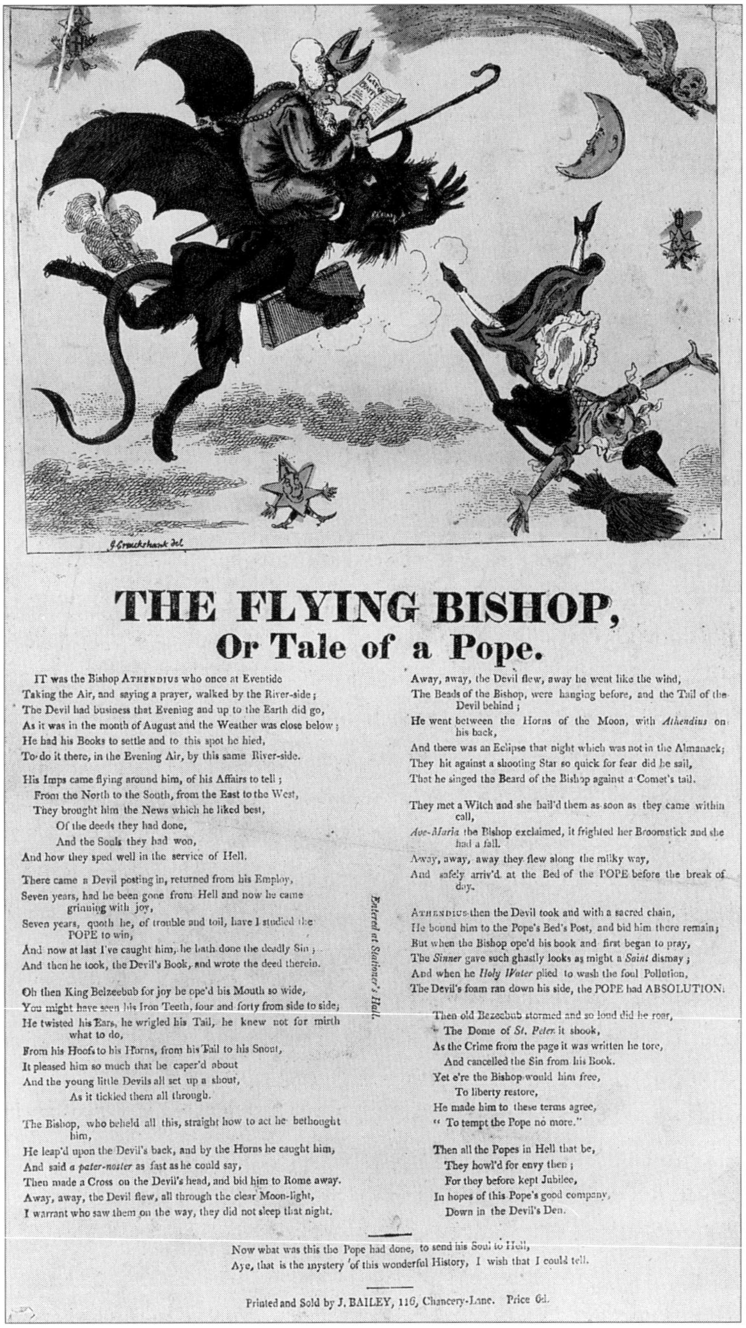

THE FLYING BISHOP,
Or Tale of a Pope.

IT was the Bishop ATHENDIUS who once at Eventide
Taking the Air, and saying a prayer, walked by the River-side;
The Devil had business that Evening and up to the Earth did go,
As it was in the month of August and the Weather was close below;
He had his Books to settle and to this spot he hied,
To do it there, in the Evening Air, by this same River-side.

His Imps came flying around him, of his Affairs to tell;
From the North to the South, from the East to the West,
They brought him the News which he liked best,
Of the deeds they had done,
And the Souls they had won,
And how they sped well in the service of Hell.

There came a Devil posting in, returned from his Employ,
Seven years, had he been gone from Hell and now he came grinning with joy,
Seven years, quoth he, of trouble and toil, have I studied the POPE to win,
And now at last I've caught him, he hath done the deadly Sin;
And then he took, the Devil's Book, and wrote the deed therein.

Oh then King Belzeebub for joy he ope'd his Mouth so wide,
You might have seen his Iron Teeth, four and forty from side to side;
He twisted his Ears, he wrigled his Tail, he knew not for mirth what to do,
From his Hoofs to his Horns, from his Tail to his Snout,
It pleased him so much that he caper'd about
And the young little Devils all set up a shout,
As it tickled them all through.

The Bishop, who beheld all this, straight how to act he bethought him,
He leap'd upon the Devil's back, and by the Horns he caught him,
And said *a pater-noster* as fast as he could say,
Then made a Cross on the Devil's head, and bid him to Rome away.
Away, away, the Devil flew, all through the clear Moon-light,
I warrant who saw them on the way, they did not sleep that night.

Entered at Stationer's Hall.

Away, away, the Devil flew, away he went like the wind,
The Beads of the Bishop, were hanging before, and the Tail of the Devil behind;
He went between the Horns of the Moon, with *Athendius* on his back,
And there was an Eclipse that night which was not in the Almanack;
They hit against a shooting Star so quick for fear did he sail,
That he singed the Beard of the Bishop against a Comet's tail.

They met a Witch and she hail'd them as soon as they came within call,
Ave-Maria the Bishop exclaimed, it frighted her Broomstick and she had a fall.
Away, away, away they flew along the milky way,
And safely arriv'd at the Bed of the POPE before the break of day.

ATHENDIUS then the Devil took and with a sacred chain,
He bound him to the Pope's Bed's Post, and bid him there remain;
But when the Bishop ope'd his book and first began to pray,
The *Sinner* gave such ghastly looks as might a *Saint* dismay;
And when he *Holy Water* plied to wash the foul Pollution,
The Devil's foam ran down his side, the POPE had ABSOLUTION.

Then old Bezeebub stormed and so loud did he roar,
The Dome of *St. Peter* it shook,
As the Crime from the page it was written he tore,
And cancelled the Sin from his Book.
Yet e're the Bishop would him free,
To liberty restore,
He made him to these terms agree,
" To tempt the Pope no more."

Then all the Popes in Hell that be,
They howl'd for envy then;
For they before kept Jubilee,
In hopes of this Pope's good company,
Down in the Devil's Den.

Now what was this the Pope had done, to send his Soul to Hell,
Aye, that is the mystery of this wonderful History, I wish that I could tell.

Printed and Sold by J. BAILEY, 116, Chancery-Lane. Price 6d.

Figure 68. Isaac Cruikshank, *The Flying Bishop Or Tale of a Pope*, c. 1790-1810, hand-colored etching, 380 × 221 mm (paper), The Pierpont Morgan Library, Peel Collection, New York.

reflecting centuries of antipapist English propaganda, relates the peregrinations of a popish bishop, on his way to Rome on the back of the devil, who singes his beard on a comet, runs amuck a shooting star, and collides with a stereotypical witch.[48]

The demand for satirical prints in the late eighteenth and early nineteenth centuries spawned talented caricaturists, some of them amateurs, who produced for the bawdy, exuberant market. With the creation of the Royal Academy in 1768, they were encouraged to improve the quality of English engraving while simultaneously being discriminated against and barred from membership. In consequence, shops sprang up that dealt solely with satirical prints. The average price for a print was sixpence plain and one shilling colored, until the end of the eighteenth century when the price doubled. There were exceptions like that of Thomas Tegg, who in 1807 started a business that still sold crudely colored prints for one shilling. Even so, many people could not afford that price and were reduced to looking in print shops' windows. The number of impressions from each plate varied from a few dozen to thousands. The color was usually mass produced by assistants, sometimes children, who applied the same color in assembly line fashion, although some have more carefully applied color. The market was fed by current political upheavals and controversies, which provided grist for satirists' mills. While these prints were the height of fashion, gentlemen hired portfolios of them to provide entertainment for parties and social events. Naturally, therefore, they were influential on public taste and opinion. Occasionally, politicians used the caricaturists for their own ends; for example, the Tory Pitt paid the caricaturist James Sayers, who was really a political bravo, for visual attacks on his Whig rival in Parliament, Charles James Fox.

It is not surprising, therefore, that these illustrators quickly latched onto the ancient visual and symbolic traditions associated with comets and meteors, incorporating them into their political attacks, while simultaneously lampooning their superstitious content. Because the vogue for caricature and satire was also fed by the British xenophobic streak, it was natural that in the early nineteenth century one of the favorite targets for British satirists was Napoleon, who looked to the tradition associating comets with rulers and victory to legitimize his reign and conquests. Thus, British artists produced scathing prints that mock both Napoleon and his rhetorical infatuation with comets.

Toward 1830, caricature shops were no longer the center of pictorial journalism. A phenomenal increase in printed media and illustrated periodicals started to eclipse them, beginning with *The Scourge* and continuing with the appearance in 1841 of *Punch*, subtitled the "*London Charivari*" (a word meaning a raucous cacophony). There even was a periodical entitled "*The Meteor*," which, true to its name, existed briefly. These developments coincided with the increasing size of the reading public and the advent of steam-driven presses, providing cheaper and quicker methods of reproduction to feed the demand for images in this age of demographic change. Technology continued to influence art with the perfection of lithography (c. 1798) and wood engraving (1780-1800).

The English political, social, and economic climate during the period encompassed in this chapter lent itself expeditiously to satire, for it was a time of sweeping changes that threatened accepted values. It was marked by the political consolidation of Great Britain, by great population increases, and by a gradual erosion of old agrarian systems which were replaced by urban centers and industrialization. During the Georgian and Regency periods even the king's authority was questioned, and there were constitutional and parliamentary reforms.[49] The French Revolution, as well as the menace of Napoleon, impacted heavily on the insular nation. All the while, the problems of England's colonies and expanding imperial structure had to be addressed. Just as writers dealt with these themes in literature, artists who rendered comet and meteor images expressed related ideas. Since comets and meteors were symbols of change, they lent themselves to the expressions of the age.

James Sayers (active 1748-1825) seems to have produced the first British political satire with a comet in 1789 [Fig. 69].[50] Before turning to caricature, Sayers was trained as an attorney, preparing him for a rhetorical presentation of ideas in his prints. Afterward he gained favor with Prime Minister Pitt, and his prints brutally feature opposition members Fox and Lord North. He was rewarded by Pitt with a courtesy post, Marshal of the Court of the Exchequer. While Sayers's draftsmanship is shaky, his ideas reveal a keen political insight, as exemplified in his aquatint with the large comet, which was influential on later artists.[51] It is signed *J S f* (James Sayers *fecit)* and was published by Thomas Cornell on 18 February 1789, a date crucial for its interpretation. The comet's tail

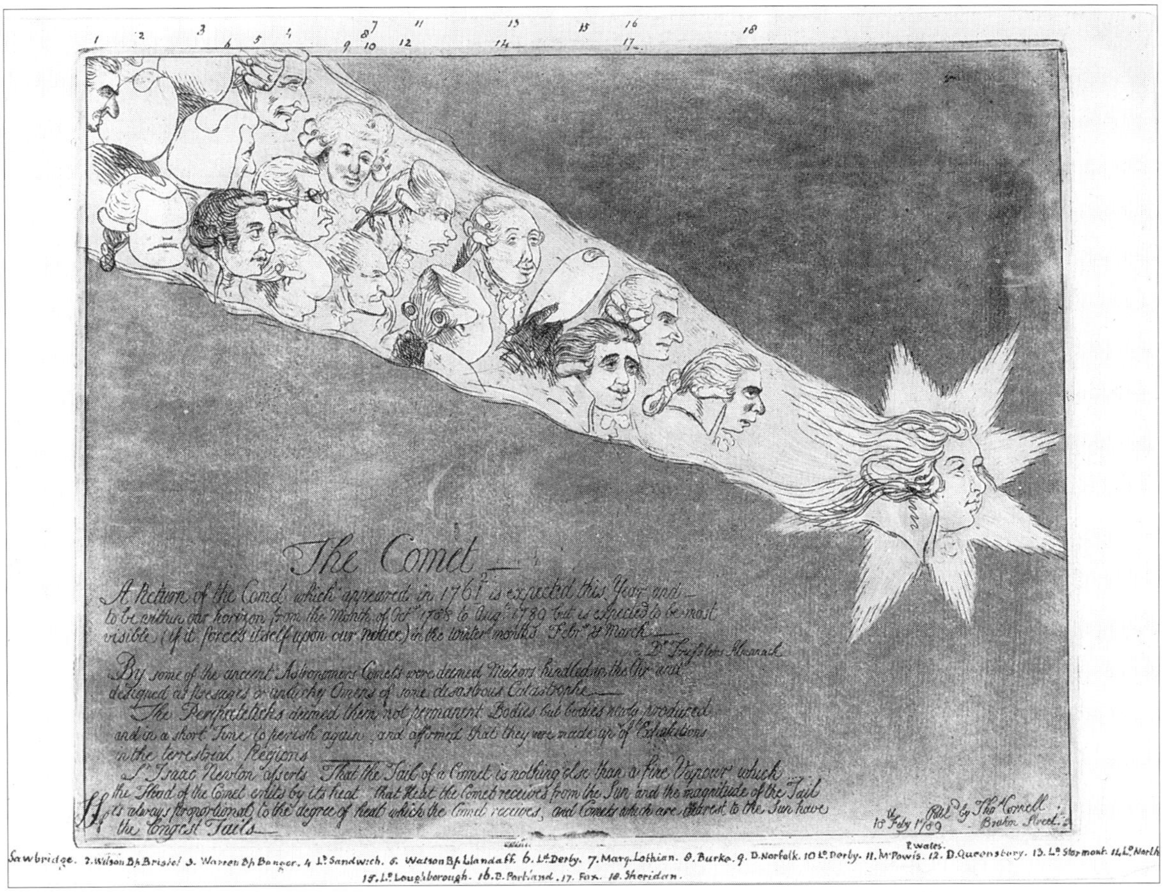

Figure 69. James Sayers, *The Comet/ A Return of the Comet which Appeared in 1761…*, 1789, aquatint, 248 × 344 mm, The British Museum, London.

traverses the design diagonally downward. Its head, an eight-pointed star (the eighth masked by its tail), encloses the head of the Prince of Wales (the future George IV), shown nearly in profile with his wig converting into the comet's tail. Sayers placed the portraits of his disappointed followers in the tail.[52] The inscription in the image can be read either as scientific theory or an allegory alluding to a conceit similar to that of Louis XIV, the "Sun King." In this etching, George III, who had manifested bouts of madness in 1788 and 1789, is the unseen sun, while his antagonistic opponent and eldest son, George Augustus Frederick, the Prince of Wales, and his followers comprise the comet, the prescient symbol of change or of impending doom. The latter are disappointed in their support of the hopeful Prince Regent, because the King's recovery was announced to the House of Lords on 19 February 1789, after Fox had written from Bath, confident that a Regency

Figure 70. Attributed to William Dent (Uncle Toby), *The Terrifying Comet or Modern Planetary System*, 1792, aquatint, 247 × 345 mm, The Pierpont Morgan Library, Peel Collection, New York.

would begin.[53] However, by 1811 (and its great comet) George III became permanently insane and blind, living in seclusion until his death in 1820, while the dissolute Prince of Wales ruled as the Prince Regent. The scandals and debaucheries of his private life led the desperate Prince to give large sums of money to artists to suppress prints that caricatured him in immoral circumstances.

Sayers's image influenced another quirky amateurish aquatint that again satirizes the monarchy [Fig. 70]. Inscribed "Uncle Toby" and dated 24 July 1792, it is attributed to the watercolorist and engraver William Dent (active c. 1782-93).[54] Its target is the madness of George III and the political situation vis-a-vis the monarchy via topical astronomical conceits. Its comet consists of an eight-pointed star, formed by the heads of George III facing to the left, his grotesque head with its wig worn backwards signaling his mental illness, and the Duke of Richmond, from whose mouth issues the comet's tail. The comet's long tail is covered by tiny figures of soldiers, artillery, and their encampments. A rising sun at the lower left has rays similar to the emblematic feathers of the Prince of Wales, eager to assume his father's place. The Prince's image is surrounded by six small stars with the heads of Fox, Sheridan, and supporters of his Regency in Parliament. A large star, inscribed *Venus*, represents the Queen. Two smaller stars above the comet's tail, inscribed *Jupiter* (the head of Pitt) and *Georgium Sidus* (the original name of Uranus given in 1781 by Herschel in honor of George III, a portrait of Burke), regard the king. This aquatint, like that of Sayers, sets an amusing precedent, which would be mined by artists in the early nineteenth century.

While meteors appear in satirical prints, they did not enjoy the same cachet as comet imagery. Two exceptions occur in figures 71 and 72. *Patriotick Meteors* of c. 1771 is an anonymous work featuring three heads of political figures personified as meteors wearing civic chains and heading for *THE GULF OF OBLIVION*.[55] At the left and right are two sheriffs (the right, Frederick Bull as a bull's head), and in the middle is the outgoing Lord Mayor of London. This humorous comment on quickly passing (meteoric) political fame employs the witty conceit of personified, caricatured meteors. The other print of 1790, *The Female Philosopher Smelling Out the Comet* [Fig. 72], is executed in a Rowlandson-like style.[56] The word "philosopher" should be read in the eighteenth-century sense of a natural scientist or astronomer and helps satirize the popular interest in astronomy. The female astronomer looks through a telescope at the comet, personified as a child within the comet's head. A "tail" issuing from the child's bottom is inscribed *Catch*. The woman clasps her hands stating: *What a strong sulphurous scent proseeds from this meteor*. From the inscriptions *meteor* and *Comet*, the artist either mistakenly used the terms interchangeably, or was launching a barb at the woman's amateurish attempt to view the comet, by "smelling it out" and calling it a "meteor." The crescent moon laughs in commentary on the scene.

As in the case of Blake, the art of James Gillray (1757-1815) is on the cusp between the eighteenth and nineteenth centuries, although the introverted Gillray's style seems to be rooted in eighteenth-century stylistic characteristics. The artist was one of the most gifted and amusing of the British satirical printmakers. His

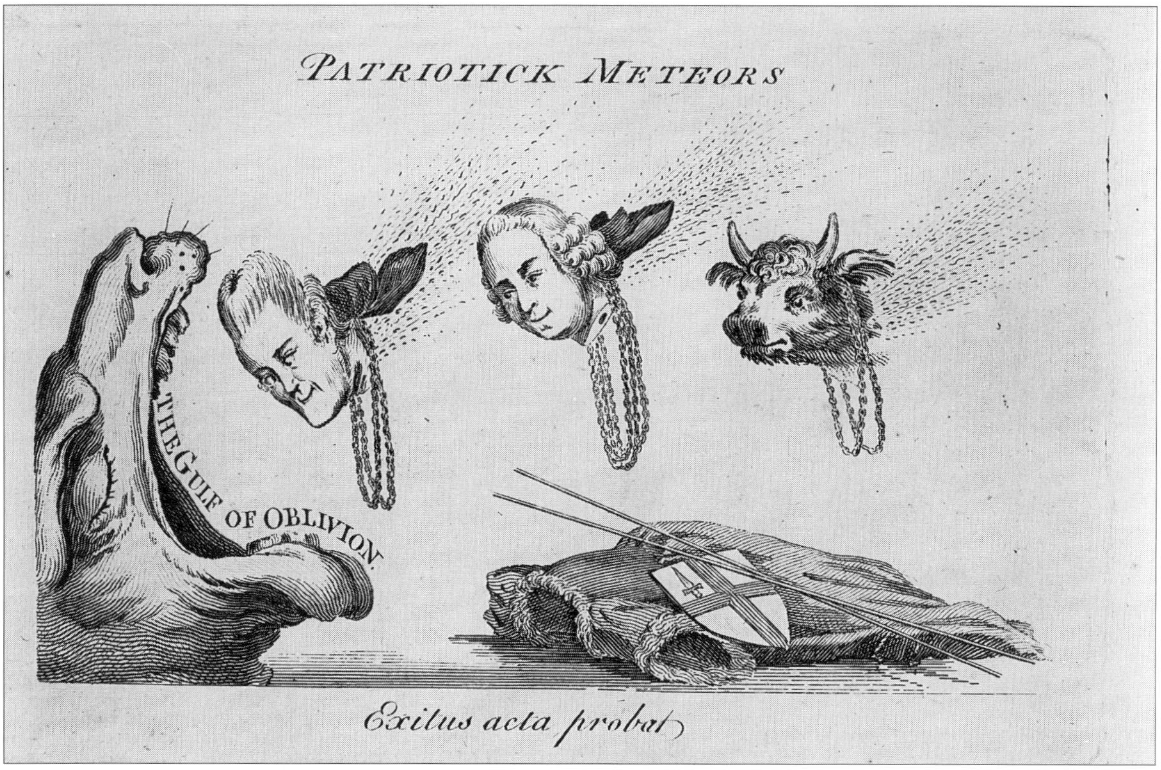

Figure 71. Anonymous, *Patriotick Meteors*, 1771, etching, 126 × 195 mm, The Pierpont Morgan Library, Peel Collection, New York.

wonderfully expressive distortions and kindled political wit gained him much notoriety. In spite of Gillray's venomous attacks, many of his victims considered it a mark of acclaim to be featured by him, as many individuals today consider tabloid fame desirable. Eventually, after being employed by various London print-sellers, he worked exclusively with Hannah Humphrey, living in her shop for the rest of his life, which ended with the hypersensitive artist lapsing into insanity. Gillray holds the distinction of being the first truly professional British caricaturist. He devoted his talents entirely to this end establishing England as the leading center for political and social caricature. With this background, it comes as no surprise that Gillray's prints frequently feature comet/meteor imagery and target Napoleon.[57]

Gillray's brilliant etching and aquatint of 1797, *Titianus Redivivus – or – The Seven Wise Men Consulting the New Venetian Oracle* [Fig. 73], is an elaborate satire of the painterly Venetian style vogue that was the chief topic of the Royal Academy's opening in 1797. The print includes comet/meteor-like forms at the upper right.[58]

Figure 72. R. Hawkins (?), *The Female Philosopher Smelling Out the Comet*, 1790, etching, 258 × 185 mm, The Pierpont Morgan Library, Peel Collection, New York.

The Venetian manner was not a new controversy in England, but the issue became topical in 1797 with an incident known as the "Venetian secret," inspiring Gillray's ridicule. The name for this incident refers to the recipe for Titian's coloring that Mary Anne Provis, a young painter and a protégé of Benjamin West, claimed to have discovered. She agreed to impart its secret to a limited number of subscribers on payment of 10 guineas. In the top half of his composition, Gillray positioned the twenty-year-old art student, who was supposed to have given Titian's secret to West. She is dressed in tatters with a long peacock train, held up by the Three

Ah! ha! _ Ah! ha! Mess.rs VAN-BUTCHELL! _ IRELAND & CHARLES! _ LANE & LACKINGTON! _ what are you now?
redeunt Titianica regna, jam nova progenies coelo demittitur alto
TITIANUS REDIVIVUS; _ or _ The Seven-Wise-Men consulting the new Venetian Oracle, _ a Scene in ye Academic Grove.

Figure 73. James Gillray, *Titianus Redivivus...*, 1797, hand-colored etching, 542 × 410 mm, The British Museum, London.

Graces, painting Titian's portrait on top of a rainbow. Above them is a phoenix holding the precious manuscript with the secret, as it rises surrounded by flames. Across the middle are seated the Royal Academicians – the seven wise men – associated with the ruse and the dupes of the secret, lined up with canvases and suitable slogans about "beginning" and "finishing" and hiding faults with paint. Behind them are artists clambering after Miss Provis and changing into monkeys (a play on the adage that art is the ape of nature). The ghost of Reynolds (d. 1792) rises from a grave in the foreground. Thomas Macklin, John Boydell, and West, President of the Royal Academy, head off to the right in the shadows, leaving a trail of cash.[59] Gillray's satire was no doubt sharpened by a resentment of the Academy's condescension towards engravers and by his own unsuccessful attempts to find employment with Boydell and West. In this etching he ridicules all those taken in by the hoax in which the follies of the Academy are revealed. By contrast, the portfolios of the artists who remained aloof of it are leaned up against a statue of Apollo – among them Fuseli, Paul Sandby, and the eleven-year-old Turner – and are spurned by an imp who relieves himself on their art. In the left background is the Temple of Fame, either incomplete or in restoration, and on the right the cracked facade of the Royal Academy at Somerset House. To the left of this facade, Fame cries through a trumpet: "*You little stars, hide your diminished heads.*" In the tradition wherein comets or shooting stars are associated with Fame, as in Blake's illustration [Fig. 43], there are five comets or shooting stars, whose tails are inscribed with the names of illustrious painters of the past. The first and fourth (Parmigianino and Correggio) have either been rubbed out on the plate or diminished in this impression, but three can still be read: *RAPHAEL*; *MICHELANGELO*; and *RUBENS*.

Then, on 1 January 1805, Gillray's famous parody of Napoleon's pretensions, *The Grand Coronation Procession of Napoleon the First, Emperor of France, from the Church of Notre-Dame, Dec. 2nd, 1804*, was published.[60] Herein, Napoleon's entourage is lampooned, and the line-up led by a figure holding a fringed banner with a huge comet, inscribed with the letter *N* and suspended above half a scored globe ablaze.[61] This comet represents a two-pronged attack, on the ill-fated nature of the French Emperor's reign and on Napoleon's pompous adoption of comet imagery.

Before the initial years of the nineteenth century, the diminutive Bonaparte had cast his shadow over European civilization. Never a man to doubt his powers, and a master at self-promotion, he cannily adopted various comets as his protecting *genii*. He was assuredly aware of the tradition wherein comets were associated with kings and great rulers. By adopting comet symbolism, Napoleon lent his reign a traditional resonance and legitimacy. The first historical comet linked with Napoleon was the Great Comet of 1769 (frequently termed "Napoleon's Comet"), which had an unusual red luster and a tail spanning more than 60°. Since portents can be read in various ways, it was later interpreted by his enemies as foreshadowing the bloodshed and devastating war brought by Napoleon. Nevertheless, his supporters viewed it as a triumphant sign of his glorious reign. When the Great Comet of 1811 (C/1811 F1, formerly 1811 I) appeared, it was also referred to as "Napoleon's Comet." Bonaparte greeted it enthusiastically as his guiding star and the controller of his destiny. What fun the English caricaturists had with this situation! However, the comet, which was visible for seventeen months, proved to be ominous for Napoleon. As his *grand armée* marched into Russian territory, the comet burned with an intensely brilliant light and its tail performed frightening acrobatics by splitting into two parts. Around October, the tail reached a maximum length of about one hundred million miles and a breadth of fifty million miles. Then, a second comet (C/1811 W1, formerly 1811 II) appeared in November and was visible until January 1812. As Napoleon's troops took Moscow in September 1812, neither of the two earlier comets was visible, but between 20 July and 21 September, the Pons-Brooks Comet (12P/1812 O1) made its first recorded appearance and became naked-eye.[62]

Perhaps Gillray's finest conception with comets/meteors is his *Phaeton Alarm'd!* [Fig. 74], which was inspired by another political figure, George Canning. It was published on 22 March 1808 inscribed with a quotation from Ovid's *Metamorphoses*.[63] Canning, who concluded one of his earliest Poetical Exercises at Eton with an ardent desire "To live in a blaze, and in a blaze expire," was active in the overthrow of the Old Ministry. Gillray's print seems to exhibit a consummation of his wishes. Herein, Canning appears as the foolish Phaeton of the political firmament,

Figure 74. James Gillray, *Phaeton Alarm'd!*, 1808, hand-colored etching, 326 × 369 mm, The British Museum, London.

startled by menacing constellations on his reckless ride. Again, every component of Gillray's incendiary satire has an identity. For example, Napoleon is the monkey in the thick of the flames riding Ursa Major, the Russian bear, and chasing the double-headed eagle of Austria.[64] Shooting stars, or perhaps a meteor shower, at the right signal disaster.

A nearly contemporary representation, influenced by Gillray and no doubt engendered by the comet of 1807, occurs in the engraving by Charles Williams (active 1797-1830), *Malignant Aspects...a new planetary system* [Fig. 75].[65] Much of Williams's

Figure 75. Charles Williams, *Malignant Aspects. . . a new planetary system*, 1807, hand-colored etching, 341 × 245 mm, The British Museum, London.

work, like this astronomical satire, was published anonymously, although some prints appeared under pseudonyms.[66] The center of its planetary system is a roundel enclosing a drinking and smoking John Bull (invented in 1712 by Dr. John Arbuthnot to represent the nation of Great Britain) with his dog. The frame of this circle is inscribed: *GOD SAVE THE KING – OLD ENGLANDS WOODEN WALLS. – BRITONS STRIKE HOME – BRITANNIA RULE THE WAVES* – around whose circumference ships sail. It is surrounded by the "malignant" aspects of the title, chief among

Figure 76. Thomas Rowlandson, *John Bull Making Observations on the Comet*, 1807, hand-colored etching, 246 × 343 mm, The British Museum, London.

them a comet with the head of Napoleon. The comet's tail is his ponytail, although Bonaparte did not wear his hair in that fashion. It is inscribed derisively: *A Corsican Comet Frenchified*. Other countries representing the international powers which either threatened England directly or waged commercial war through a blockade are personified in the clouds. (By the Berlin Decree of 21 November 1806, Bonaparte forbade all trade between the British dominions and countries under French control.)

In the same year Thomas Rowlandson (1757-1827), another artist with an essentially eighteenth-century style whose activity spans both centuries, also cast Napoleon in the guise of a comet in *John Bull Making Observations on the Comet* [Fig. 76], on the design of George Moutard Woodward.[67] Rowlandson, who had studied at the Royal Academy as well as in France, was a brilliant but sometimes facile draftsman. His facility with watercolor was often necessitated by his passion for gambling, which resulted in a

Figure 78. Thomas Rowlandson, *Looking at the Comet till You Get a Criek in the Neck*, 1811, pen and ink and watercolor, 406 × 308 mm, whereabouts unknown, formerly H. Schickman Gallery, New York.

need to execute drawings in payment of debts. The rococo style and delicate coloring of his watercolors is not evident in his political engravings, although their light-hearted wit is less intense and vitriolic than that of Gillray.[68] In this engraving, Rowlandson personified the comet's head as Napoleon wearing a plumed bicorne hat. It rushes towards the sun, the profile of George III, who by this time was acknowledged as suffering from bouts of insanity. A carbuncled John Bull crouches as he looks through a telescope towards the comet. The two slices of foreground land represent those flanking the English Channel; the one with a frog climbing its bank is France. The bubble issuing from John Bull's head is inscribed with his thoughts: *Aye – Aye – Master Comet you may attempt your Perihelioms – or your Devil heliums for what I care but take the word of an Old Man you'll never reach the Sun depend upon it.* Rowlandson possessed at least a smattering of scientific knowledge about comets, because he cleverly represented the comet as reflecting the sun!

It is fascinating that in the same year Robert Southey (1774-1848) in this revelatory commentary on Georgian English life (in three volumes entitled "Letters from England") made pertinent comments on science, the arts, and comets.

> To end the list of fashions.... The arts and sciences are now taught in lectures to fashionable audiences of both sexes; and there is a Royal Institution for the purpose, where some of the most scientific men in the kingdom are thus unworthily employed.... Above all stands the man who comprehends the mystiques of astronomy for not only does he hold sway on the proposed actions of people but should he perchance foretell the appearance of a natural phenomenon then his reputation is great indeed.[69]

Comet 1811 I Redux: Satire, Caricature, and Drinking

In 1811, coinciding with the brilliant comet 1811 I, Rowlandson and others produced amusing prints that comment not only on the current politics but also on the public's fascination with comets. *Looking at the Comet Till you Get a Criek in the Neck* [Fig. 77 color] was one of these.[70] Here, Rowlandson drew an older man in night-cap and dressing gown gazing through a small spyglass at the comet. He is so intent on his quarry that he misses the dalliance nearby: the young woman at his side, who also looks at the comet while slipping her hand into that of the young man beside her and pressing her foot on his. Clearly the comet is a diversion that allows flirtation to prosper. A preparatory watercolor testifies to the subject's popularity [Fig. 78].[71]

Napoleon's linkage to the comet of 1811-12 provided much fodder for the English satirists. For example, William Elmes's (active 1811-1820) print *The Gallic Magi Led by the Imperial Comet* [Fig. 79][72] attacks with scatological humor Napoleon's exploitation of the traditions wherein the Star of Bethlehem and

Figure 79. William Elmes, *The Gallic Magi Led by the Imperial Comet*, 1811, hand-colored etching, 240 × 341 mm, The British Museum, London.

the births of rulers were linked to comets.[73] Napoleon and his second wife, Marie Louise, are watching a comet, whose head encloses a representation of their infant son, the King of Rome, propelled by his flatulence forming the comet's tail. (Certainly, this motif is dependent on the image in figure 72.) The child's governess, Madame de Montesquiou, proudly holds aloft the earthly child, wearing a plumed bicorne hat, to gaze at the apparition through a telescope. The infant in the comet's head wears an identical hat and holds a bowl; a gallic cock stands on his posterior. Three grotesque and elderly Frenchmen (the Magi) kneel in adoration. Beside them is a plan showing the child-comet surrounded by stars and a waning moon. Elmes's etching satirizes the pomp surrounding the infant King of Rome and the imperial dynasty.

In the same year, Elmes etched a related, more explicit *The Imperial Comet Shedding its Baneful Influence* [Fig. 80].[74] Herein,

Figure 80. William Elmes, *The Imperial Comet Sheding its Baneful Influence*, 1811, hand-colored etching, 254 × 349 mm, Yale Center for British Art, Paul Mellon Collection, New Haven.

the King of Rome, a naked infant wearing a cocked hat, crouches on his hands and knees within the comet's head, its tail formed by a flatulent blast from his buttocks replete with emblems of Popery. The child (resembling Bonaparte) spoons up pap, which is smeared across his face, from a bowl. Around the comet are fixed stars and planets. For example, the *Georgium Sidus* in the lower left encloses the head of George III, looking up at the comet. At the lower right on the earth are John Bull gazing at the comet from England and Napoleon looking at it, while Madame de Montesquiou lifts up the King of Rome, from France.

In a similar spirit, Williams's *Devils Among the Flats or Boney Getting into Hot Water* [Fig. 81] uses cometary symbolism to taunt England's enemy.[75] Above Napoleon soars a comet whose head is the profile of Admiral Horatio Nelson, the British naval hero (d. 1805). Its tail, tinted in orange and yellow watercolor, takes the form of barbed rays slanting in a threatening manner toward Bonaparte.

Figure 81. Charles Williams, *Devils Among the Flats or Boney Getting into Hot Water*, 1811, hand-colored etching, 248 × 353 mm, The British Museum, London.

Another print that has been attributed to Gillray, *The Comet of 1811*, depends on Sayers's political satire of 1789. It features a portrait of the Prince of Wales in the comet's head [Fig. 82].[76] In this later etching, the comet's direction is reversed from Sayers's print [Fig. 69], while the identities and order of the portraits are changed.

Charles Williams employed the same comet device in the even more amusing and highly original print of the same year *THE COMET!!!* [Fig. 83].[77] The comet's head is the visage of the Tory Pitt in which his hair forms its tail, a conceit derived from the ancient "hairy star" appellation for comets. It streams across the entire design covered with the words: *War* and *Taxes*, each repeated alternately across the tresses six times. A caricatured Fox (ironically a Whig), floating above the comet's tail, leans forward holding a pair of snuffers, inscribed *PATENT*, which he is about to

Figure 82. Attributed to James Gillray, *The Comet of 1811*, 1811, hand-colored etching with an aquatint background, second state, 247 × 300 mm, The British Museum, London.

Figure 83. Charles Williams, *THE COMET!!!*, 1811, hand-colored etching, 247 × 354 mm, The British Museum, London.

place on the comet's head. Below this comet is a five-pointed, smaller star containing the profile of Spencer Perceval, Prime Minister of England from 1809 to 1812; the points of the star are inscribed: *WAR* (three times) and *TAXES* (twice). Below is a military officer, Baron de Geramb (whose eccentricities and adventures in England in 1811 attracted much attention). He looks up angrily in fear of his enormous moustache saying, "*Cot dam! this*

Figure 84. Rebecca Emes and Edward Barnard I, *Tankard with the Comet of 1811*, 1811/12, silver with a gilt interior, 149 mm tall, Sterling and Francine Clark Art Institute, Williamstown, MA.

fellow will burn my whiskers!" His presence adds a whimsical touch to an otherwise biting criticism of Perceval, who carried out the war policy of Prime Minister Pitt (who had died in 1806), ruthlessly imposing taxes, though, it would seem, on a smaller scale, against the wishes of the opposition (inspired by Fox who also died in 1806, but who had been appointed Secretary of Foreign Affairs at Pitt's death).

A gauge of the impact of comet 1811 I on other individuals is encountered in a handsome, hallmarked silver tankard by Rebecca Emes and Edward Barnard I, who worked in a London partnership between 1808 and 1829 [Fig. 84].[78] This late Georgian, slightly tapered tankard for beer is engraved with illustrations that point to a highly specialized commemorative commission. It is ornamented with two bands of horizontal rings that suggest the old Ptolemaic rings of heaven as well as planetary orbits in spite of the fact that they appear on many other tankards of the period. A third decorates the top of the lid surrounding the twelve-point star finial positioned on a base of clouds. (These concentric rings have also been compared to the metal hoops of beer barrels.) Instead of the more customary armorial decoration, the tankard's cylinder is engraved with depictions of planets and stars. The planet Saturn with its rings and seven satellites (the number known in 1811) is flanked by two other planets, whose identifications are complicated by four satellites orbiting one and six the other. Next to Saturn is Jupiter with its four satellites discovered by Galileo. It is represented sideways with respect to the orientation of planetary orbits. The third celestial object is presumably Uranus, discovered in 1781, which is featureless in a ground-based telescope even today and which showed surface features only to the Voyager II spacecraft in 1986 and the Hubble Space Telescope starting in 1994. We do not know why it is shown with six moons, because only two (discovered by William Herschel) were then known (fifteen are now known, five of them discovered from the spacecraft). However, the heavenly body that steals the show is the wonderful comet, whose diaphanous tail sweeps over fixed stars. Its differentiated coma and nucleus are also amazingly captured in lightly engraved lines, and its distinctive profile accords with astronomers' drawings of comet 1811 I.[79] If the tankard were held by a right-handed individual, the side with the comet would be visible to his/her companion(s), and thus would function as a conversation piece.[80] Because of the inscribed *F* and the engraved crest, it has been suggested that the elegant tankard was once owned by a member of the Frend family. A likely candidate is William Frend (1757-1841), a Fellow of Jesus College, Cambridge, who was elected to the Royal Astronomical Society in 1832 and served on its Council in 1835-36.[81]

It would seem that comet 1811 I was the first nascent media comet of the nineteenth century, because it started to receive attention in different art forms and what was then the popular press, satiric prints. It also engendered a wider consideration of historical comets in pamphlet form.[82] This topically oriented attitude about comets would snowball by mid-century with the virtual explosion of illustrated journalism.

Already by 1813 the monthly periodical *THE METEOR, OR, Monthly Censor* had appeared. True to its title, it was brilliant, short-lived, and topical. Only seven issues were produced (between November 1813 and May 1814). George Cruikshank (1792-1878) – considered the leading caricaturist of the Regency and the

Figure 85. George Cruikshank, *Cover of "THE METEOR,"* 1 November, 1813, woodcut, 208 × 123 mm, The British Museum, London.

successor of Gillray as well as a celebrated Victorian illustrator – designed the initial cover (1 November 1813) with a number of amusing personifications, presided over by a comical little fellow with a meteor on his forehead carrying a lantern [Fig. 85].[83] Below at the right in Roman armor is the hero holding the shield of the Free Press and armed with the sword of Truth and Justice. He

Figure 86. George Cruikshank, *The Man in the Moon*, 1820, woodcut, 140 × 102 mm, The British Museum, London.

protects himself against the three-headed serpent of Licentiousness, Fraud, and Hypocrisy. At the left is Punch, wearing the cap and bells of Folly, standing on a pedestal and regarding his reflection in a mirror held up by a laughing satyr with a goat's head ("satire" is from Greek satyr plays). These loaded images are printed on mauve-pink colored paper. Quite obviously, the periodical's title was selected for its impact.

Another Cruikshank satire adorns the title page of *The Man in the Moon*, published in 1820 [Fig. 86]. Herein the Prince Regent, inside the moon, wearing a cocked hat with three peacocks' feathers, attempts to put out the sun which encloses a printing press, an allusion to his censorship. The comet, a symbol of change, whose head is in the shape of a cap of liberty hat and whose tail is labelled *REFORM*, rushes toward the Prince Regent. By 1821, the print was in its 51st edition and was followed by a series of smaller woodcuts emulating it.[84]

Figure 88. Thomas Rowlandson, *The Observation of a Comet*, c. 1821?, pen and brown and gray ink and watercolor over graphite, 311 × 245 mm, Museum of London.

Later Cometomania and Rowlandson

Rowlandson was the British caricaturist most involved with astronomical subjects. He must have kept a watch on the heavens, where with few exceptions at least one comet appeared every year during the first three decades of the nineteenth century. Because of demand, he executed at least four versions of a bawdy composition that demonstrates how his contemporaries were smitten with cometomania [Figs. 87 color, 88].[85] One of the four is signed and dated 1821, although the stylistic sequence of the group is not clear, and only one other sheet bears the artist's name. What is patently obvious in all of them, however, is Rowlandson's understanding of human foibles and follies, such as the urban public's unquenchable curiosity about new celestial phenomena. Rowlandson shows his audience that the comet functioned rather like a side show with amateur astronomers turning their spy glasses and portable telescopes, possibly comet-seekers, to view the spectacle. A few, like the crouching opportunist in the right foreground who cups the robust bottom of a fellow watcher, uses the celestial apparition as a diversion. Small variations among the four suggest that Rowlandson kept inventing in a minor fashion, but must have kept one version from which to work. The one in the Wiggin Collection is clearly unfinished, while the signed and dated one seems to be the most carefully detailed. It is probably safe to conclude that it was executed during the time when the comet of 1821 (C/1821 B1) was visible. This comet was discovered simultaneously with a telescope by Jean Nicolas Nicollet in

Rowlandson

Paris and Jean-Louis Pons in Marseilles. By the second half of February it became naked-eye, the time when Rowlandson and others in London might have witnessed highjinks similar to those depicted in the watercolors.[86] As reported by John Herschel, the comet had a 5° tail, which on the first of March reached 7°, and after perihelion on 1 April was recorded as measuring 8°.[87]

Rowlandson also exhibited an awareness of astronomy in other works; for example, in a wonderfully mordant drawing known in versions in which an aged astronomer gazes through his telescope, only to be startled by the appearance of Death. The works are related to Rowlandson's 1814-16 series *The English Dance of Death.* One of these is inscribed in Rowlandson's hand on a separate sheet mounted with the drawing: *The Astronomer/Why I was looking at the Bear/But what strange Planet see I there.*[88]

Rowlandson frequented the Royal Observatory at Greenwich and its surrounding park, although we have not been able to connect him with any astronomer because of the almost complete lack of correspondence either from or to the artist. However, his drawings and watercolors serve as a good indication of his movements and thoughts because he was afflicted with graphomania.[89] From them, we know that Rowlandson was in Greenwich a number of times and at the Royal Observatory. He discusses the latter in his preface to the Dent Sketchbook, which also has a drawing of the Observatory, before which stand three men, one looking through a telescope.[90]

The Royal Observatory, The Royal Institution, and the Royal Astronomical Society

Among the British organizations and locations for both professional and popular astronomical activity during the early nineteenth century were: the Royal Observatory, The Royal Institution, and the Royal Astronomical Society. They will be discussed in the following paragraphs.

The Royal Observatory was founded to provide an astronomical solution to the problem of finding longitude for ships at sea. In 1675, King Charles II appointed an Astronomer Royal "...to apply himself with the utmost care and diligence to the rectifying the tables of motions of the heavens and the places of the fixed stars, or as to find out the so much-desired longitude of places for perfecting the art of navigation."[91] The original observatory building, whose main part was designed by Sir Christopher Wren, was built at the Royal Park at Greenwich, a short boat ride along the Thames from London. Telescopes and accurate clocks were installed in it to measure the positions of the stars.[92] Comets, by nature irregular in their timing, however, were of no use for marine navigation, although Astronomers Royal observed them from this observatory.[93] As successive Astronomers Royal constructed additional buildings at Greenwich, they were aligned to the east of the original, and a meridian (north-south line) was defined by the telescopes in them. The meridians of Halley, Bradley, and Airy superseded that of Flamsteed, and it is the Airy meridian that was adopted in 1844 as the world's Prime Meridian, the 0° of longitude.[94] During the early twentieth century, parts of the Observatory were moved out of Greenwich, because of the magnetic disturbances caused by the power station and the railways and because of light pollution. In 1948, the rest was moved to Herstmonceaux Castle in Sussex and took the name of Royal Greenwich Observatory.[95] Some observation continued at Greenwich until 1954, although the old buildings are now a museum known as the Old Royal Observatory. A comet weathervane of c. 1899 [Fig. 89 color] still decorates one of the subsidiary buildings of the Old Royal Observatory, perhaps to mark the British interest in cometary studies.[96] Weathervanes in the shapes of comets have been noticed in the United States (in New England) as well as in Italy. The directional thrust and unpredictability of comets suits this form and function.

In 1799, The Royal Institution of Great Britain for the Dissemination of Useful Knowledge was founded in London by Benjamin Thompson (Count Rumford). Soon the Royal Institution, which stressed practical science, became a fashionable place for the public to attend lectures, including those on comets and meteors.[97] It housed and still contains research laboratories, where important discoveries were made.[98]

As part of the general interest in astronomy and the desire to establish its autonomy and the status of the discipline, the Astronomical Society was founded in 1820, some 160 years after the Royal Society, by a group of professional and amateur astronomers. John Herschel was active in the arrangements and gave the inaugural address. Francis Baily, who later discovered the "Baily's beads" phenomena of total solar eclipses, became Secretary. At the time of its founding, the President of the Royal Society, Sir Joseph Banks, was antagonistic to the formation of this specialized society, as he was to the new Geological Society. He did not want the authority of the Royal Society to be diluted.[99] He saw that the first president of the Astronomical Society, the Duke of Somerset, resigned his post and membership.[100] These circumstances made it difficult for the fledgling organization to find a successor. The aged William Herschel, then eighty-three, was asked to accept the post. Though he refused, when Banks died later that year, Herschel was prevailed upon to accept on the condition that it was an honorary one without any duties.[101] Thereafter, the Astronomical Society flourished. In 1831 it was given a Royal Charter and after several homes headquartered in Burlington House (1874), which it shares with, among others, the Royal Academy of Art, the Geological Society, and the Society of Antiquaries.

Ashworth has made a case that the interests of many of the founders of the Royal Astronomical Society were connected to business, and, in particular, to accounting. Just as accountants and actuaries made accurate tables of the earth, Ashworth argues that the purpose of the Astronomical Society was to make accurate tables of positions and motions in the heavens.[102] The publications of the Society about comets included many tables of positions, whereas the earlier publications of the Royal Society had included letters from a wide range of professionals and nonprofessionals describing in written words the comets and meteors that they had seen. This change in approach supports Ashworth's thesis that the Astronomical Society was not established to expand astronomy, as is popularly supposed, but rather to narrow and define the science's boundaries. It set out to remove astronomical speculation and place it on a solid calculating base.[103]

While the number of papers about comets in the *Philosophical Transactions of the Royal Society* dwindled after c. 1810, not even covering the brilliant comet of 1811, the publications of the Royal Astronomical Society and other astronomical publications on the Continent that started in the 1820s began to feature comet and meteor articles. Whereas only a couple of papers per decade about comets were published in the *Philosophical Transactions* through 1833, papers about comets appeared in publications of the Royal Astronomical Society at a good pace. (Since the 1850s, no papers on comets and meteors appeared in the *Philosophical Transactions*.[104]) Even the Royal Society's new *Proceedings of the Royal Society*, which did not start publication until 1840, had nothing on comets or meteors and only two brief notes about meteorites in its issues from 1854 through 1865, despite the fact that these years included brilliant comets, notably Donati's of 1858. Clearly, the Royal Astronomical Society's *Memoirs*, which began publication in 1822, and the *Monthly Notices*, which began in 1827, had taken over the role of publishing these papers. It should be noted that further specialization continued because while the *Memoirs* regularly included articles about comets (though none on meteors) through 1838, these observations gradually began to appear in publications of observatories – Greenwich, Cambridge, and Edinburgh.[105] The *Monthly Notices*, which started as meeting reports, expanded rapidly and its format changed.[106] Starting in the 1830s, it even began to publish some papers that did not appear in the *Memoirs*.[107] The issues contained reports of comets with some regularity. For decades, they were bound in with the *Memoirs*, but by 1858, the year of Comet Donati, the *Monthly Notices* was recognized as a totally separate journal.[108] The *Monthly Notices* survives to this day as a major astronomical journal publishing full scientific papers, while the last issue of the *Memoirs* appeared in 1978 with the catalogue by J.A. Bennett of manuscripts in the collection of the Royal Astronomical Society.

The Comets and Meteors of Later Visionaries

Although at the beginning of this chapter we discussed Blake and three of his close friends, Linnell and the Varleys, his fourth and youngest friend, Samuel Palmer (1805-1881), who also represented comets, needs to be introduced at this juncture.[109] The two artists met in 1824, and like Linnell, whose eldest daughter Hannah he married in 1839, Palmer had an interest in both the pastoral and spiritual. However, both he and Blake were more visionary than Linnell. Blake was one of the most potent forces in

Figure 90. Samuel Palmer, *A Shepherd and His Flock under the Moon and Stars*, c. 1827, pen and brown ink with brown and grey wash and white heightening, 108 × 132 mm, Yale Center for British Art, Paul Mellon Collection, New Haven.

Palmer's life and art. Palmer's 1827 drawing – *A Shepherd and His Flock under the Moon and Stars* [Fig. 90] – reveals him to be a precocious visionary already when he moved to rustic Shoreham. Because this drawing was executed at the height of Blake's influence on him, it is not surprising that it features a star with a tail – a shooting star/meteor or a comet – to the left of its crescent moon.[110] The hallucinatory quality of its technique gives the illusion that the stars are twinkling, while a two-pronged tail attached to one large star and rendered in dry brush has a wispy vaporous quality of a shooting star or meteor.

Around the same time (1825-28), Francis Danby,[111] another visionary artist, painted his gigantic *An Attempt to Illustrate the Opening of the Sixth Seal* [Fig. 91], a theme traditionally associated with meteors.[112] Its scale is the opposite of Palmer's intimate sketch and demonstrates a change in public taste. Danby's canvas embodies Romanticism in the grand manner, wherein the Sublime and Picturesque, espoused by Edmund Burke and others, is taken to a rhetorical extreme. The work was exhibited at the Royal Academy in 1828 and was an immediate success, receiving rave reviews that mark the most successful moment of Danby's career. In fact, the Irish artist's apocalyptic canvas was the most

Figure 91. Francis Danby, *An Attempt to Illustrate the Opening of the Sixth Seal*, 1828, oil on canvas, 185 × 255 cm, The National Gallery of Ireland, Dublin.

popular painting exhibited and had to be moved to a distant gallery because of the crowds it attracted.

Danby's literary source was the traditional passage related in Revelation 6:12 ("And I beheld when he had opened the sixth seal, and lo, there was a great earthquake; and the sun became black as sackcloth of hair, and the moon became as blood; And the stars of heaven fell unto the earth...."). Since the passage's visual tradition included stars falling from heaven, it is understandable why Danby represented a meteorite or fireball crashing into a cliff with great impact at the right. His depiction contrasts so markedly with earlier stylized depictions of the theme, as well as with his preparatory study, that one wonders whether he actually observed a similar phenomenon.[113]

John Martin (1789-1854)[114] designed his earlier (1824-25) *The Creation of Light* [Fig. 92] in a similar Romantic spirit.[115]

Figure 92. John Martin, *"The Creation of Light" from "Paradise Lost," Book 7, I, 339*, 1824-25, mezzotint, 191 × 278 cm, The British Museum, London.

In contrast to Danby's canvas, however, Martin's mezzotint concerns creation and illustrates Book 7, I, 339ff, of Milton's *Paradise Lost.*

It is significant that while Milton does not mention a comet in this passage, Martin felt it necessary and fitting to include one. He may have been inspired either by other images in Milton or the current speculation about comets.

The vaporous tail of Martin's comet hovers above the crescent moon to the right of the specter of the Creator and neatly echoes the trajectory of the sun's rays. Its presence in the firmament testifies to the importance of comets in Martin's day and to their positive symbolism. It may reflect the artist's memories of comet 1811 I or another one of the brighter comets in the early decades of the century (see Appendix II). Its presence lends a primordial note to his dramatic composition that rings true to the scientific knowledge about the origin and chemical composition of comets: that they are made from matter left over from the genesis of the universe. Fifteen years later, Martin would render an even more spectacular comet in his *Eve of the Deluge* [Fig. 110 color], in which he painted a visually accurate and convincing comet in a meteorologically correct sky.

Ephemera, et cetera

Ephemera of the period also testify to a deep cultural fascination with comets in all strata of British civilization. For example, another trade card, later than that of Benjamin Martin, was made for Egbert Storer, who identifies himself as an horologist. It depicts four comets, labelled with their discoverers' names, from *Hooke 1656* to *Haley* [sic].[116] It demonstrates that comets lent a glamor to everything and that clocks and clock-making were still intimately related to astronomy. A second example of early nineteenth-century ephemera is a large engraved board game entitled: *Science in Sport or the Pleasures of Astronomy.* It is dated 1804 and features a large comet at number 23.[117] A later set of game cards, identified as *Astronomia* on the box where they are dated 1829, is accompanied by a pamphlet entitled "Explanation of the Celestial Cards, and of Two of the Games in which they are Used....A Brief Introduction to Astronomy," to demonstrate that it was a didactic game.[118] *The Comet* card [Fig. 93] features the large Comet of 1680 surrealistically appearing against a starry sky between two columns and below a yellow curtain. Its reverse contains information about the comet's tail, its distance from the sun, etc. Other astronomical card games were also produced in the nineteenth century. One from c. 1825 has perforations so that when the viewer held it up to the light, the positions of the stars were illuminated as points.[119] We have also seen two examples of a papier-mâché snuffbox with a print showing the Great Comet of 1811 viewed from the Pont-Neuf in Paris. However, while the print may record the 1811 event, the boxes appear to date from a slightly later time and seem to be English.[120]

Figure 93. Anonymous, *The Comet of 1680*, from a Game of Celestial Cards called "*Astronomia*," 1829-31, engraving on board with watercolor, 98 × 65 mm, The Science Museum, London.

Finally, to close our consideration of the early nineteenth century, we will turn to the comets or shooting stars by another Romantic artist taken with such imagery, but in the imaginative mode of an illustrator and with connections to the eighteenth century: Henry Howard (1769-1847). Although a respected pro-

Figure 94. Henry Howard, *"The Mermaid on the Dolphin," from Shakespeare's "A Midsummer Night's Dream," Act II, Scene 1*, watercolor, 153 × 229 mm, whereabouts unknown, formerly Christie's, London.

fessor at the Royal Academy as well as its Secretary for a time, Howard's reputation has disappeared into artistic oblivion. He specialized in paintings of poetical subjects and illustrated writings by poets, many of whom, such as Milton and Spenser, were employers of comet/meteor symbolism. Howard painted a number of astronomical subjects, including *The Pleiades* for the Duke of Sutherland, a replica of which reportedly ornamented the Leicester Galleries. For the Duke he also painted a large version of the solar system.[121]

His watercolor [Fig. 94], identified in a Christie's sale catalogue as *The Rising of the Pleiades*, may be related to the Duke of Sutherland's commission of the same subject but that is doubtful.[122] According to the Christie's sale catalogue, it supposedly illustrates Act II, scene 3, from Shakespeare's *A Midsummer Night's Dream*, although this scene does not exist. Furthermore, the depiction seems to have nothing to do with the Pleiades. Rather, the work shows a female figure with a lyre seated on a large dolphin in the water. Above her is a full moon, against which is silhouetted a bust-length female figure, probably the personification of the moon or Diana/Artemis. From the left fly in two figures above whom hover two stars with tails. The first figure is a neoclassical nude female, followed by a male figure who can be identified by his caduceus and winged helmet as the god Mercury. Their aerial postures are reminiscent of the voluptuous lover wind gods, a motif taken from an ancient cameo, in Botticelli's *Birth of Venus*. Certainly the woman on the dolphin derives from Shakespeare's *A Midsummer Night's Dream*, Act II, scene 1, a passage that Walter Crane and Arthur Rackham later illustrated with similar images [Figs. 150, 151]. Shakespeare's text reads:

> Since once I sat upon a promontory,
> And heard a mermaid on a dolphin's back
> Uttering such dulcet and harmonious breath,
> That the rude sea grew civil at her song.
> And certain stars shot madly from their spheres.
> To hear the sea-maid's music.
>
> That very time I saw – but thou couldst not –
> Flying between the cold moon and the earth,
> Cupid all armed; a certain aim he took
> At a fair vestal throned by the west.
> And loosed his love-shaft smartly from his bow
> As it should pierce a hundred thousand hearts:
> But I might see young Cupid's fiery shaft
> Quenched in the chaste beams of the watery moon,
> And the imperial votaress passed on,
> In maiden meditation, fancy-free....

According to Frank Howard, Henry Howard is recorded as exhibiting in 1801 "The Mermaid sitting on a Dolphin's back" – "And certain stars shot madly from their spheres/To hear the sea-maid's music." – "one of his most beautiful compositions, which was purchased by the poet William Hayley [1745-1820] out of the exhibition."[123] Hayley, who was a patron and friend of Blake, was also a biographer of Milton. Howard repeated the subject again in 1841 on a large scale.[124] If this watercolor is indeed the same work or related to it, which we think may be the case, the stellar forms

Figure 95. Henry Howard, *Mythological Scene, Echo Flying from Narcissus (?)*, medium unknown, 137.5 × 117 mm, whereabouts unknown, formerly Christie's, London.

and their human companions perhaps were meant to be the planets Mercury and Venus, which the artist associated with the line "And certain stars shot madly."

Howard rendered a similar stellar form with a tail in another work depicting a flying and veiled female figure, dressed in vaguely neoclassical garb; it was described in another Christie's sale catalogue as "Echo Flying from Narcissus" [Fig. 95].[125] The star with its tail partly masked is inside the central break in the clouds. The subject cannot be identified securely but seems to have more in common with other works Howard is known to have executed: in 1830 an *Aurora ushering in the May Morning* or in 1829 *Night*.[126]

It is fitting that Howard's Romantic works close our consideration of the early nineteenth century, because in contrast to his somewhat anachronistic works, the fourth decade of the nineteenth century witnessed a new emphasis on realism, which was accompanied by a change in the subjects of works of art and an even more pronounced emphasis on landscape painting.[127] Some of these changes were mitigated by shifting factors of patronage, social reorganization in urban centers, and the monarchy. This period, which will be examined in the subsequent chapter, experienced several major comet apparitions that elicited the contributions of British astronomers, a number of whom can be directly linked to artists who represented comets and meteors.

Notes

* Brilliant comets of the nineteenth century seem bright as well as frequent. It had seemed that bright comets are not appearing with the same frequency today until the apparitions of C/1996 B2 (Hyakutake) in 1996 and C/1995 O1 (Hale-Bopp) in 1997. Certainly all comets and meteors appear less bright today because of electricity and light pollution. Comet Hale-Bopp demonstrated, however, how magnificent a brilliant comet appears even with urban sprawl. See also M.R. Kidger, "The Frequency of Appearances of Bright Naked Eye Comets since 1750: Are bright comets less frequent in the 20th Century?," *Earth, Moon, and Planets*, 66, 1994, 231-238; M. Gingerich, "Great Comets, Novae, and Lady Luck," *Sky and Telescope*, 89:6 1995, 86-88.

1 I. Morus, S. Schaffer, and J. Secord, "Scientific London," in *London – World City – 1800-1840*, ed. C. Fox, exh. cat., New Haven and London, 129-142.

2 See P. Dunbar, *William Blake's Illustration to the Poetry of Milton*, Oxford, 1980, as well as W. Ingram and K. Swain, *A Concordance to Milton's English Poetry*, Oxford, 1972, 96, 148, 490; L. Sterne and H.H. Kollmeier, *A Concordance to the English Prose of John Milton*, Binghamton, 1985, 268, 854.

3 Inv. no. 1949.4:8 in the Pierpont Morgan Library, New York, in pen and black ink and watercolor measuring 161 × 123 mm. Signed *WBlake inv.* See Butlin, 1981, 398-399, no. 543 8, 394-395; Dunbar, 146-149. For the series, executed for Thomas Butts, see also Butlin, 1978, 112, 119-123, nos. 239-250.

4 Dunbar, 146-148.

5 J.M.Q. Davies, *Blake's Milton Designs: The Dynamics of Meaning*, West Cornwall, Conn., 1993, 136-137, interprets them as stellar scourges. We had allied them to the Bard's whip prior to 1986.

6 See R.N. Essick and J. Viscomi, eds., *Blake's Final Illustrated Works*, 5: *William Blake. Milton a Poem and the Final Illustrated Works*, ed. D. Bindman, London, 1993, 20-21, 23-24, 27-28. See also J. Viscomi, *Blake and the Idea of the Book*, Princeton, 1993, 315-329, for the various copies. He believes that copy C in the New York Public Library was restructured by Blake himself at least three times (and rebound incorrectly by someone else). He thinks it was printed c. 1818 after copies A (the British Museum) and B (Huntington Library, San Marino) and before D (the Rosenwald Collection, Library of Congress, Washington, D.C., the only one with 50 plates). See also Bindman, 1982, 149. The size of each image varies, although the measurements of the sheets in copy C are 230 × 149 mm.

7 See Bindman, 1978, nos. 415-464. They are permeated with celestial forms; the relief-etchings with watercolor and pen and ink all measure c. 210 × 155 mm (plate).

8 For this very rare book and its complicated history, see: Butlin, 1978, 126-132, nos. 257-276; M.D. Paley, *The Continuing City. William Blake's "Jerusalem,"* Oxford, 1983; *idem.*, ed., *Blake's Illuminated Books, 1: William Blake. "Jerusalem,"* Princeton, 1991; Viscomi, 338-361. In Blake's own time only one copy was ever finished, copy E in the Yale Center for British Art, Paul Mellon Collection, New Haven (inv. no. B1992.8.1, 1-100, referred to as Bentley Copy E); Noon, 78. In addition, there are separate plates and at least three copies that were printed without the addition of the watercolor after Blake's death. The book's subtitle, "The Emanation of the Giant Albion," suggests that the protagonists are Albion or Mankind and his female emanation, Jerusalem, with whom he is reunited as part of Man's Redemption through Christ. In her fallen state, she becomes Vala or Nature, and Albion's struggle with her is paralleled by that of Los, the poetic principle.

9 Paley, 1983, 243.

10 Inv. no. N 01110/B 651 in the Tate Gallery, London, in tempera heightened with gold on canvas measuring 74 × 62.7 cm. It is signed *WBlake 1805*[?last digit illegible]. An old label, seemingly inscribed by Samuel Palmer, supplies the date of 1805, which is obscured in the inscription. In Blake's catalogue, the work had an extended title: "...; he is that Angel who, pleased to perform the almighty's orders, rides on the whirlwind, directing the storms of war: He is ordering the Reaper to reap the Vine of the Earth, and the Plowman to plow up the Cities and Towers."
See Butlin, 1990, 148-150, no. 59, as well as R. Lister, *The Paintings of William Blake*, Cambridge, 1986, no. 36.

11 Inv. no. N 03006/B 649 in the Tate Gallery, London, in tempera on canvas measuring 76.2 × 64.9 cm. See Butlin, 1990, 146-147, no. 58. The 1809 exhibition at his brother James's house at 28 Broad St., Golden Square opened in mid-May. The paintings of Pitt and Nelson are apocalyptic visions of war that parallel Blake's writings of the period. The reaper and the plowman accompanying Pitt are the angels who, in Revelation 14:14-19, prepare "the great winepress of the wrath of god," which Blake introduced into *Milton* as the "Wine-press of Los." The biblical monsters Leviathan and Behemoth appear in *Jerusalem* as the "War by Sea enormous and the War by Land astounding."

12 See *ibid.*

13 Olson, 1985b, 83.

14 Inv. no. N 05889/B 750 in the Tate Gallery, London, in tempera heightened with gold on panel measuring 21.4 × 16.2 cm, with a provenance from John Varley's collection. It is signed at the lower right: *WBlake Fresco*. See Butlin, 1990, 159-160, no. 64.

15 Butlin (*ibid.*, 160), who adds that Varley's account, while being an accurate description of the painting, may be a fanciful one. Other scholars have not been as skeptical. See also W. Klonsky, *William Blake. The Seer and His Visions*, New York, 1977, 124-125. Cunningham's interview was for his *Lives*, published in 1830.

16 Inv. no. N 05184/B 692 97-8 in the Tate Gallery, London, in graphite measuring 189 × 153 mm. The side of the sheet with the head of the flea, p. 98 of the sketchbook considered the recto, is inscribed faintly by Varley: *Original [?] W.Blake*. See Butlin, 1990, 157-158, no. 63. The standing flea also resembles Blake's depiction of the scaly personification of the plague in his watercolor *Death of the First Born* (c. 1805) in the Museum of Fine Arts, Boston, and studies for it (see Butlin, 1981, 2, fig. 518). For John and Cornelius Varley and their circle, see C. Klonk, *Science and the Perception of Nature: British Landscape Art in the Late Eighteenth and Early Nineteenth Centuries*, New Haven and London, 1996, 101-147.

17 See M. Butlin, *The Blake-Varley Sketchbook of 1819 in the Collection of M.D.E. Clayton-Stamm*, 2, London, 1969, 27-28, as p. 94 of the sketchbook (illustrated in vol. 1); Christie's, London, sale cat., 15/VI/71, no. 141, for an illustration after it was removed.

18 It is on pl. 34 of Hooke's treatise. Klonsky, 124-125, illustrates the entire flea while most other sources merely reproduce its head. The connection

with Hooke was first offered by C. Singer, "The First English microscopist: Robert Hooke 1635-1703," *Endeavour*, 14, 1955, 12-18. See also G. Keynes, "Blake's Visionary Heads and the Ghost of a Flea," *Bulletin of the New York Public Library*, 64, 1960, 567-572; *idem.*, *Blake Studies. Essays on his life and work,* Oxford, 1971, 139-135, pl. 33. Furthermore, Keynes points out that Hooke's plate of the flea was reprinted in two other eighteenth-century publications, although be believes that there is some reason to believe that Blake saw the original.

19 Klonsky, 124-125.

20 Inv. nos. 1907-102; 1891-12; 1921-814; 1921-818; 1981-74 (which belonged to Varley) in the Science Museum, London. See Parris, 1973, 125, no. 305, as well as M. Pidgley, "Cornelius Varley, Cotman, and the graphic telescope," *The Burlington Magazine*, 114, 1972, 781-786; M. Butlin, "Blake, the Varleys, and the Graphic Telescope," in *William Blake: Essays in honour of Sir Geoffrey Keynes*, ed. M.D. Paley and M. Phillips, Oxford, 1973, 294-304; J.H. Hammond and J. Austin, *The Camera Lucida in Art and Science*, Bristol, 1987, 57-77, on Varley's far-ranging scientific pursuits. Cornelius Varley was acquainted with lenses and worked on the manufacture of the Graphic Telescope himself. He had trained under Samuel Varley, his uncle, an instrument maker.

21 Olson, 1985b, 86.

22 K. Crouan, *John Linnell: A centennial exhibition*, exh. cat., Cambridge, 1982, xiii.

23 Inv. no. 1967-12-9-3 in the British Museum, London.

24 See R.J.M. Olson and J.M. Pasachoff, "The 1816 Solar Eclipse and Comet 1811 I in John Linnell's Astronomical Album," *Journal for the History of Astronomy*, 12, 1992, 121-133.

25 Crouan, ix, quoting articles in the *Art Journal* and the *Atheneum*.

26 *Ibid.*, xi.

27 Inv. no. 1967-12-9-3 (1) in the British Museum, London, in pen and brown ink on white paper measuring 197 × 165 mm. The paper is irregular with deckled edges. The sheet has a verso with unrelated sketches in graphite.

28 In the Fitzwilliam Museum, Cambridge, *Journals*, folios 31v and 32r, 1811, in pen and brownish gray ink on pages c. 254 × 212 mm (irregular). Before working with the original *Journals*, Joan Linnell Burton had sent us xeroxes of a copy made by an unknown family member. She and we assumed that it was a literal copy, so that the following differences with the original were not suspected. Presumably, these additions were annotated by the unknown copyist. Therefore, our article on Linnell's astronomical drawings in the 1992 *Journal for the History of Astronomy* (n. 24 above) needs to be amended with the following information. In the copy to the left of the inscription *Comet-9-* and the sketch of the comet is the following note: *The sketch is/This size and angle.* Above is: *The Comet referred to. which was "remarkably conspicuous" was had a length/according to W Herschel on Oct 15. of a hundred million miles./ greatest breadth fifteen million. Vis [visible] all autum. to the naked eye/ Phil Trans. Royal Soc for 1812.* At a later date, some Linnell relative diligently researched the information and annotated this page referring to the article by William Herschel in volume 99 of the *Philosophical Transactions.* Linnell's *Journals* are not only spartan but also dispassionate. No other comets are noted and there is only one other sketch that may have some astronomical import, what appears to be a sketch of the solar system, also made in 1811.

29 In July 1811, Linnell had taken a second floor flat at number eleven Queen Street Edgeware Road (A.T. Story, *The life of John Linnell*, 1, London, 1892, 80).

30 Inv. no. 1967-12-9-3 (2) in the British Museum, London, in white chalk on gray paper measuring 304 × 221 mm.

31 Thanks to Lindsay Stainton for her perceptive interpretation of the long "s."

32 Inv. no. 1967-12-9-3 (3) in the British Museum, London, in black and white chalk on gray paper measuring 295 × 221 mm.

33 D. Linnell, *Blake, Palmer, Linnell and Co.*, Lewes, East Sussex, 1994, 25, refers to these drawings of comet 1811 I with knowledge of the article by Olson and Pasachoff, but without citing it or the location of the drawings.

34 It was one of two that year; comet 1811 II was quite faint by comparison. See Vsekhsvyatskii, 144-145; Kronk, 1984, 27; Olson, 1985b, 74-77.

35 Crouan, xvi.

36 Quoted from W3/1.12, p. 37 in the Herschel Archive, Royal Astronomical Society Library, London, in pen and brown ink measuring 334 × 210 mm. William Herschel's observational notes belong to his log *Comets and Asteroids*, No. 1 (21 November 1781 to 9 November 1811; 16 September 1801 to 3 June 1807), listed in Bennett, 1978, 65. For Caroline Herschel's observations, see chap. II, nn. 119, 121.

37 In the collection of Gerald H. Morris, London, an engraving measuring 100 × 170 mm. It is inscribed at center: *The Comet of 1811/As seen at Day break the 15th of Oct.r from Otterbourne Hill, Winchester.* At left: *Engraved by H.R. Cook from a Drawing by Pether.* At right: *From the Gallery of Nature and Art.* It was published by R. Wilks, 89 Chancery Lane, on 15 August 1814.

38 Inv. no. B1975.4.1867 in the Yale Center for British Art, Paul Mellon Collection, New Haven, in pen and black ink and gray wash over graphite measuring 305 (311 uneven) × 557 mm. Its verso has a partially legible graphite inscription in a modern hand: *By Cornelius Varley (or John Varley)/ From Folio bought at Sucklings/ y Colnaghi 195?/Folio inscribed by C. Varley.* At the lower left in graphite it is also inscribed: *By C. Varley.* Cornelius Varley used his Graphic Telescope to render an architectural structure – John Heathcoat's Lace Factory at Tiverton, Devon – in a watercolor dated 1824, which could be compared with this sheet; see Parris, 1973, 101, no. 238, ill.

39 Vsekhsvyatskii, 141; Kronk, 1984, 25.

40 Inv. no. M.5-1956 in the Victoria and Albert Museum, London, with silver mounts in walnut and metal with gold, measuring 155 cm in length. It has a 1801 London hallmark, is signed *D. Egg*, and has the mark of Moses Brent, who made the silver mounts. See J.F. Hayward, *European Firearms*, London, 1969, 62-63, no. 98, wherein the forms with tails inlaid in silver on the stock are termed "rosettes."

41 For Egg, see A. M. Carey, *English, Irish, and Scottish Firearms Makers*, London and Edinburgh, 1954, 28. For Georgian firearms, J.F. Hayward, *The Art of the Gunmaker*, 2 vols., London, 1962 (for Egg, vol. 2, 287); H.L. Peterson, ed., *Encyclopedia of Firearms*, London, 1964, 138-140; W.K. Neal and D.H.L. Back, *Great British Gunmakers 1740-1790*, London, 1975.

42 See chap. II, n. 121; Vsekhsvyatskii, 140-143. W. Herschel, "Observations of a Comet, made with a View to investigate its Magnitude and the Nature of its Illumination," *Philosophical Transactions*, 97, 1807, 265,

notes that toward evening on January 27, he gave his sister the location where the comet might be and "between flying clouds that same evening about 6h 49' she saw it just long enough to make a short sketch of its situation." William Herschel calculated the diameter of the "well-defined disk" inside the coma to be 538 miles. Another manuscript in the Royal Astronomical Society Library, London, records this comet visible for three months, when in fact it was visible much longer; Ms. Madras 6, "An account of the Comet, which appeared in the Months of September, October and November 1807," written by Captain J. Warner. See Bennett, 1976, 42. For this comet, Vsekhvyatskii, 142; Kronk, 1984, 26.

43 It is listed as whereabouts unknown in Getty Art History Information Program, Documents for the History of Collecting on CD-ROM of the Provenance Documentation Collaborative, Malibu, 1995, which notes the sale of the painting on 6 April 1813 at an anonymous auction house in London. The engraving by G. Garrard (31 May 1813) was also included in *A Description of the different varieties of oxen, common in the British Isles; embellished with engravings*, London, 1800, opposite p. 2 of the section "Tees Water Cattle." We thank Marty Schlabach for checking the information. Another painting of this bull by Thomas Weaver is in the Beamish Open Air Museum, Chester le Street, County Durham, measuring 81.2 × 71.1 cm (according to Colin Pillinger).

44 W. Feaver, *Masters of Caricature from Hogarth and Gillray to Scarfe and Levine*, ed. A. Gould, New York, 1981, 5.

45 R. Godfrey, "Introduction," in London, Victoria and Albert Museum, *English Caricature 1620 to the Present*, exh. cat., London, 1984, 10-25, who believes that "The art of caricature was considered as British as roast beef and beer, a manifestation of Liberty denied to the beggarly French."

46 M.D. George, *Hogarth to Cruikshank: Social Change in Graphic Satire*, London, 1967, 13.

47 *Ibid.*, 17. As George III lay perilously close to death during his first bout of insanity, which was due to a genetic condition known as porphyria, a new comet discovered by Messier was reported in the pages of Topham's newspaper (*The World*, 6 December 1788). The fact that Herschel was also searching for the return of a periodic comet had been reported by the paper a couple of weeks earlier.

48 Vol. 7, p. 36v, no. 114, in the Peel Collection, Pierpont Morgan Library, New York, a hand-colored etching measuring 380 × 221 mm. It is included neither in George, nor in the standard monograph on the artist – E.B. Krumbhaar, *Isaac Cruikshank. A Catalogue Raisonné with a Sketch of his Life at Work*, Philadelphia, 1966 – nor in J. Nygren, ed., *Isaac Cruikshank and the Politics of Parody*, exh. cat., Los Angeles and San Marino, CA, 1994. It was printed and sold by J. Bailey, 116 Chancery Lane for 6 d. Its verse reads:

> It was the Bishop Athendius who once at Eventide....
> He [the Devil] went between the Horns of the Moon, with Athendius on his back...
> And there was an Eclipse that night which was not in the Almanack:
> They hit against a shooting Star so quick for fear did he sail
> That he singed the Beard of the Bishop against a Comet's tail....
> Now what was this the Pope had done, to send his Soul to Hell,
> Aye, that is the mystery of this wonderful History, I wish that I could tell.

49 See F.M.L. Thompson, ed., *The Cambridge Social History of Britain 1750-1950*, 1, Cambridge, 1990; C. Fox, ed., *London – World City – 1800-1840*, exh. cat., New Haven and London, 1992.

50 Inv. no. 1868-8-85840 in the British Museum, London, an aquatint measuring 248 × 344 mm. See George, 6, 590-591, no. 7508. The British Museum has two impressions of the same state. Above the margin of one are numbers lettered in brownish black ink to correspond to the identities of those portrayed, while below is the inscription in brownish black ink: *1. Sawbridge. 2. Wilson Bp Bristol 3. Warren Bp Bangor. 4. L*^{D}*. Sandwich. 5. Watson Bp. Llandaff. 6. Ld. Derby. 7. Marq. Lothian. 8. Burke. 9. D. Norfolk. 10. L*^{D}*. Derby. 11. M*^{r}*. Powis.* $^{\text{P Wales}}$ *12. D. Queensbury. 13. L*^{D}*. Stormont. 14. L*^{D}*. North. 15. L*^{D}*. Loughborough. 16. D. Portland. 17. Fox. 18. Sheridan.* The image is inscribed:

> *The Comet – /A Return of the Comet which appeared in 1761 is expected this Year and/to be within our horizon from the month of Oct.r 1788 to Aug.t 1789 but is expected to be most /visible (if it forces itself upon our Notice) in the Winter months Feb.y & March/—vide Dr Trusslers Almanack/By some of the ancient Astronomers Comets were deemed Meteors kindled in the Air and/designed as Presages or unlucky Omens of some disastrous Catastrophe—/The Peripateticks deemed them not permanent Bodies but bodies newly produced/and in a short Time to perish again, and affirmed that they were made up of Exhalations in the terrestrial Regions—/Sr. Isaac Newton asserts That the Tail of a Comet is nothing else than a fine Vapour which/the Head of the Comet emits by its heat that Heat the Comet receives from the Sun and the magnitude of the Tail /is always proportional to the degree of heat which the Comet receives, and Comets which are nearest to the Sun have /the longest Tails –*

51 See Feaver, 1981, 58; M. Bryant and S. Henage, *Dictionary of British Cartoonists and Caricaturists 1730-1980*, Aldershot, Hants, 1994, 192.

52 Immediately after the Prince is Richard Brinsley Sheridan, Fox, and Portland. See George, 6, 591, for all the identities.

53 *Ibid.*, 6, 591. Later prints of 1811 [Figs. 82, 83] depend on it. See nn. 76, 77.

54 Vol. 12, p. 77, no. 257, in the Peel Collection, Pierpont Morgan Library, New York, an aquatint, measuring 247 × 345 mm. It is inscribed: *Executed by Uncle Toby* and dated: *July 24 1792*. Above it is faintly inscribed: *Pub. by W. Dent*. George, 6, 923-924, no. 8115, notes that it was published several times in the *London Chronicle*. Uncle Toby is a reference to a character in Laurence Sterne's *Tristam Shandy*. For Dent, see Feaver, 1981, 58; Bryant and Henage, 57.

55 Vol. 4, p. 21, no. 101, in the Peel Collection, Pierpont Morgan Library, New York, an etching measuring 126 × 195 mm. It is inscribed: *Exitus acta probat*. George, 5, 28-29, no. 4887, notes that it was published on 1 November 1771 in the *London Magazine* and on 1 January 1772 in the *Hibernian Magazine*.

56 Vol. 9, p. 77v, no. 245, in the Peel Collection, Pierpont Morgan Library, New York, an etching measuring 258 × 185 mm. The inscription is difficult to decipher: *Pub. Feb. 11/2(?) 1790 by R. Hawkins...No 53. Old Comton St. Soho*. It is not in George's volumes.

57 For the artist, see: T. Wright., ed., *The Works of James Gillray, The Caricaturist*, London, n.d. (1970 reprint); *idem.* and R.H. Evans, *Historical and Descriptive Account of the Caricatures of James Gillray,*

London, 1831; D. Hill, *Mr. Gillray, The Caricaturist*, London, 1965; *idem.*, *Fashionable Contrasts. Caricatures by James Gillray*, London, 1966; London, Arts Council of Great Britain, *James Gillray 1756-1815*, exh. cat., London, 1967; H. Guratzsch and M. Hasse, *James Gillray 1757-1815. Meisterwerke der Karikatur*, exh. cat., Stuttgart, 1986; K.W. Hart, *James Gillray. Prints by the Eighteenth-century Master of Caricature*, exh. cat., Hanover, NH, 1994.

58 Inv. no. 1851-9-1-881 in the British Museum, London, a hand-colored etching measuring 542 × 410 mm. It is signed: *J s Gy inv & fect* and dated 2 November 1797. George, 7, 387-390, no. 9085, classifies most of the satirical prints as engravings, when many are in the freer technique of etching. See also London, Arts Council of Great Britain, 34, no. 55, where it is entitled: *Titianus Redivius; – or – The Seven-Wise-Men consulting the new Venetian Oracle – A Scene in ye Academic Grove No. 1. November 2, 1797*, as well as Wright, 231.

59 George, 1967, 126.

60 Inv. no. 1868-8-8-7512 in the British Museum, London, a hand-colored etching measuring 232 × 763 mm. See George, 8, 318-321, no. 10362; M. Duffy, *The Englishman and the Foreigner*, ed. M. Duffy, Cambridge, 1986, 328-329. A first uncolored state without inscriptions in the British Museum (inv. no. 1863-1-10-141) and an anonymous copy in the Metropolitan Museum of Art, New York (inv. no. 56.57), attest to its popularity.

61 See R. Olson, "Representations of Pope Pius VII: The First Risorgimento Hero," *Art Bulletin*, 68:1, 1986, 77-93, as well as Wright, 314-315; Wright and Evans, 239-40, no. 294; London, Arts Council of Great Britain, 42, no. 79, which notes that Armand Drayot (*Napoléon Raconté*, Paris, 1895, 171) reported that it "enjoyed a resounding success...Napoleon could not see it without entering into the most violent anger."

62 Since it was confused with the earlier comets by nonscientists, it was therefore logical that Leo Tolstoy exercised a bit of artistic license in *War and Peace* when, as Napoleon's troops invaded the Russian capital, a comet blazed not only as a sign of war, but also as a transcendental metaphor of the renewing power of love between his protagonists. For the three comets, see Vsekhvyatskii, 130, 144-145; Kronk, 1984, 27-28; B. Marsden and G.V. Williams, *Catalogue of Cometary Orbits 1996*, Cambridge, MA, 1996, 16-17, 48-49. For Napoleon's linkage with comet symbolism in art and literature, see Olson, 1985b, 75-81. His debacle at the Battle of Waterloo was also linked with a comet, the first appearance of the P/Olbers comet (13P/1815 E1); Vsekhyvatskii, 147.

63 Inv. no. 1868-8-8-10394 in the British Museum, London, a hand-colored etching measuring 326 × 369 mm. See George, 8, 635-636, no. 10972; it was published by H. Humphrey. There are two other impressions in the British Museum, inv. nos. 1868-8-8-13678 (a proof without inscriptions) and 1831-9-1-1249.

64 See Wright, 154; R. Evans and M. Evans, *The Life and Art of George Cruikshank 1792-1878: The Man who Drew the Drunkard's Daughter*, New York, 1978, 307-308, no. 340.

65 Inv. no. 1868-8-8-7595 in the British Museum, London, a hand-colored etching measuring 341 × 245 mm. George, 8, 560-561, no. 10768, notes it was published by Thomas Tegg on 21 October 1807 (sold for one shilling colored).

66 Feaver, 1981, 62; Bryant and Henage, 238-239.

67 Inv. no. 1868-8-8-7597 in the British Museum, London, a hand-colored etching measuring 246 × 343 mm. See George, 8, 561-56, no. 10769. Rowlandson engraved and designed the print published by Thomas Tegg III on 10 November 1807 (sold colored for one shilling). It was also bound in vol. 2 of *Caricature Magazine or Hudibrastic Mirror* as no. 274. For the magazine, see George, 8, 599-600; 610-611; 723-725; 889-891; five volumes (vols. III and 3 are different) were published by Tegg from 1807 to 1821 (Krumbhaar, 174, states that it began in 1803). See also J. Grego, *Rowlandson the Caricaturist*, 2, New York, 1880, 116.

68 For Rowlandson, see Grego, 2 vols.; A.P. Oppé, *Thomas Rowlandson. His Drawings and Water-colours*, ed. G. Holme, London, 1923; J. Hayes, *Rowlandson. Watercolours and Drawings*, London, 1972; R.R. Wark, *Drawings by Thomas Rowlandson in the Huntington Collection*, San Marino, CA, 1975; J. Baskett and D. Snelgrove, *The Drawings of Thomas Rowlandson in the Paul Mellon Collection*, New York, 1977; R. Southey, *Mr. Rowlandson's England*, ed. J. Steel, Woodbridge, Suffolk, 1985; J. Hayes, *The Art of Thomas Rowlandson*, exh. cat., Alexandria, VA, 1990.

69 Southey, 93-94. Another eccentric, Southey, who married his wife in Bristol and left on a visit to Spain and Portugal without her the same day, was witness to a meteorite fall on his trip. He was an associate of Humphrey Davy who made significant comments about the acceptance of meteorites as extraterrestrial objects on taking over the Presidency of the Royal Society after Banks.

70 In a private collection, New York, a hand-colored etching measuring 345 × 247 mm. See George, 9, 67, no. 11810. It was published on 20 September 1811 by Thomas Tegg (sold colored for one shilling) as well as in vol. I of *The Caricature Magazine or Hudibrastic Mirror* as no. 91.

71 Whereabouts unknown, in pen and ink and watercolor measuring 406 × 308 mm (larger than the print); see New York, H. Schickman Gallery, *Exhibition of Old Master Drawings at the H. Schickman Gallery*, sale cat., New York, 1968, no. 53, where it is identified incorrectly as "The Astrologer."

72 Inv. no. 1863-8-8-7985 in the British Museum, London, a hand-colored etching measuring 240 × 341 mm. See George, 9, 30-31, no. 11737. It was published by Thomas Tegg on 24 September 1811 (sold colored for one shilling) as well as in vol. 1 of *Caricature Magazine* as no. 92.

73 See Olson, 1985b, 15-17; R.J.M. Olson and J.M. Pasachoff, "New Information on comet P/Halley as depicted by Giotto di Bondone and other Western artists," *Astronomy and Astrophysics*, 187, no. 1/2, November (II) 1987, 1-11 (reprinted in *Exploration of Halley's Comet*, ed. M. Grewing, F. Praderie, and R. Reinhard, Berlin/Heidelberg, 1988); R.J.M. Olson, "Much Ado About Giotto's Comet," *Quarterly Journal of the Royal Astronomical Society*, 35, 1994, 145-148.

74 Inv. no. L.220.64 (f°) in *Caricature Magazine*, 1, no. 93, in the Yale Center for British Art, Paul Mellon Collection, New Haven, a hand-colored etching measuring 254 × 349 mm. See George, 9, 31, no. 11738. It was published on 25 September 1811 by Thomas Tegg (sold colored for one shilling).

75 Inv. no. 1868-8-8-7986 in the British Museum, London, a hand-colored etching measuring 248 × 353 mm. See *ibid.*, 9, 31-32, no. 11739. It was published in September 1811 by Walker and Knight. It is inscribed: *The First Glorious Exploit of the Invincible Boulogne-Flotilla. Sept. 20, 1811.*

76 Inv. nos. 1851-9-1-1358 (first state) and 1868-8-18-7971 (second state) in the British Museum, London, two hand-colored etchings with aquatint backgrounds, the first (without inscriptions) measuring 231 × 288 mm and the second 247 × 300 mm. Both were published in January 1811 by T. Moon. See *ibid.*, 9, 2-3, no. 11705, who notes that they are sometimes attributed to Gillray, although no name is inscribed. She also points out that the heads in the comet's tail seem to be copied from heads by Gillray. In the second state, Williams portrayed various figures in the Prince's wake; eight of their names are inscribed below (*Moira, Sheridan, Erskine, Derby/ Norfolk, Grenville, Buckingham, Temple*), while four others are inscribed along the right edge, reading from the lower right up the margin (*Temple, Buckingham, Grenville, Grey*).

77 Inv. no. 1868-8-8-7387 in the British Museum, London, a hand-colored etching measuring 247 × 354 mm. See *ibid.*, 9, 32-33, no. 11740. It was published in October 1811 by William Holland.

78 Inv. no. 114 in the Sterling and Francine Clark Institute, Williamstown, MA, in silver with a gilt interior, measuring c. 149 × 90 mm, and weighing 427.7 gm. Engraving on the handle, the script initial *F* and buck's head cabossed, probably crest of Frend, of county Limerick, or Friend. The mechanical regularity of its banding, as well as the plain treatment of the foot-ring and the D-shaped handle, suggest that the vessel was a mass-produced item, as does the finial, comprised of a stamped and chased mass of clouds and a cast and applied twelve-point star. A vertical seam under the handle confirms that the cylinder was not manufactured through the traditional process of hand raising. Other technical information is in the museum's files; R. Fernandez, J.M. Pasachoff, and R. Volz, "*Urania Observed*, an Exhibition of Works of Art with Astronomical Content," in *The Teaching of Astronomy*, eds. J.M. Pasachoff and J.R. Percy, Cambridge, 1989, 430, 433, no. 11. For its marks, see J. Banister, "A Postscript to the Barnard Ledgers," *Proceedings of the Silver Society*, 3, 1983, 39; C. Culme, *The Directory of Gold and Silversmiths, Jewellers and allied Traders, 1838-1914*, 1, Woodbridge, Suffolk, 1987, 29-30; A.G. Grimwade, *London Goldsmiths, 1697-1837: Their Marks and Lives*, London, 1990, 431.

79 See, for example, the drawing of the Great Comet of 1811 in the observations of Wilhelm Olbers and Lindenau; Vsekhsvyatskii, 146, pl. XVII.

80 It has been suggested in the museum files that the inclusion of Uranus, Saturn, and Jupiter may refer to the Greek cosmological myths recounting creation as a succession of the sky god Uranus, his son Chronos (Saturn in Latin), and his grandson Zeus (Jupiter in Latin), a sequence that was used in naming Uranus. However, if the makers had intended this reference, the planets would have been positioned in that sequence.

81 This suggestion was made by P.D. Hingley in a letter of 11 October 1991.

82 For example, inv. no. CR.C2.130 in the Crawford Collection of the Royal Observatory Edinburgh, a pamphlet by J. Williams entitled "A description of the Nature and Number of Comets seen Within Five Hundred Years," published in Birmingham in 1811. Its frontispiece features a quotation from Proverbs 3:17 and celestial bodies including a prominent comet.

83 Inv. no. 1862-12-17-432 in the British Museum, London, a woodcut measuring 208 × 123 mm. See George, 9, 332, no. 12170; G.W. Reid, *A Descriptive Catalogue of the Works of George Cruikshank*, 1, London, 1871, 205, no. 2800; F.G. Stephens, *A Memoir of George Cruikshank*, New York, 1891, 132. For the artist, see: R.L. Patten, *George Cruikshank: A Revaluation*, Princeton, 1974; Evans and Evans; J. Buchanan-Brown, *The Book Illustrations of George Cruikshank*, North Pomfret, Vermont, 1980; U. Bode et al., *George Cruikshank 1792-1878*, exh. cat., Stuttgart, 1983; R.L. Patten, *George Cruikshank's Life, Times, and Art*, 2 vols., New Brunswick, 1992-96.

84 Inv. no. 13508 in the British Museum, London, a woodcut measuring 140 × 102 mm. See George, 10, 5, no. 13508; H.T. Dickson, *The Caricature and the Constitution 1760-1832*, ed. M. Duffy, Cambridge, 1986, 268-269; Fox, 1992, 218. It was published first by William Hone in 1820 (sold for one shilling).

85 (1) The first example, whereabouts unknown, is in pen and brown ink and watercolor over traces of graphite measuring 330 × 270 mm. It is signed and dated at the lower left *Rowlandson 1821*. See Christie's, New York, sale cats., 25/X/84, no. 439, 9/VII/91, no. 70 (where the dimensions are reported as 430 × 270 mm); R.J.M. Olson, "The Draftsman's Comet," *Drawing*, 7:3, 1985, 52-53. (2) The second, in a private collection, is in pen and brown and gray inks and sanguine and grey washes, measuring around 302 × 245 mm with uneven edges. Its provenance is listed as from Eugen Hollander, Berlin, in Sotheby's, New York, sale cat., 20/I/82, no. 108. (3) The third is in the Albert H. Wiggin Collection of the Boston Public Library, where it is unnumbered, in graphite, pen and brown ink and watercolor, measuring 307 × 248 mm. (4) The fourth is inv. no. 49-47 in the Museum of London, in pen and brown and gray ink and watercolor over graphite measuring 311 × 245 mm. See Southey, 94; J. Hayes, *Thomas Rowlandson*, London, 1960, 20, no. 24.

86 See Vsekhsvyatskii, 150-151; Kronk, 1984, 29.

87 See also J.N. Nicollett, "On the Comet discovered in the Constellation Pegasus in 1821...," and W. Olbers, "On the Comet discovered in the Constellation Pegasus in 1821...," *Memoirs of the Royal Astronomical Society*, 1:1, 1822, 154-158; J.N. Nicollett, "The Right Ascension and the Declination of the Comet of January 1821," *Memoirs of the Royal Astronomical Society*, 1:2, 1825, 271-272.

88 Inv. no. Rare Book 144954, f. 313, in the Huntington Library, San Marino, CA, in pen and ink and watercolor over graphite measuring 226 × 127 mm. Signed in Rowlandson's hand. See Wark, 90, no. 297. Other versions are in the Widener Collection at Harvard University and the Daragnès Collection, reproduced in the Witt Library.

89 Hayes, 1972, 23.

90 For the Dent Sketchbook, see Christie's, London, sale cat., 10/VII/84, no. 39, with 150 drawings on 91 leaves of an album. Noteworthy are p. 47, no. 74, *The Observatory, Greenwich Park.*, and p. 51, no. 80, *On the Lookout: Greenwich Pensioners with a Telescope*. The prefatory note reads:

> GREENWICH – a town in Kent 4 Miles... was the birthplace of Queen Mary & Queen Elizabeth & here/Edward the VI died. A Palace Erected here....He [Charles II] also enlarged/ the Park, walled it around, planted it, & erected a Royal Observatory/ on the top of the hill, for the use of the celebrated Flamstead [mis-transcribed in the Christie's catalogue as Hamstead],/ whose name the hill retains. He likewise furnished it with/ mathematical instruments for astronomical observations/ and a deep dry well for observing the stars in the day time.

91 K. Lippincott, *The Story of Time and Space: The Old Royal Observatory at Greenwich, Past, Present and Future*, Tonbridge, Kent, 1994, 7.

92 D.H. Sadler, "Place in Space," in *Greenwich Observatory: 300 years of astronomy*, ed. C.A. Ronan, London, 1975, 17-22.

93 The Astronomer Royal at Greenwich for a long time was the only professional scientist in the pay of the King (for a list of the Astronomers Royal, see Appendix III). When other scientists became useful to the Nation, no similar posts were created. Conversation with Sir Martin Rees, Astronomer Royal, 17 April 1995.

94 Lippincott, 15-27.

95 There were problems with observing in that site because of overcast skies. Though a l00-inch Isaac Newton Telescope (since moved to La Palma) was built at Herstmonceaux, the Royal Greenwich Observatory was moved to Cambridge in 1990. A building was constructed adjacent to the existing Observatory and the Institute of Astronomy, and the archives transferred to the Cambridge University Library. The title of Director of the Royal Greenwich Observatory was separated from that of Astronomer Royal with the appointment of E. Margaret Burbidge to the former in 1972. See D. Howse, *Greenwich Observatory: the Royal Observatory at Greenwich and Herstmonceaux, 1675-1975,* 3: *The Buildings and Instruments*, London, 1975, 13-15. The other volumes are: E.G. Forbes, 1: *Origins and Early History (1675-1835)*, London, 1975; A.J. Meadows, 2: *Recent History (1836-1975)*, London, 1975.

96 In apparent gilded brass (unknown), measuring 54 × 84 cm. It is on the Altazimuth Pavilion, built in 1899, and was patterned after the depiction of Halley's Comet in the Bayeux Tapestry. See Howse, 1975, 56, fig. 51 caption. In the material about the Altazimuth Pavilion from the archives at the Cambridge University Library, inv. no. RGO 7/35, there is no reference to the weathervane, although it appears in the unsigned architect's diagram labelled: *Royal Observatory Greenwich Building for New Altazimuth/rt side: East Elevation.*

97 B. Lowell, ed., *Astronomy: The Royal Institution Library of Science*, 2 vols., Barking Cross, 1970, 34, notes that of the Friday Evening Discourses between 1851 and 1939, two (W. Huggins in 1882 and H.H. Turner in 1910) were about comets, while eight concerned meteors or meteorites. The list of Christmas Juvenile lectures includes ten on astronomy between 1826 and 1911; one was J. Dewar's on meteorites (1885-86).

98 D. Rosen and S. Rosen, *London Science*, London, 1994, 123-124.

99 H.B. Carter, *Sir Joseph Banks, 1743-1820,* London, 1988, 194-202, 540.

100 J.L.E. Dreyer and H.H. Turner, eds., *History of the Royal Astronomical Society, 1820-1920,* London, 1923 (reprint Oxford, 1987), 7-9.

101 G. Buttmann, *The Shadow of the Telescope: A Biography of John Herschel*, New York, 1970 (reprint London, 1974).

102 W.J. Ashworth, "The calculating eye: Baily, Herschel, Babbage, and the business of astronomy," *British Journal for the History of Science*, 27, 1994, 411.

103 *Ibid.*, 412.

104 Dreyer and Turner, 127

105 *Ibid.*, 79.

106 *Ibid.*, 38-42.

107 *Ibid.*, 79-80.

108 *Ibid.*, 239-240. For example, it published the important articles of T. Maclear, "Observations of Donati's Comet, made at the Royal Observatory, Cape of Good Hope, between October 11th, 1858, and March 4th, 1859," *Memoirs of the Royal Astronomical Society*, 29, 1861, 59-83; W. Lassell et al., "Observations on Donati's Comet," *Memoirs of the Royal Astronomical Society*, 30, 1862, 58-84.

109 For Palmer, see A.H. Palmer, *The Life and Letters of Samuel Palmer*, London, 1892; G. Grigson, *Samuel Palmer the Visionary Years*, London, 1947; D. Cecil, *Visionary and Dreamer. Two Poetic Painters: Samuel Palmer and Edward Burne-Jones*, Princeton, 1969; R. Lister, *Samuel Palmer: A Biography*, London, 1974; *idem.*, ed., *The Letters of Samuel Palmer*, 2 vols., Cambridge, 1974; D.B. Brown, *Samuel Palmer 1805-1881. Loan Exhibition from the Ashmolean Museum Oxford,* exh. cat., London, 1982; R. Lister, *The Paintings of Samuel Palmer*, Cambridge, 1985; A. Abley, ed., *The Parting Light: Selected Writings of Samuel Palmer.*, Manchester, 1985; R. Lister, *Samuel Palmer, his life and art*, Cambridge, 1987; *idem.*, *Catalogue Raisonné of the works of Samuel Palmer*, Cambridge, 1988.

110 Inv. no. B1981.25.2655 in the Yale Center for British Art, Paul Mellon Collection, New Haven, in pen and brown ink with brown and grey wash and white heightening measuring 108 × 132 mm. See White, 119-120, no. 215. Lister, 1987, 20, believes that c. 1824 Palmer, like Blake, discerned in the English landscape the spiritual Jerusalem, an idea that finds reflection in this watercolor.

111 For Danby, see E. Adams, *Francis Danby – Varieties of a Poetic Landscape*, New Haven and London, 1973; F. Greenacre, *Francis Danby 1793-1861*, exh. cat., London, 1988.

112 Inv. no. 162 in the National Gallery of Ireland, Dublin, in oil on canvas measuring 185 × 255 cm. See Adams, no. 26; J. Hutchinson, *James Arthur O'Connor*, exh. cat., Dublin, 1985, 57, no. 17; Greenacre, 99-100, no. 24, wherein he entitles it "*An Attempt to Illustrate the Opening of the Sixth Seal*" and mentions that in 1826 Danby abandoned it, after the exhibition of Martin's *The Deluge* (see chap. IV), privately accusing Martin of plagiarism. Greenacre, 26-28, also discusses this situation, and on p. 98, no. 23, illustrates a probable study, which is dated c. 1825-28 and measures 63.5 × 76.1 cm in the Victoria Art Gallery, Bath City Council. Here, the form hitting the cliff looks more like a bolt of lightning, whereas in the painting it seems a meteorite. For the mezzotint by G.H. Phillips of 1830, see *ibid.*, 150-151, no. 134.

113 As noted in chap. II, there was a great awareness of meteorites at this time.

114 For Martin, see T. Balston, *John Martin 1789-1854. His Life and Works,* London, 1947; W. Feaver, *The Art of John Martin,* Oxford, 1975; J.D. Wees, *"Darkness Visible." The Prints of John Martin*, exh. cat., Williamstown, MA, 1986.

115 Inv. no. 193a.25 1978-5-20-11(16) in the British Museum, London, a mezzotint measuring 191 × 278 mm (image) from the bound version. A separate impression in the same institution is inv. no. 1895-10-15-13. Martin's *Creation of Light* was one of the 24 mezzotint illustrations of *Paradise Lost* commissioned in 1823 by the publisher Septimus Prowett, together with a smaller set of plates. Prowett published *Paradise Lost* in four versions – folio and octavo and two quarto sets – between 1825 and 1827. Individual prints could be bought as well. The success of the project encouraged Martin to set up his own printing shop. For Martin's illustrations to Milton, see Balston, 94-105; Feaver, 1975, 72-113; M.J. Campbell, *John Martin Visionary Printmaker*, exh. cat., York, 1992, 81.

116 Inv. no. 1951-685/78 in the Science Museum, London, an engraving measuring 127 × 95 mm.

117 Inv. no. 1978-160/2 in the Science Museum, London, the engraving is only a fragment of the whole, a xerox of which is in their collection. The entire engraving measures 428 × 553 mm and was published by John Wallis on 17 December 1804 in London. It is inscribed: *..of whom may be had Science in Sport or the Pleasures of NATURAL PHILOSOPHY, an entertaining game.* We thank Wendy Sheridan for bringing this and the previous example to our attention.

118 Inv. no. 1979-351 in the Science Museum, London, the cards are engravings on board with watercolor measuring roughly 98 × 65 mm.

119 It is entitled *Urania's Mirror*, Urania being the muse of Astronomy. The cards measure 203 × 140 mm and were published in several editions by Samuel Leigh and later M.A. Leigh at 421 Strand. The set was often accompanied by one of the editions of *A Familiar Treatise on Astronomy* by Jehoshaphat Aspin. The identity of the author of the game, the Reverend Richard Rouse Bloxam, who became a member of the Royal Astronomical Society in 1830, was discovered by P.D. Hingley, "*Urania's Mirror* – a 170-year old mystery solved?," *Journal of the British Astronomical Association*, 104:5, 1994, 238-240. *Urania's Mirror* is also included in C. Tennant, *The Box of Stars*, Boston, Toronto, and London, 1993.

120 One is inv. no. 1988-123 in the Science Museum, London, measuring 100 (length) × 60 (width) × 30 (height) mm. Another identical example was advertised in a Tesseract, Croton-on-Hudson, sale cat., 1994.

121 *Bryan's Dictionary of Painters and Engravers*, 3, London, 1904, 78-79. See also H. Howard, *A Course of Lectures on Painting delivered at the Royal Academy of Fine Art*, ed. F. Howard, London, 1848. His astronomical subjects are varied. In 1795 he illustrated the "Planets drawing Light from the Sun," quoting Milton's *Paradise Lost*, Book VI, which was subsequently modified into *The Solar System* and exhibited at the Royal Academy in 1823. It was duplicated and a circular adaptation was painted for the boudoir of the Duchess of Sutherland. The sketch ultimately adopted, in which had been introduced the subsequent "astronomical discoveries of the planet Herschel, and the asteroides [sic], Juno, Pallas, Ceres, and Vesta," was purchased by the painter Sir Thomas Lawrence. In 1797 Howard exhibited at the Academy another small sketch from Milton, "Sin and Death passing through the Constellations." F. Howard adds "astronomical personification being evidently the ruling passion, which, it will be seen, at various periods in his after-life...." (*Ibid.*, lxii-lxiii).

122 Whereabouts unknown, in watercolor measuring 153 × 229 mm. See Christie's, London, sale cat., 13/VII/65, no. 117.

123 H. Howard, lxii.

124 *Ibid.*, lxxv.

125 Whereabouts and medium unknown, measuring 137.5 × 117 mm. See Christie's, London, sale cats., 16/III/84, no. 82, as attributed to the circle of Richard Cosway, 1/III/91, no. 77, as "Attributed to Henry Howard."

126 H. Howard, lxxiii-lxxiv.

127 See Wilton, 1992, 167-186.

IV

The Triumph of Realism

By the fourth decade of the nineteenth century, new astronomical discoveries helped to weaken further the Romantic approach to comets and meteors in favor of a more empirical one, leading to more down-to-earth attitudes. After 1820 there was a general increase in comet discoveries, a trend that peaked between 1840 and 1890. While these celestial vagabonds still elicited feelings of awe, they were more frequently observed not out of fear but rather for their inherent beauty and unusual presence in the heavens.

Improving technology continued to aid astronomers in their investigations of not only bright comets but also fainter ones so that the century proved to be *the* comet century. (There were over thirteen spectacular comets during the hundred-year span [Appendix II], an unprecedented record that has not been equaled to date.) By 1825, telescopic observation of comets had improved so much that four comets were spotted in that year, followed by five in 1826.[1] As the technology of optics grew more sophisticated, larger refracting telescopes aided celestial sleuthing. They facilitated examining comets in detail, although their fields of view were usually too small for discovery. Better for discovery was a compact type called "the comet-watcher" or "comet-seeker." Similarly, most comets found today are discovered with wide-field binoculars or wide-field telescopes. During this period, comet hunting became as much a practice of amateurs as of professionals due to the long, often fruitless hours that it requires (on average it takes a few hundred hours of searching to discover a comet). Because of the intense scrutiny of the skies, much of the remaining mystery and lore surrounding comets not already dispelled by the momentous discoveries of the previous two hundred years disappeared for natural scientists.

Nevertheless, the nineteenth century maintained an erratically dualistic attitude toward comets and meteors. For example, Samuel Taylor Coleridge, the great Romantic poet, looked at nature in a mystical manner while attempting to understand comets in scientific fashion. He failed in the latter, however, as shown by a letter in which Coleridge argues against Newton's theory about comets.[2] Segments of the public seemed periodically unaffected by the new scientific understanding, with some individuals clinging to age-old superstitions that billed comets as instant doomsday machines. This situation follows from the fact that the era was prone to attacks of excess and journalistic-induced hysteria whenever comets appeared, both of which tended to exacerbate recurring bouts of virulent comet fever. However, these attacks neither lasted as long as in previous centuries nor afflicted everyone. Other individuals, freed by the new scientific knowledge about the nature of comets and meteors, invented fantasies based on old themes with a novel detachment. Thus, by comparison with earlier times, the predominant attitude toward them in the middle decades of the century was more objective.

This chapter will examine images of comets and meteors produced in Great Britain, together with the historical apparitions and the astronomers who studied them, from the fourth decade of the century until Donati's Comet in 1858. It will explore the strong connection between individuals in the visual arts and astronomers during a time that witnessed the advent of photography and a more realistic art.

Halley's Comet 1835-36 and Astronomical Improvements

When Halley's Comet returned in 1835 (1P/1835 P1, formerly 1835 III) in its twenty-eighth recorded appearance, many artists were already pursuing naturalistic studies. A similar attitude is also present in the increasingly detailed observational drawings by astronomers, who no longer simply drew thumbnail sketches. Instead they produced analytical studies of the physical configurations of these phenomena made possible by ever more sophisticated instruments. Needless to say, this comet's return caused a great flurry of activity by astronomers and artists alike.

Figure 96. Thomas Maclear, detail of a drawing of *Halley's Comet on 1 February 1836 (Ms. 15/149, fol. 43v)*, pen and brown ink, 70 × 80 mm (image), Royal Greenwich Observatory Archives, Cambridge University Library.

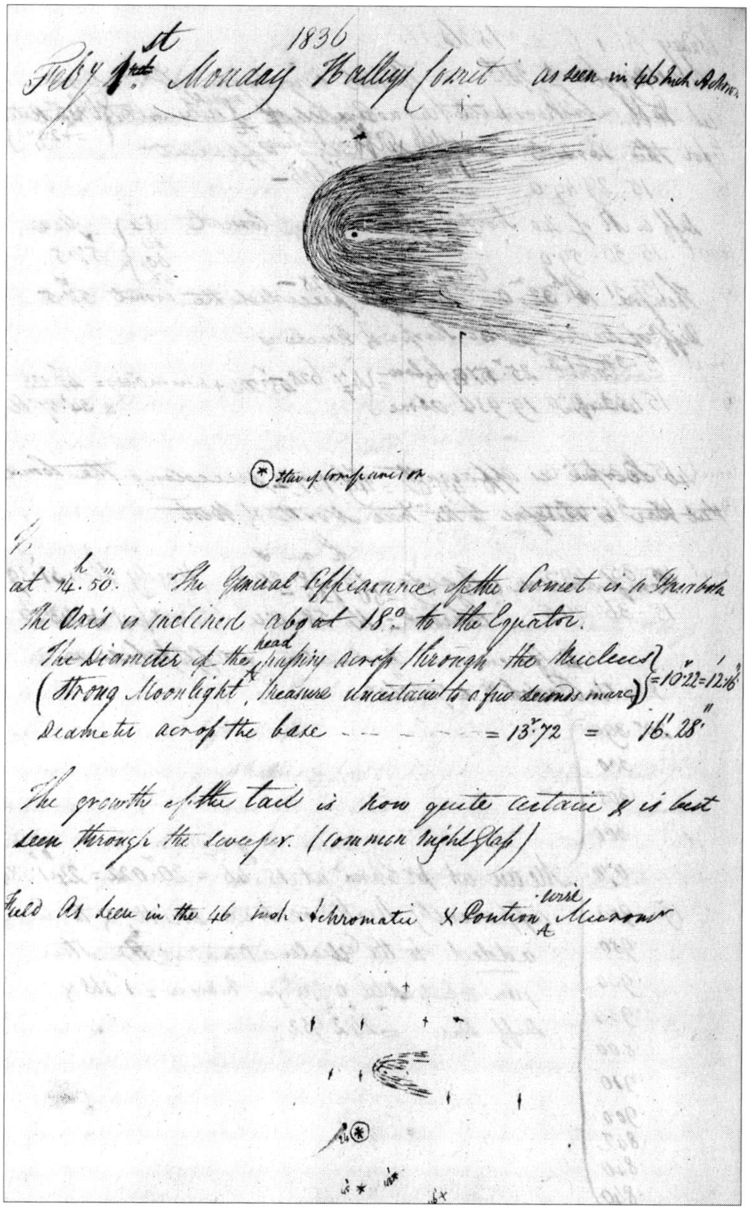

Halley's Comet was rediscovered on 5 August 1835. It was naked-eye by the end of September and observed until May 1836.[3] Both Thomas Maclear (1794-1879) and John Herschel (1792-1871) observed and drew it from the southern hemisphere at the Cape of Good Hope [Figs. 96-98], as in a pen and ink drawing by Maclear [Fig. 96], and published articles on it.[4] They recorded daily changes in its head that are preserved in prints after both Herschel's

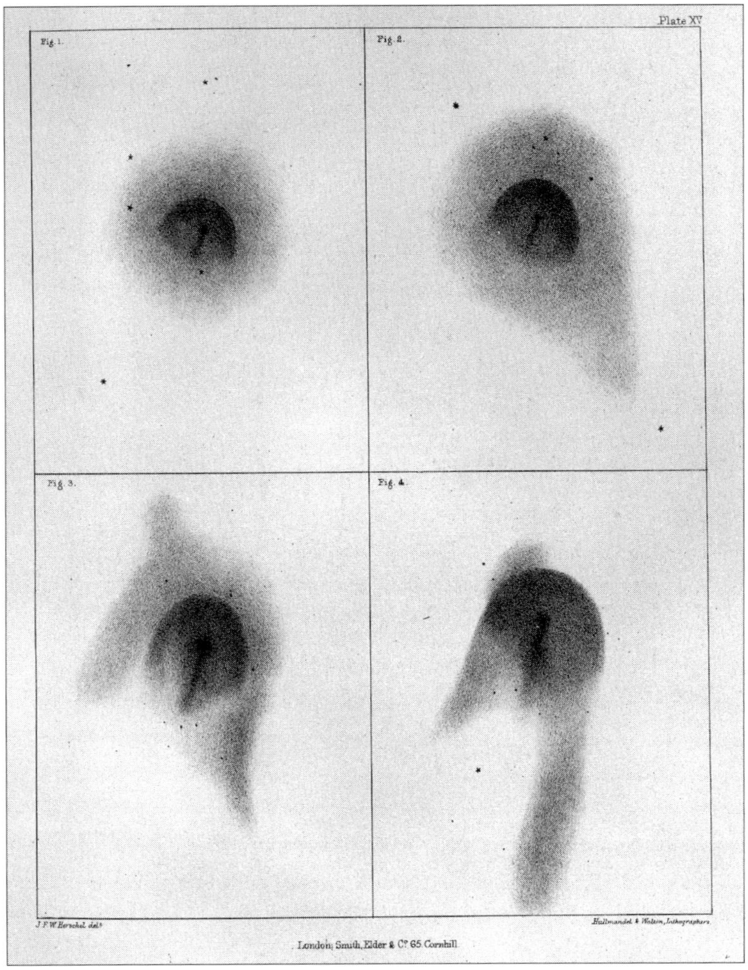

Figure 97. John Herschel, *Halley's Comet of 1835-36*, plate XV from *Observations at the Cape*, 238 × 185 mm (image), Royal Astronomical Society Library, London.

[Fig. 97][5] and Maclear's sketches.[6] These plates reveal a desire to capture with great verisimilitude the physical properties of the internal structure of its volatile head and temporal changes, such as the gas and dust flowing out of it.

John Herschel, who observed this comet with great care, played a major role in early nineteenth-century British astronomy.[7] His father William was already a famous astronomer when he was born, and during his youth, he was close to his Aunt Caroline, the indefatigable comet-hunter. His brilliance brought him early recognition at the University of Cambridge and a precocious election to the Royal Society at the age of 21. While his father may be the more original astronomer, John was more significant in the political interplay between science and the government and in the

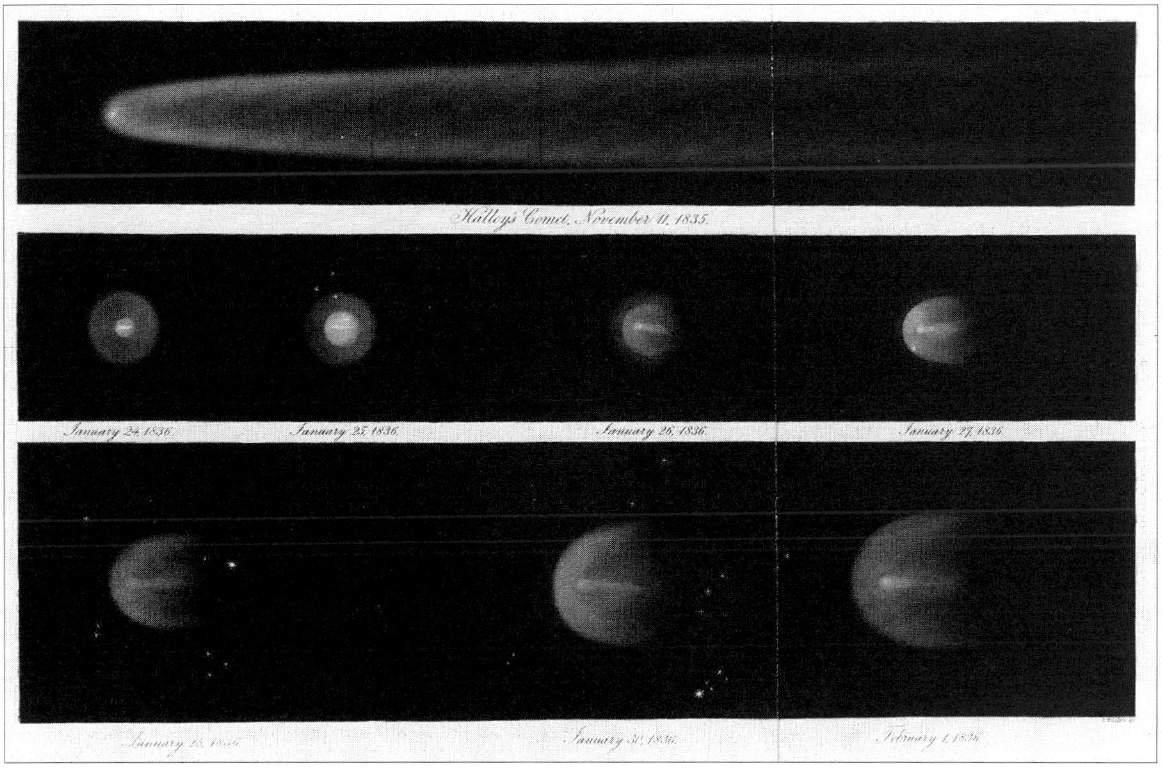

Figure 98. Thomas Maclear (with the assistance of Charles Piazzi Smyth), *Halley's Comet of 1835-36*, plate II from volume 10, *Memoirs of the Royal Astronomical Society*, 192 × 310 mm (image), Royal Astronomical Society Library, London.

professional organization of science. John continued the astronomical sweeping of the skies his father had pioneered. In 1833, he published a catalogue of thousands of nebulae and star clusters, based on both their observations at Slough, and later catalogues of double stars. Herschel also researched problems involving physics and mathematics and worked on the chemistry of nascent photography, searching for appropriate materials. In the process, he discovered the basis of what we call "developer," and the terms "positive" and "negative," still in use in photography, were his invention. He also took the first photograph on glass, a view of his 40-foot telescope on 9 September 1839.[8] His book *Results of Astronomical Observations Made...at the Cape of Good Hope, Being a Completion of a Telescope Survey of the Whole Surface of the Visible Heavens* (London, 1847) was a tour de force. He received three medals from the Royal Society, served a term as its Secretary, and was a founder of the Royal Astronomical Society and its President for three different intervals. Herschel went on to government service, taking Newton's former position of Master of the Mint.[9]

John Herschel also refurbished William's 20-foot telescope and used it at Slough from 1825 to 1833. Subsequently he took it to the Cape of Good Hope, South Africa, using it from 1834 to 1838 in collaboration with the observatory's director, Maclear to view the different set of stars seen from the southern latitudes. Because of the speed with which the mirror tarnished, he brought with him his polishing machine and three specula (each mirror was usable for only two or three months and sometimes only for a week).[10] He also used this telescope for his exceptional 1835-36 observations of Halley's Comet.

Herschel's telescopes were surpassed in size in the nineteenth century only by those of William Parsons, Lord Rosse. Rosse's giant telescope, the Leviathan of Parsonstown, on his Birr Castle estate in Ireland was cast in 1842 and became operative in 1845. Its mirror measures 1.8 meters (72 inches) in diameter, weighs four tons, and remained the largest telescope in the world for 70 years.[11] However, Rosse's viewing was plagued by the bad weather of Ireland and by the tarnishing of its mirror.[12]

Herschel's and Rosse's telescope mirrors were made of speculum metal, an alloy of tin and copper. When new this alloy gave good images, although it reflected only sixty percent of the incident light at best and soon tarnished. To restore its reflectivity, it had to be polished, which damaged the accuracy of the shaped reflecting surface. The problem was overcome only when astronomers began in 1857 to employ glass coated with a reflecting surface of silver.[13] This surface started with a reflectivity of ninety percent, and when it tarnished, the silver could be recoated without damage. This technique was replaced in the 1930s by the currently used non-tarnishing aluminum coating.

Another problematic area of telescopic observation that needed improvement was chromatic aberration in refracting telescopes – the focusing of light of different colors at diverse points. It had first been solved by a discovery of Chester Moore Hall of Essex, but ignored until its rediscovery by John Dollond in 1758. Both men found that stacked layers of flint glass and crown glass would give images that were more-or-less color free, known as "achromatic," something that Newton had thought impossible. For a time, these lenses could not be made larger than about 12 centimeters in diameter, a limitation which led to an emphasis on reflectors,

Figure 99. Anonymous, *Comet Almanack for 1836 (Ms. Add. 180)*, 1836, wood engraving in reddish brown and black ink, 334 × 258 mm, Royal Astronomical Society Library, London.

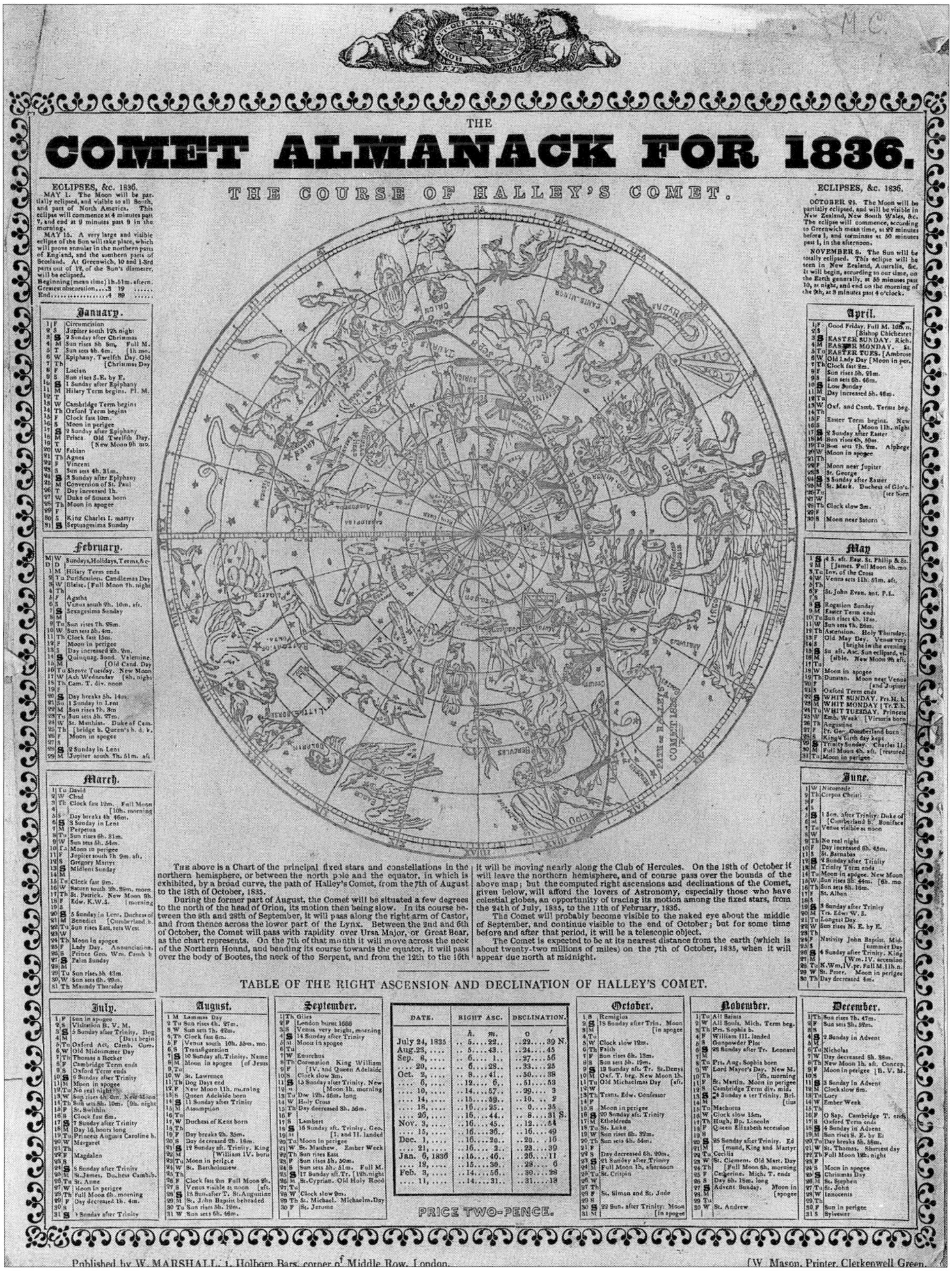

THE

COMET ALMANACK FOR 1836.

THE COURSE OF HALLEY'S COMET.

ECLIPSES, &c. 1836.

MAY 1. The Moon will be partially eclipsed, and visible to all South, and part of North America. This eclipse will commence at 4 minutes past 7, and end at 9 minutes past 9 in the morning.

MAY 15. A very large and visible eclipse of the Sun will take place, which will prove annular in the northern parts of England, and the southern parts of Scotland. At Greenwich, 10 and 1-3rd parts out of 12, of the Sun's diameter, will be eclipsed.

Beginning (mean time) 1h. 51m. aftern.
Greatest obscuration....3 19 ......
End....................4 30 ......

ECLIPSES, &c. 1836.

OCTOBER 24. The Moon will be partially eclipsed, and will be visible in New Zealand, New South Wales, &c. The eclipse will commence, according to Greenwich mean time, at 22 minutes before 1, and terminate at 50 minutes past 1, in the afternoon.

NOVEMBER 8. The Sun will be totally eclipsed. This eclipse will be seen in New Zealand, Australia, &c. It will begin, according to our time, on the Earth generally, at 55 minutes past 10, at night, and end on the morning of the 9th, at 8 minutes past 4 o'clock.

The above is a Chart of the principal fixed stars and constellations in the northern hemisphere, or between the north pole and the equator, in which is exhibited, by a broad curve, the path of Halley's Comet, from the 7th of August to the 18th of October, 1835.

During the former part of August, the Comet will be situated a few degrees to the north of the head of Orion, its motion then being slow. In its course between the 8th and 28th of September, it will pass along the right arm of Castor, and from thence across the lower part of the Lynx. Between the 2nd and 6th of October, the Comet will pass with rapidity over Ursa Major, or Great Bear, as the chart represents. On the 7th of that month it will move across the neck of the Northern Hound, and bending its course towards the equator, it will pass over the body of Bootes, the neck of the Serpent, and from the 12th to the 16th it will be moving nearly along the Club of Hercules. On the 18th of October it will leave the northern hemisphere, and of course pass over the bounds of the above map; but the computed right ascensions and declinations of the Comet, given below, will afford the lovers of Astronomy, especially those who have celestial globes, an opportunity of tracing its motion among the fixed stars, from the 24th of July, 1835, to the 11th of February, 1836.

The Comet will probably become visible to the naked eye about the middle of September, and continue visible to the end of October; but for some time before and after that period, it will be a telescopic object.

The Comet is expected to be at its nearest distance from the earth (which is about twenty-two millions of miles) on the 7th of October, 1835, when it will appear due north at midnight.

TABLE OF THE RIGHT ASCENSION AND DECLINATION OF HALLEY'S COMET.

DATE.	RIGHT ASC.	DECLINATION.
	h. m.	° ′
July 24, 1835	.. 5....22..	..22....39 N.
Aug. 23, ...	.. 5....43..	..24....45
Sep. 8, ...	.. 6.... 1..	..27....56
.... 20, ...	.. 6....28..	..33....25
Oct. 2, ...	.. 8....41..	..50....38
.... 6, ...	..12.... 6..	..51....53
.... 10, ...	..14....57..	..29.... 3
.... 14, ...	..15....59..	..10. .. 2
.... 18, ...	..16....25..	.. 0....35
.... 26, ...	..16....44..	.. 8....31 S.
Nov. 3, ...	..16....45..	..12....54
.... 15, ...	..16....36..	..16....49
Dec. 1, ...	..16....20..	..20.. .16
.... 21, ...	..16.... 2..	..23....39
Jan. 6, 1836	..15....46..	..26....11
.... 18, ...	..15....30..	..28.... 6
Feb. 3, ...	..14....56..	..30....28
.... 11, ...	..14....31..	..31....18

PRICE TWO-PENCE.

Published by W. MARSHALL, 1, Holborn Bars, corner of Middle Row, London.

[W. Mason, Printer, Clerkenwell Green.

whose mirrors could be much larger. By the early nineteenth century, techniques became available for making lenses that were larger, and by mid-century similar lenses 50 centimeters in diameter were made. Only in the last half of the century were larger refractors made, up to the 100-centimeter (40-inch) one at the Yerkes Observatory in Williams Bay, Wisconsin, which is still the largest. Today's grand telescopes are all reflectors with the 2.4-meter Hubble Space Telescope in orbit and the two 10-meter Keck telescopes on Mauna Kea in Hawaii as prime examples.

Because people knew in 1835 that Halley's Comet would to be visible for an extended period, *The Comet Almanack for 1836* [Fig. 99] was issued.[14] It is an example of the connection between comets and calendars/almanacs that dates back several centuries.[15] Its dominating feature is a circular diagram of the principal fixed stars and constellations of the northern hemisphere through which is drawn the path of Halley's Comet between 7 August and 18 October 1835. Directly below is the text about the comet's predicted behavior. The lateral and lower boxes provide additional almanac material. Thus, Halley's Comet proved to be not only the central device of the print, but also the selling gimmick for its publisher.

This comet's appearance engendered comet valentines, echoing the old link between comets and sexual love. At the time British valentines occasionally incorporated current events into their imagery, so that it is not at all surprising that Halley's 1835-36 apparition was included in at least one of them [Fig. 100].[16] The valentine features Venus in her chariot poised over seated lovers, bathed in the light of a comet. They are accompanied by two putti – Cupid with a quiver and a bow holding two flaming hearts and Hymen brandishing a burning torch and holding a flower garland. The inscription *The COMET changed to a VALENTINE, for the 14th of February* crowns its composition.

Other ephemera, many no doubt engendered by the return of Halley's Comet in 1835-36, survive as well. One example is a carnelian intaglio watch fob engraved with a profile portrait of Edmond Halley with a comet behind his cranium [Fig. 101 color].[17] From c. 1830 through the 1870s many so-called "comet pins," set with semiprecious stones in the heads and tails, were produced in quantities that prove comets were both topical and

Figure 100. Anonymous, *Comet Valentine Inspired by Halley's Comet of 1835*, 1835, engraving, measurements unknown, whereabouts unknown.

fashionable.[18] They are stylized descendants of an earlier example in cut steel, a material popular in the mid-eighteenth century, with a tail that trembles as the wearer moves [Fig. 102 color]; it indisputably represents a comet and may date as early as the 1759 apparition of Halley's Comet.[19] Since so many people wore these pins, we conclude that the sting of superstition about comets had been largely replaced by an excitement about them.

Halley's Comet of 1835-36 attracted a number of English artists. John Constable (1776-1837), one of the major figures of the time, was strangely not among them. It is all the more puzzling that he was not attracted to the comet apparitions which

Figure 104. John Doyle (HB), *The Alarming Comet of 1835*, lithograph, 234 × 343 mm, The Science Museum, London.

illuminated the skies of his lifetime, because Constable claimed that the sky is the "chief organ of sentiment" and made many breathtaking studies of clouds and the English atmosphere. But Constable rarely rendered nocturnal scenes, and when he did, he usually did not depict specific astronomical details, even shying away from the moon. A devoted family man, Constable preferred daylight studies of meteorological conditions. However, he knew the Chalons (Alfred Edward and John James) during the 1830s in London, and John James Chalon rendered a wonderful watercolor of a crowd of people viewing Halley's Comet with the aid of a telescope [Fig. 103 color].[20] The work has an authentic feeling as though the artist quickly recorded an actual event, wherein those gathered discuss and point at the awesome comet. The correspondence between Constable and the Chalons demonstrates that they were in contact during the very year of Halley's appearance.[21] At the time, comets were still used in political satires, albeit somewhat diluted ones, like the lithograph by John Doyle, who worked under the initials HB, that was occasioned by Halley's Comet in 1835 [Fig. 104].[22] In it, Doyle has employed the same device of the personified comet as earlier satirists [Figs. 69, 82, 83] but in a less forceful manner. The difference in tone is due partly to artistic conception and partly to technique; its light lithographic crayon is less abrasive than deeply etched lines.

Turner's Comets and Meteors and an Obsession with Light

Joseph Mallord William Turner (1775-1851), who must also have observed Halley's Comet in 1835-36, was fascinated by the effects of light. One could even say that he was obsessed with them, in substance and metaphor. He painted rainbows, lightning, sunsets, moonlight through clouds, fireworks, exploding shells, light shining through atmosphere as well as fog in oil paintings and seemingly endless watercolor sketches. While it may be argued that he never painted an historical comet – such as those he is most likely to have observed in 1811 and 1835-36 – in a contemporary setting, for reasons about which we can only speculate, he did depict at least one comet in a context calling for it. He also represented meteors, again when a literary source had established a tradition. This information implies that Turner preferred a suggestive light – that is, he was partial to intense, poetic but slightly mysterious evocations, rather than to more journalistic recordings or highly symbolic depictions.[23]

Two of Turner's 1835 vignette illustrations for Milton's *Paradise Lost* feature meteors or shooting stars [Figs. 105, 106]. Needless to say, the passages which they illustrate have an apocalyptic tenor. Both were engraved after watercolors by the artist, an avid reader of Milton's poetry.

The first depicts the Mustering of the Warrior Angels from Book V [Fig. 105].[24] Turner located the scene, where the Warrior Angels gather in response to God's order to drive the Rebel Angels through Chaos into Hell, among celestial spheres, which reflect light in ways reminiscent of the Ptolemaic system's crystalline spheres as well as Linnell's drawing of a sphere [Fig. 64]. Turner condensed his image from a number of lines (in Book V, 384ff). Gage remarks that the scene reveals Turner's interest in cosmolo-

gy and that the globes may be related to the artist's own thoughts on the structure of the universe. He views the illustration as depicting a triumph of light over darkness and believes that it combines elements of mood and form from Milton's primordial "Creation of Light Scene" (which, we might add, was illustrated by John Martin with a comet – Fig. 92).[25] Moreover, there is a small watercolor in the Tate Gallery, *Galileo's Villa*, planned but rejected as an illustration for Rogers' *Italy*, which is related to Turner's thinking about cosmic forms in the Milton illustrations. On this sheet, Turner drew a telescope, a celestial globe, and a celestial map with orbits of spherical bodies before the architectural villa of the title.[26]

The second of these engravings features a prominent meteor and illustrates the Fall of the Rebel Angels to the bottomless pit from Book VI [Fig. 106].[27] Turner supplemented the text from *Paradise Lost*, Book VI, 824-837, with an earlier passage narrating the original expulsion of Satan and the rebel angels.[28] In addition, his illustration depends on Revelation 12:7-9 and its meteor (there are no specific meteors mentioned in Milton's text which,

Figure 105. *(Top left)* J.W.M. Turner, *Mustering of the Rebel Angels*, vignette for *Paradise Lost,* Book V (*Milton's Poetical Works*), 1835, engraved by Robert Brandard, c. 90 × 83 mm (image) and 217 × 152 mm (plate), Yale Center for British Art, Paul Mellon Collection, New Haven.

Figure 106. *(Top right)* J.M.W. Turner, *Fall of the Rebel Angels*, vignette for *Paradise Lost,* Book VI (*Milton's Poetical Works*), 1835, engraved by Edward Goodall, c. 120 × 83 mm (image) and 204 × 150 mm (plate), Yale Center for British Art, Paul Mellon Collection, New Haven.

however, does refer to a feature of Revelation 9:1-3, the bottomless pit). Piggott believes that "the falling star," the prominent meteor depicted in Turner's illustration, represents "the phosphorescent Lucifer" or possibly "eternal wrath."[29]

Gage theorizes that Turner acquired his knowledge of cosmology through conversations with Mrs. Somerville [Mary Fairfax], whose 1831 treatise on astronomy – *Mechanisms of the Heavens* – he had acquired.[30] Somerville was herself an amateur painter and had long admired Turner's work. In turn, Turner referred to her early experiments on the magnetizing properties of colors, which for him were a confirmation of the influence of light. Gage believes that Turner met Somerville through the sculptor Francis Legatt Chantrey, who was a close friend of this distinguished, well-connected scientist of the day.[31] Somerville, a member of the Royal Astronomical Society and a close friend of John Herschel as well as the godmother of his daughter Rosa, knew many scientists, among them William Herschel, Lord Rosse, George Airy, and Michael Faraday. In her memoirs, Somerville writes about Turner:

> I frequently went to Turner's studio, and was always welcomed. No one could imagine that so much poetical feelings existed in so rough an exterior....[32]

Because Somerville followed one of the fashions of the day – not saving correspondence but simply preserving one of Turner's signatures from a letter as a specimen – we have been deprived of knowing more about their relationship.[33]

Somerville's experimental work was of particular interest to Turner because it concerned the solar spectrum.[34] Further, Gage believes that she influenced the vignette of the Mustering of the Warrior Angels and that Galileo's discoveries, which are discussed by Somerville in her book, were linked to Milton (since Milton had visited Galileo in Florence, Gage believes that the two can be connected).[35] Rather than linking the forms in the Mustering of the Warrior Angels scene to Galileo, we believe that it is more accurate to state that they echo concepts belonging to the Ptolemaic universe and that they amplify ideas in Turner's *Galileo's Villa* as well as the "orbs" cited by Milton. Interest in these astronomical ideas may have been kindled by Turner's contact with Somerville.[36]

Returning to our discussion of Turner's second illustration for *Paradise Lost* [Fig. 106], we note that it is one of Turner's rare

Figure 108. J.M.W. Turner (after), *Napoleon's Abdication at Fontainebleau*, vignette for Sir Walter Scott, *Prose Works*, vol. 15, engraved on steel by William Miller, 1834-36, 96 × 75 mm (image), The British Museum, London.

depictions of heaven as the realm of the Deity.[37] Herein, Turner has arranged his angels in rings referring to the old hierarchical, concentric Ptolemaic circles. The composition's prominent meteor with its zigzag, fizzling tail, cuts across the center of the composition to increase the aura of cosmic doom associated with Revelation while echoing the trajectories of the falling host.

At the time of Halley's apparition in 1835-36, Turner's career as an illustrator was flourishing, and between 1834 and 1836 he executed vignette watercolors for the *Prose Works* of Sir Walter Scott. For volume 15 and the life of Napoleon, Turner created a watercolor with a comet to illustrate the Emperor's abdication at Fontainebleau [Fig. 107 color]. Instead of representing the event inside the palace, where Scott placed it, Turner exercised artistic license and set his scene outdoors before the main entrance, where there could be greater spatial and dramatic light effects. Turner, deeply aware of Scott's admiration of Napoleon, has represented the Emperor atop the grand staircase, where he is dwarfed by the huge palace. In comparison with Turner's studies of Fontainebleau in his sketchbooks, it is apparent that he has exaggerated the size and especially the height of the architecture to overshadow the former Emperor. An heroic but tragic figure, Napoleon stands alone, apart from his staff, probably in contemplation of his destiny.[38] Above, a celestial light breaks through the clouds. It is telling that in the final steel engraving, this celestial form is unclear, resembling a crescent moon bisecting a planet or star [Fig. 108], evidently the result of the engraver's intervention, whereas in the preparatory watercolor the form is clearly a comet [Fig. 107 color].[39]

It is important to recognize that no such luminary body – either a comet or meteor – is present in Scott's account. Turner's inclusion of it indicates that he knew about Napoleon's infatuation with comets as well as the association of comets with doom and the fall of rulers. Evidence found in the engraver's proof, which was annotated by the artist himself, in the Rawlinson Collection of the Yale Center for British Art sheds further light both on the comet of the watercolor and the artist's intention. The proof contains a blank sky in which Turner wrote boldly in graphite on a diagonal: *falling Star*. We know, therefore, that he corrected his engraver, William Miller, and emphasized that this element was important for his composition. However, he used the non-specific term "falling star," which usually refers to a meteor but was also used by individuals when they really meant a comet. By using "falling star," he stresses the downward course of Bonaparte's career. Because of the context, Turner most likely conflated a falling star with a comet and intended a reference to the traditional linkage of Napoleon with comets. Turner's use of the comet also suggests that he viewed Halley's Comet in 1835-36, which appeared during the very time that these illustrations were conceived. However, the artist's terminology proves that he was not here concerned with scientific nomenclature. As was often the case in Turner's tempestuous relationships with the individuals who engraved his designs, Miller did not translate the comet form correctly when he engraved, and thus the image is not successful. Because Miller did not follow Turner's instructions with care, the apparition in the sky is ambiguous and lacks power. It could easily be passed over by a viewer wondering what it represents. Hence, the symbolic resonance of the scene is diminished in the engraving, while it remains intact in the watercolor.

That Turner knew about comets and probably had paid attention to comet 1811 I is apparent in a letter/poem to John Taylor written on 16 January 1811:

> Thanks gentle Sir for what you sent...
>
> Where lines so round about applied
> At last gives the parabolide
> Commixt perplexing and obscure
> 'Fitter to puzzle than allure'
> While on each point your Comet shines
> It gleams to truth with truth defines
> Unclothes at once the Magic shell
> The Donjon Keep and Elfin yell.[40]

In this poem Turner is quoting from Taylor's satire on Walter Scott, *The Caledonian Comet* (1810).[41]

Turner painted more generalized, sublime light effects in several other works involving Napoleon, in contexts that could have also included a comet, but do not. One example is his *Field of Waterloo from Hougoumont*, which was exhibited at the Royal Academy in 1818.[42] Turner returned to the theme again in the 1830s when he engraved the episode for Scott's *Prose Works* and rendered it in a watercolor vignette.[43] In all of these, the blinding central light above the carnage of battle has been compared to several unusual light effects – lightning, exploding shells, or rockets.[44] The painting has been allied to Lord Byron's poem *Childe Harold's Pilgrimage*; and Byron was, of course, a passionate employer of comet/meteor symbolism.[45] In preparation for his canvas, Turner visited Waterloo on 16 August 1817 and devoted several pages in his sketchbooks to the battle. A note also reveals that he had in hand *The Traveller's Complete Guide* by Charles Campbell, who quotes the "Waterloo Stanzas" of Byron's *Childe Harold*. That the artist was fascinated by Bonaparte is apparent in another painting depicting the defeated Emperor, *The War – The Exile and the Rock Limpet*, which Turner executed in 1842.[46] Turner's canvas represents a bitter meditation on Napoleon's former glory showing him before a brilliant sunset, allegorical of the mighty Emperor's fall. Throughout his life, Turner was obsessed by the fickleness of fortune, and Napoleon's career provided a concrete example of fortune's rise and fall. In each case, Turner employed cosmic symbolism that could be read poetically as a comment on that destiny.

Thus, his illustrations of Napoleon's abdication at Fontainebleau [Figs. 107 color, 108] take on new resonance. Moreover, the falling star in the watercolor and the intense light of *Waterloo* become even more provocative when examined in the context of Byron's poems about Napoleon that employ comet and meteor images. For example, in "Napoleon's Farewell," written 25 July 1815, Byron has Napoleon remark:

> ...I have warr'd with a world which van-
> quish'd me only
> When the meteor of conquest allured me
> too far....

Then in "Ode From the French," first published in 1816, Byron allies the Star Wormwood of Revelation, usually illustrated as a meteor, with the Battle of Waterloo:

> ...we do not curse thee, Waterloo!...
> Never yet was seen such lightning
> As o'er heaven shall then be bright'ning!
> Like the Wormwood Star foretold
> By the sainted Seer of old,
> Show'ring down a fiery flood,
> Turning rivers into blood.[47]

Revelation 8:10-11 reads: "And the third angel sounded and there fell a great star from heaven burning as it were a lamp.... And the name of the star is Wormwood."[48] In light of the Byronic evidence, the intense central illumination in Turner's *Waterloo*, which has been interpreted as lightning or a bursting shell, should be rethought. It may represent a conflation of one of these forms with an exploding meteor, a bolide, which would have connoted the Star Wormwood, to lend an apocalyptic resonance to the scene.

Furthermore, in the very same year, Byron first published his ironic poem "On the Star of the Legion of Honour," which may also be related to Turner's blinding light in *Waterloo*. It further suggests that Turner represented a meteoric phenomenon therein.

> Star of the brave! – whose beam hath
> shed
> Such glory o'er the quick and dead –

Thou radiant and adored deceit,
Which millions rush'd in arms to greet!
Wild meteor of immortal birth
Why rise in Heaven to set on Earth?

Certainly, it is conceivable that Turner also knew Byron's "Ode of Napoleon Bonaparte" of 10 April 1814, which can be ironically related to the Battle of Waterloo and to Turner's painting *The War – The Exile and the Rock Limpet*:

...And now thou art a nameless thing:
Since he, miscalled the Morning Star....

Comets Commercialized

By the 1830s and 1840s the word "comet" had already entered into the vernacular lexicon for naming items. To whatever it was attached, the word lent a certain cachet of glamour and speed. Already, for example, in 1809, there was a boat named the *Smack Comet*, known today from a rigged model copied from a contemporary model in the possession of the London and Edinburgh Shipping Company.[49] Slightly later, in 1812, just after the Great Comet of 1811, Henry Bell constructed a paddle steamer named "Comet."[50] And publications – daily newspapers, literary magazines, weekly publications – named "The Comet" began to proliferate during mid-century in many cities of the British Isles. One of the scandals of early steamboat travel in Great Britain revolved around a steam Packet named "*The Comet*." The boat, bound from Glasgow with sixty to seventy passengers on board, was rammed 300 yards from shore by another ship, the Ayr, on 21 October 1825 at 2:00 a.m. during its return to Greenock. *The Comet* had no light on her bow, and the Ayr fled the scene without rescuing any passengers, fearing it would capsize if additional ones were taken on. Only eleven survived to tell *The Comet*'s story. An etching by William Turner de Lond (active 1767-1826) commemorates this maritime incident, which the press vociferously reported.[51]

As we noticed earlier, the 1835-36 apparition of Halley's Comet proved to be a commercial watershed for the manufacture of comet items. In addition, a locomotive named the "Comet" was made for the Newcastle and Carlisle Railway.[52] It was followed by another c. 1845, known in a model built on a scale of 1:12 for the Stockton and Darlington Railway.[53] Soon thereafter, c. 1850, another steamboat named the "Comet" was built by Henry Bell, today preserved only in a lithograph.[54] Comets were also used humorously or inventively, such as in a whimsical brass celestial letter opener with a comet and a jolly full moon that dates from c. 1830-40 [Fig. 109 color].[55]

John Martin's Apocalyptic Comet

Certainly comet 1811 I or Halley's 1835-36 apparition inspired John Martin, who had used comets in several earlier works, when painting his operatic *Eve of the Deluge* (1840) [Fig. 110 color]. The painting is based on Martin's 1833 mezzotint with etching frontispiece, *The Eve of the Deluge*, which he executed for the first volume of John Galt's *The Ouranoulogos or, The Celestial Volume*, whose comet resembles the spectacular comet of 1811.[56] The painting belongs to a trilogy on the Biblical flood. According to Martin's son, Prince Albert – husband of Queen Victoria (reigned 1837-1901), herself an accomplished watercolorist – was partially responsible for the genesis of this canvas during a visit to Martin's studio. The Prince saw a painted version of the artist's *The Deluge* (c. 1826, exhibited at the British Institution and now lost; another version of 1834 is in the Yale Center for British Art) and suggested that Martin paint two related themes (The Eve of the Deluge and The Assuaging of the Waters) for a series on a grand scale (although the artist had already thought of that idea). Martin subsequently exhibited the latter two at the Royal Academy in 1840 and *The Eve* at the British Institution in 1841. *The Eve* was purchased by Prince Albert in 1841. This epic Romantic subject was popular at the time, with Turner painting his version c. 1841 and Linnell his c. 1843.

The cataclysmic *The Eve of the Deluge* is based on Genesis 6:5-8, in which God decides to destroy the world grown weary with sin, with the exception of Noah's family and two of every species. The seated white bearded figure on the craggy promontory at the left is Methuselah, who instructs Noah to open the scroll written by his father, Enoch, so that he can compare the signs in the sky (the sun, buried in the dusty clouds of sunset, the moon, and a mesmerizing comet) with those on the scroll. The fact that these astrological (in the pre-zodiacal, Babylonian sense meaning astronomy) signs match ordains the end of the world. In the middle distance under the trees at the right, a group of hedonistic revelers of the antediluvian age heed neither the comet nor the circling ravens, another ill omen.

Martin published his own account of the three paintings on the theme of the Old Testament flood in a pamphlet (1840), *The Deluge*, which reveals that his sources were fairly recondite, extending beyond historical and religious texts to cosmological and poetical writings, including Byron's *Heaven and Earth*. Byron's Sublime poem is the poetic equivalent of the sentiments in Martin's canvas:

> The earth's grown wicked
> And many signs and portents have proclaimed
> A change at hand, and an overwhelming doom
> To perishable beings.

However, while neither Byron nor the Bible mention a comet, Martin felt compelled to include one to help establish an apocalyptic tenor. In fact, Martin wrote in his pamphlet that he endeavored to represent "the near conjunction of the Sun, Moon, and a Comet, as one of the warning signs of the approaching doom," echoing Galt's statement in his *Ouranoulogos* that "the comet whose fiery visage nightly scatters disasters through the air...and celestial orbs are moving into terrible conjunctions."[57] Further, it is probable that Martin witnessed the Great Comet of 1811 as well as Halley's in 1835-36. In addition to these spectacular comets, in 1838 the tenth appearance of Comet Encke occurred (2P, formerly 1845 IV), at which time that comet, although faint, was naked-eye a short while. Then, there were four comets recorded in 1840. The first was visible to the naked eye, although none of them was especially memorable.[58] However, as we have noticed, the fact that comets were observed made the subject topical and kept them in the public's attention. Because Martin was such a literary painter, he was conversant about comet traditions, which may have fueled his interest in actual apparitions.

The stupendous comet of Martin's *The Eve of the Deluge* plunges downward toward the mist-enshrouded sun and rising moon, a vivid physical prophecy of the impending flood. (The twentieth-century quack Immanuel Velikovsky, among others, has shown that comet superstitions persist into this century by crediting the biblical flood to a comet.[59]) The convincing nature of the comet's vaporous tail and its center of condensation in Martin's canvas suggests that he was reflecting his own observations of historical comets. This realistic interpretation is also argued by the nature of the cometary light and its appearance in the sky at sunset.

The mood established in Martin's painting seems the antithesis of a concurrent pamphlet – "THE COMET OF MANY TAILS. A COMIC KALENDAR FROM 1841 with sketches of Kings and Commoners." The lighthearted ephemeris by W.H. Harrison is really not about comets but features on its cover a comet consisting of a bowl forming the head and a tail consisting of spoons, knives, and forks.[60] The contrast between it and Martin's painting demonstrates the range of attitudes about comets at mid-century.

Shortly after the second exhibition of Martin's *The Eve of the Deluge* in 1841 an intensely bright comet appeared in 1843 (C/1843 D1, formerly 1843 I). It was the most brilliant of the three that year and was in every sense remarkable. Not only was it a daylight comet, but also it was first detected during daylight, indicating its extreme brilliancy. It boasted the longest tail ever recorded, an estimated 200 to 300 million miles in length, two or three times greater than the distance between the sun and earth. Besides, it was a sungrazer, passing at perihelion within one hundred thousand miles of the sun's surface with a period variously estimated at 360 and 513 years. (During the twentieth century, many sungrazers were discovered using devices called "coronagraphs" on earth-orbiting spacecraft to cut the solar glare, and it was realized that some of them actually hit the sun, never to emerge. The Solar and Heliospheric Observatory, a European Space Agency/NASA satellite, is discovering more sungrazers at a rate greater than one a month [Fig. 159 color].) Comet 1843 I was first visible in the southern hemisphere, where its magnitude reached −8, over ten times brighter than Venus. By the time it could be observed in the northern hemisphere, around 12 March, its brilliance was diminished to around zero magnitude, equivalent to the brightest stars in the northern sky, save Sirius. Its tail remained over 30° in length through the month, by which time

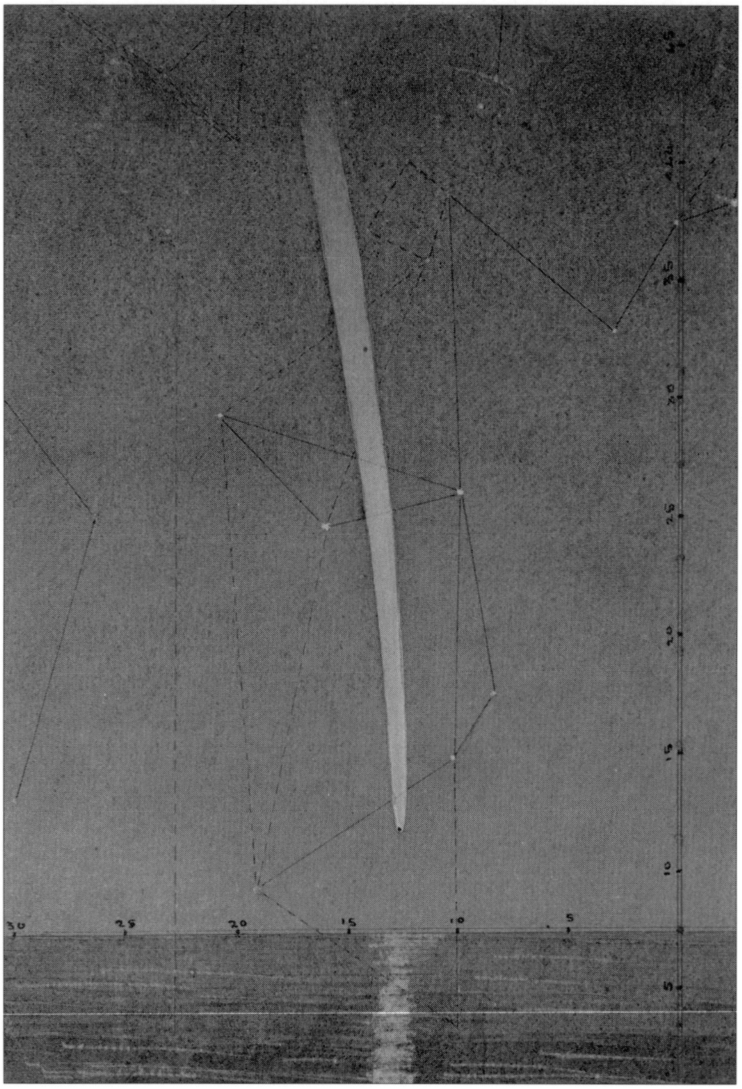

Figure 111. William Clerihew, *Observational Drawing of the Comet of 1843 (Ms. Add. 183/3)*, 1843, pen and red and black ink with watercolor and white heightening, 186 × 135 mm, Royal Astronomical Society Library, London.

it had faded from naked-eye view.[61] As the astronomer John Herschel wrote discerningly in 1849:

> By far the most remarkable comet, however, which has been seen during the present century, is that which appeared in the spring of 1843, and whose tail ... a great beam of nebulous light. ... the head and nucleus appeared with extraordinary splendour, exciting in every country where it was seen the greatest astonishment and admiration. Indeed, all descriptions agree in representing it as a stupendous spectacle, such as in superstitious ages would not fail to have carried into every bosom.[62]

Clerke notes an important result from the observation of this comet: the behavior of its tail was regarded as an *experimentum crucis* in the nature of comet tails. Clearly no fixed appendage many millions of miles in length could be whirled like a brandished sabre, as comet lore would have it, from one side of the sun to the other in 131 minutes. Thus, cometary trains or tails, were proved not to be solid. As Clerke described in 1902, Olbers rightly viewed comet tails to be emanations, rather than appendages that consist of outflows of highly rarefied matter, the greater part, if not all, of which becomes permanently detached from the nucleus.[63]

William Clerihew, an artist and amateur astronomer, produced four fascinating studies of this apparition that reveal its visually spectacular nature.[64] Although Clerihew trained as an architect and became a member of the Institute of British Architects (now the R.I.B.A.) in 1839, his name neither appears in members' lists after 1849 nor is found among the members of the Royal Astronomical Society. He was thus truly an amateur astronomer. He exhibited two architectural subjects at the Royal Academy in 1840 and 1841 and, in addition to being an artist of some merit, was also something of a surveyor, traits evident in the third drawing of the series of comet 1843 I [Fig. 111].[65] The reflections of the comet's light on the water and the measurements of the altitude and the azimuth noted were made by Clerihew in the waters in the southern Bay of Bengal. Clerihew, like Piazzi Smyth, observed this arresting comet when it was at its brightest in the southern hemisphere.

Piazzi Smyth: Astronomer, Artist, Photographer

Charles Piazzi Smyth (1801-1900), who also rendered comet C/1843 D1 a number of times [Figs. 112-114] and studied it assiduously, was one of the more enigmatic individuals of nineteenth-century British astronomy. Because he has acquired a reputation as an artist as well as an astronomer, he is another linking figure in our study. Known in his earlier years as a man of widespread interests – including astronomy, painting, and photography – in his later years he became highly independent and was known more for eccentricity than astronomical or artistic accomplishment.[66]

His middle name was the surname of his godfather, the Sicilian monk Abbé Giuseppe Piazzi, who was court astronomer at

Palermo and famous for discovering the first known asteroid, Ceres, in 1801. Piazzi Smyth's father, William Henry Smyth, an officer in the Royal Navy, was stationed in Italy and became a friend of Piazzi. After all, name is destiny, but Charles Piazzi Smyth's passion for Urania can also be explained by astronomy in his blood. When the boy was six years old, the family returned to England, where his father set up a well-equipped observatory. The elder Smyth, who was well known in European scientific circles and a member of both the Royal Society and the Royal Astronomical Society, took an interest in comets and published an article on Halley's apparition of 1835-36.[67] A friend of John Herschel, he also built a well-equipped observatory at Bedford, referred to as his "Uranian Temple." The elder Smyth's book, *A Cycle of Celestial Objects*, won the Royal Astronomical Society gold medal in 1845 and exerted great influence at the start of the finest period of British amateur astronomy.[68]

In 1835 when Piazzi Smyth was 16, he went to South Africa to assist Maclear at the Royal Observatory, the Cape of Good Hope. At that time he began to mix some of his astronomical observations with drawing and painting in a series of sketchbooks, a practice he would continue throughout his life and one that was more or less expected of a Victorian gentleman. The majority of these, in fact 38 of them, are on deposit in the Library of the Royal Observatory Edinburgh. Piazzi Smyth seems to have been both a self-taught artist and an autodidactic astronomer, and his artistic endeavors have a naive quality to them, especially his figures. It is in the skies of his landscapes where Piazzi Smyth's two interests meet and where he was able convey some of his best observations.

Piazzi Smyth's astronomical observations at the Cape included two bright comets, those of Halley's Comet in 1835-36, which he helped Maclear observe, and C/1843 D1. When Maclear wrote the account of his observations of the former he commissioned his assistant to work up his sketches for the engraver, resulting in the mezzotints published in the *Memoirs of the Royal Astronomical Society* [Fig. 98].[69] Piazzi Smyth observed the comet between 3 and 25 March, while he was in charge of the observatory in Maclear's absence. He used a 3½-inch portable telescope, the largest then at the Cape for observations other than on the north-south meridian. Though Piazzi Smyth worked with Maclear on

Figure 112. Charles Piazzi Smyth, *The Great Comet of 1843*, watercolor and mixed media (ink, body color, and scratching out), 234 × 192 mm, Don Africana Library, Durban Metropolitan Library, South Africa.

the accurate positional observations, he was also eager to note physical characteristics, which Maclear did not think important to report. (Maclear's own observing notes included sketches, however.) Positional information was considered by the astronomical community at that time to be the most crucial information to gather for comets. Warner describes how "The difficulties he [Piazzi Smyth] experienced and a *contra temps* with Maclear, led to delay in publication of the results. When they finally appeared in 1851, only the bare positions were given, shorn of all physical description." It is immediately followed by Piazzi Smyth's article with descriptions limited to brightness and meteorological conditions.[70] The incident also involved Piazzi Smyth complaining to John Herschel about the inadequate equipment. While this resulted in Herschel having the Admiralty approve a better telescope for

Figure 113. Charles Piazzi Smyth, *The Great Comet of 1843*, oil on canvas, 91 × 61 cm, National Maritime Museum, Greenwich.

the observatory, it vexed Maclear, who claimed Piazzi Smyth was not making proper use of the equipment at hand.

Piazzi Smyth captured his observations of C/1843 D1 in several types of media. The first is an oval lithograph rendered in such a way as to simulate a wood engraving.[71] Another in watercolor and mixed media shows the comet observed by a group of people, who stand on rocks in the foreground and look out in wonder across the still water at it and its reflection in March of 1843 [Fig. 112].[72] At this time, the comet was at its greatest brilliance but still

Figure 114. Charles Piazzi Smyth, *Moonlight View over Table Bay Showing the Great Comet of 1843*, oil on canvas, 91 × 61 cm, National Maritime Museum, Greenwich.

invisible in Europe. Piazzi Smyth has captured both the star-studded sky and the particularly long, elegant tail of this spectacular comet. Another watercolor in a horizontal format in one of the Piazzi Smyth's sketchbooks is inscribed on the page below the image *SKETCH OF THE COMET OF 1843.* It features the comet streaking across the starry sky tinged by the last oranges of sunset above a campfire and a covered wagon.[73]

Two oil paintings, both in vertical formats, demonstrate how Piazzi Smyth's observations transferred to a more permanent medium in both day and night scenes [Figs. 113, 114].[74] While these paintings parallel his journals and atmospheric observations in watercolor, they were clearly intended to be works of art, not mere scientific recordings of this comet.[75] The first shows the daylight comet as it would have appeared in the late afternoon of 28 February 1843 with the artist oriented toward the west [Fig. 113]. Only a few hours before, the comet had passed perihelion and was, therefore, at its intrinsic brightest. Piazzi Smyth painted it over Table Mountain, Cape Town, in a technically difficult position to capture juxtaposed to the sun. Three sailors in a small boat point at the comet, the head of which is only 1.5° from the sun.[76] The second oil painting [Fig. 114] contains a westward view from the Cape of Good Hope and probably reflects the comet as it appeared 9 March 1843.[77] It is a more typical nocturnal scene with Table Bay illuminated by moon and starlight but with the horizon still holding the glow of sunset. Here, the comet has a bifurcated tail of about 60° – that is, concave on its south side (*pentimenti* reveal that Piazzi Smyth first drew the comet about six inches to the right of its final position). The bright star above the comet's head is Tau Ceti, and Stott and Hughes identify the other stars.[78] Again, as with the two previous works, a tiny observer in the lower foreground emphasizes the awesome scale of this large comet.

It was at this time, in the dawn of the photographic era, that Piazzi Smyth, realizing the importance of an ability to draw what he observed, published a significant paper with a fascinating, milestone account of the usefulness of astronomical drawing.[79] This practice was as a discipline still relatively new. In the article, he articulates the need for accurate visual recordings of astronomical observations, claiming that, with the exception of William Herschel, most astronomers contented themselves with hasty sketches (many of Herschel's sketches were probably by his sister Caroline). He also acknowledges the difficulty of translating observational drawings into more permanent printed media (something he had experienced) for wider circulation, a problem to this day. To that end, he reviews the strengths and weaknesses of the various print techniques. He refers to both astronomers' reproductive attempts and to the work of old masters – such as Rembrandt and Leonardo da Vinci – while tackling the limitations of conveying light in black-and-white.[80] Throughout the article, Piazzi Smyth's desideratum is faithful imitation.[81] The prescient nature of his comments and the pressing need for accurate reproductive methods for astronomical phenomena were borne out in his other activities.

1843 was a pivotal year for Piazzi Smyth, for it also marks his first successful attempt with photography. He was one of the earliest photographers, trying to make photogenic images as early as 1839, after news of John Herschel's success with representations printed on sensitized paper (within months of the announcement in 1839 of Daguerre's images on a silver-coated, polished copper plate). He adopted William Henry Fox Talbot's method utilizing paper negatives, called "talbotyping" or "calotyping," perhaps as early as 1842 (see chapter VI). In 1843 he produced a series of photographs, which appear to be the oldest known of people and scenes of South Africa. And his photograph of the Cape Observatory is most likely the oldest one of an observatory in the world. However, there is no indication that Piazzi Smyth ever made a photographic attempt to record a comet, something he reserved for watercolors, prints, and oils. Indeed, the first successful comet photograph was not taken for another fifteen years, when he was in Edinburgh, as discussed in chapter VI. It follows that Piazzi Smyth wrote a pioneering essay on the topic of photographic illustration in 1861, however.[82]

While in South Africa, Piazzi Smyth's father, who was ambitious for the youth, solicited for him a position in Edinburgh. Admiral Smyth apparently sent in the application himself because his son, who was in South Africa, could not have received a response in time. Sir John Herschel's extremely favorable letter of recommendation won the post for the young astronomer. As a result, Smyth was able to return in 1845 at the age of 26 with the distinguished titles of director of the Observatory Edinburgh, Regius Professor of Practical Astronomy at the University of Edinburgh, and Astronomer Royal for Scotland.

In Edinburgh he observed Coggia's Comet in 1874 and made watercolors of it that are preserved in one of his sketchbooks [Figs. 115, 116, both color]. Because of their freshness, together with the astronomical and meteorological quality of his observations, these works are among his best artistic efforts. In fact, they serve, perhaps better than any early photograph, to capture fully what the astronomer observed. The first is inscribed in pen and brown ink below the image: *Coggia's Comet: at midnight on Meridian below Pole/12 July, 1874./Observed transit at R.O. Ed. with Transit. Inst. Inverted head=*. The view may have been executed from Calton Hill, site of the observatory before it moved to its current site on Blackford Hill, and contains a view of the silhouetted skyline of Edinburgh. There is also a drawing of the comet's head in graphite in the right lower margin, where Piazzi Smyth was probably studying the formation of gases.[83] The following page contains no watercolors, but has notations for 13 July 1874, together with observational notes on the comet for 14 and 15 July.[84] For the latter, he notes that the nucleus was very bright, equal to a small star of the first magnitude, a fact that he visually recorded on the subsequent page in an even more dramatic watercolor with the comet's bifurcated tail longer and wider [Fig. 116].[85] It is inscribed under the image in pen and brown ink, beginning with the symbol for Wednesday (or Mercury) *15 July at 10^{h}.30^{m} P.M; Fog covered all the sky/at the Meridian Transit afterwards*. Below the middle ground hill with its romantic, silhouetted trees is Edinburgh with some of its lights glowing red. In these two watercolors Piazzi Smyth captured the full impact of the comet's apparition in both scientific and aesthetic terms.[86]

It is interesting to note that Piazzi Smyth was one of the first individuals to champion the placement of major observatories on mountains, something we take for granted. In 1856, crediting Isaac Newton for the original idea of mountain sites, he led an expedition to Tenerife in the Canary Islands, with support of the British Admiralty, to assess the advantages of a high altitude site for astronomical observation. (Typically for a Victorian gentleman and a person with an inquisitive mind, he also studied the fauna and the geology of the region.) Today, the Canary Islands host major astronomical observatories atop some of its mountains, and other major observatories around the world are also on mountain sites, all stemming in large part from Piazzi Smyth's trip.

Nearly a decade later, Piazzi Smyth's interests became much more controversial, when he went to Egypt in 1865 to study the Great Pyramid at Giza. He measured its various orientations and tried to correlate them with astronomical phenomena. Such measurements are back in vogue today, and correlations with the rising and setting positions of the sun and moon, for example, are accepted. But Piazzi Smyth went further and speculated on mysteries hidden within the Great Pyramid. In any case, his astronomical colleagues did not approve, and when a paper on his pyramidology was rejected for publication in the *Philosophical Transactions*, Piazzi Smyth resigned from the Royal Society. From today's point of view, Piazzi Smyth's speculations and his data were unreliable, and in short he was over the deep end on pyramidology.[87]

Eventually Piazzi Smyth resigned from his official posts in Scotland, ostensibly because the Observatory's equipment had not been satisfactorily kept up. Unfortunately, he had sullied his reputation with his pyramidology, which eclipsed his positive contributions. With his solid traditional astronomical work at the Cape and in Scotland, his work on photography, and his application of photography to spectroscopy, Piazzi Smyth's reputation for parts of his work, including that on comets, deserves resuscitation.

David Scott's Romantic Comets and Meteors, John Pringle Nichol, and the Earl of Rosse

Another artist with documented connections to contemporary astronomers is the Scottish painter David Scott (1806-1849). During his short life in Edinburgh, Scott associated with individuals involved with the natural sciences.[88] Already by the late 1830s and 1840s, one of his most intimate friends, in fact, was the chemist Dr. Samuel Brown, whose star-spangled nocturnal portrait Scott painted in 1844.[89] Its romantic nature, replete with a decided peppering of astronomical features, is characteristic of David Scott's paintings and graphic works. Scott was also a friend of the pioneer photographer David Octavius Hill.

David Scott was the son of a well known Edinburgh engraver, Robert Scott. He also gained some recognition as an illustrator and among his illustrations are those planned for a late, expanded edition of *Architecture of the Heavens* (first published in 1837) by his friend the astronomer John Pringle Nichol. David Scott's illustrations to supplement his friend's book are one of the artist's last works and one of his few commissions. His younger brother, William Bell Scott (1811-1890) – an artist who specialized in etching, a friend of Dante Gabriel Rossetti and the Pre-Raphaelites, and a poet, who clearly idolized his talented elder brother – describes them in his *Memoirs*.

> In the early part of the year 1848.... From thence he visited Professor Nichol, having engaged himself on a series of inventions to accompany a new edition of the Architecture of the Heavens. These designs, some of which are among his finest thoughts, employed him to the latest day he could hold a pencil.[90]

It is precisely because of William Bell's *Memoirs* and his devotion to his brother that we know so much about David, an often misunderstood artist with a pronounced imagination who suffered from severe melancholia. In his later years, David Scott became a landmark in the intellectual life of Edinburgh, its resident gloomy Romantic genius.

John Pringle Nichol (1804-1859), Regius Professor of Astronomy at the University of Glasgow, was another of David Scott's friends. He was trained as a minister but because of some unorthodox opinions became a popularizer of astronomy. Nichol, a fellow of the Royal Astronomical Society as well as the Royal Society of Edinburgh, lectured widely, eventually coming to the United States in 1848-49. He delivered his last public lecture in 1858 on the appearance of Donati's Comet.[91] The first edition of Nichol's popular but serious book was published in Edinburgh in 1837 as *Views of the Architecture of the Heavens*. It was subtitled *In a Series of Letters to a Lady*, a literary format that was customary for popularizing books in Britain. Many editions followed.[92]

It is with the 1850 edition that the philosophical aspects of Nichol's book received a visual counterpart in illustrations based on David Scott's drawings of 1848, engraved posthumously by William Bell Scott.[93] This new edition in a larger format haphazardly mixes Scott's plates with astronomical plates, some credited to the Earl of Rosse. Herein, the scientific illustrations are more lavishly reproduced. Moreover, its dedication page is inscribed "To the Right Hon./The Countess of Rosse." The Countess of Rosse, born Mary Field, had married her famous husband in 1836.

The third Earl of Rosse, who was born William Parsons (1800-1867), was the most distinguished British amateur astronomer of his time. As previously mentioned, at his estate in Ireland, he saw to the construction of telescopes larger than any yet built. The size

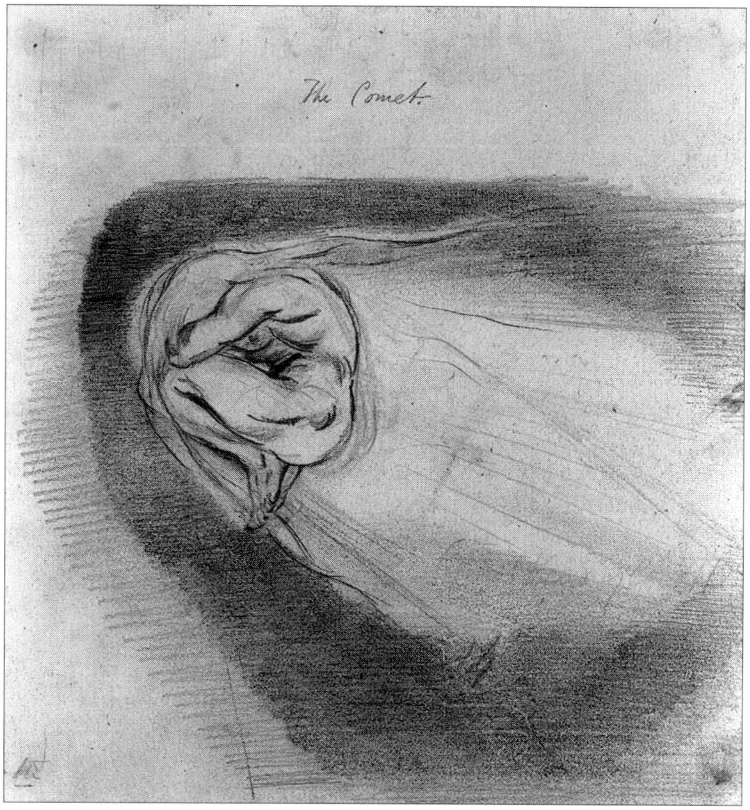

Figure 117. David Scott, *The Comet*, 1848, graphite with the possible addition of black chalk, 213 × 206 mm, The National Gallery of Scotland, Edinburgh.

of his largest, with a 1.8 meter diameter, which went into operation in 1845, allowed him to make fundamental discoveries. In particular, his examination of nebulae revealed for the first time that some had a spiral structure. He found the first in Ursa Major during the third month of the telescope's operation.[94] It was not until the 1920s that these "spiral nebulae" turned out to be "spiral galaxies," island galaxies in their own right and not merely part of our Milky Way. Rosse's discoveries are thus at the basis of our current understanding of the universe's structure.[95] It is not surprising that the 1850 edition of Nichol's book is illustrated with plates taken from his friend's nebulae observations.

David Scott's eighteen drawings for Nichol's book are mounted in a bound album compiled by William Bell Scott.[96] Only nine of these sheets, permeated with celestial imagery, were engraved by William Bell Scott for Nichol's 1850 *Architecture of the Heavens*.[97] Among the most interesting, especially for the present study, are two that were not engraved: *Creation of a Star*, which depicts a

Figure 118. David Scott, *Scandinavian Night*, before 1849, graphite and wash, 372 × 310 mm, The National Gallery of Scotland, Edinburgh.

fantastic star that looks very much like it has a tail (positioned below the creator inside the circles of Heaven) and *The Comet* [Fig. 117].[98] In both examples, we can see the influence of Michelangelo as well as of Blake, Fuseli, and Flaxman, especially in Scott's reliance on the human figure to convey abstract ideas. His image of the comet's nucleus – visualized as a nude male figure curled up in a fetal position hurtling through the darkness of space, covered only by a drape that metamorphoses into light rays – is especially evocative, personal, and, in the final analysis, spartanly powerful. Gray has remarked that: "*The Comet* seems to shadow forth a theory – one rather of the mystic than of the astronomer – that comets are embryo worlds in the process of formation."[99] Interestingly, Scott's image is close to the scientific truth that comets are made from primordial material.

Another bound album with drawings by David Scott contains a folio with a different but related type of celestial imagery: meteors.[100] This large sheet is inscribed with a long passage, beginning "*Scandinavian Night...*" [Fig. 118].[101] From the inscription it is clear that Scott's apocalyptic drawing illustrates the Ragnarök, the end of the world of gods and humans in Icelandic mythology. The word means literally "Doom of the Gods" or "Twilight of the Gods," and its texts, which contain images that evoke meteors or meteor showers, link it to a collision with a comet.[102] Scott's drawing depicts Odin (the older form is Woden), the principal god, who was also god of war and of poets (significant for Scott, who wrote poetry), on the crumbling rainbow arch of heaven. The artist has drawn two or perhaps three meteors at the left to heighten the sense of cosmic disintegration. This sheet may have been part of a series in which Scott intended to illustrate various Scandinavian myths, although we have no others to support this hypothesis. Nevertheless, this drawing reinforces the fact that Scott was a literary-oriented artist, who frequently selected esoteric subjects. It also provides additional evidence that he was in the vanguard of the later Scottish Romantic movement, and thus a significant figure.[103] Although Scott may have witnessed meteors, the nature of this representation clearly reflects a literary tradition for their presence in apocalyptic scenes rather than an actual experience.

Mid-Century Comets and Meteors

In 1850 a huge meteor appeared over England and was visible in London. It was captured dramatically by Matthew Cotes Wyatt over Paddington in a mezzotint that suggests, due to the explosion and sparks of its head, that it was a bolide [Fig. 119].[104] Its letterpress inscription reads: *A REPRESENTATION OF THE METEOR SEEN AT PADDINGTON 12 MINUTES BEFORE 11 O'CLOCK, ON THE EVENING OF THE 11TH . OF FEbY. 1850./MR. WYATT having retained that splendid object in view from its appearance until it passed away in brilliant corruscations* [sic], *made a drawing of it whilst its impression upon him /remained undiminished, and he shortly afterwards Engraved this Plate, in order that a faithfully graphical exhibition of its appearance*

Figure 119. Matthew Cotes Wyatt, *Meteor Seen at Paddington about 12 minutes ...1850*, mezzotint, 280 × 424 mm, The British Museum, London.

might be more generally diffused.... Other accounts and representations from various locations were reported in *The Illustrated London News*, which had begun publication in 1842, as well as in other periodicals. James Glaisher, the assistant to the Astronomer Royal, published an appeal for additional reports in the same issue, and consequently so many accounts were sent in that Glaisher had them published in the *Philosophical Magazine*.[105]

Meteors were newsworthy items in the popular press at mid-century and they continued to be reported, sometimes only visually without comment in periodicals such as *The Illustrated London News*. A case in point is the "meteor" reported as streaking before Westminster Abbey in the 8 April 1854 issue.[106] Its artist records two explosions near to the ground during the "meteor's" descent and shows that it invaded this famous urban space lit by gas lamps. However, if the engraving is accurate, the phenomenon could not have been an actual meteor since none has ever been seen in front of terrestrial objects so low; perhaps the artist merely

Figure 120. T.H. Jones, *H. Walker's "The Great Comet! A Fearful Tragedy,"* 1835?, lithograph, 194 × 139 mm, The Science Museum, London.

imagined the view. Nevertheless, the illustration demonstrates the topical fixation on meteoric phenomena.

Comets continued to appear after the brilliant comet of 1843, with three recorded in 1844. Of these, only comet 1844 II (4P/1843 W1) was brilliant and its observation from the Cape Observatory underscores the importance of British observation in the southern hemisphere.[107] During this time, observations were still being made with comet-seekers, such as the two fainter comets observed in 1845 (C/1844 Y2 and C/1845 D1, formerly 1845 I and II), which were followed by the bright, naked-eye comet 1845 III (C/1845 L1)

Figure 121. George Cruikshank, *Passing Events: or the Tail of the Comet of 1853, from "George Cruikshank's Magazine,"* 1854, etching, 225 × 435 mm, The British Museum, London.

and the twelfth recorded apparition of Encke's Comet (2P, formerly 1845 IV).[108] 1846 featured eight comets, and 1847, six.[109] Due to all these apparitions, the media responded in multiple ways, such as in the amusing lithographic print by T.H. Jones entitled *T.H. Walker's The Great Comet! A Fearful Tragedy Written under the Influence of Great Excitement* [Fig. 120].[110]

Comets remained on British minds throughout mid-century – and no wonder, for in 1853 four comets were recorded. In fact, comet 1853 IV (C/1853 R1) closed the year in December, and not coincidentally John Martin included another apocalyptic comet in his *Last Judgment* of 1853.[111] Comets continued to pique the public interest as witness George Cruikshank's comet that functions as a clever device to sum up his impressions of the events during the Victorian year of 1853 [Figs. 121, 122]. Cruikshank's print *Passing Events: or the Tail of the Comet of 1853* was published on 1 January 1854 in *George Cruikshank's Magazine* as an elongated foldout frontispiece. It also exists as an independent etching, several impressions of which are in the British Museum, including a first state and a signed proof.[112] The etching [Fig. 121] and its preparatory watercolor [Fig. 122] demonstrate Cruikshank's response to the public's consciousness of comets.[113] Although there are differences between the two, both feature a giddy comet

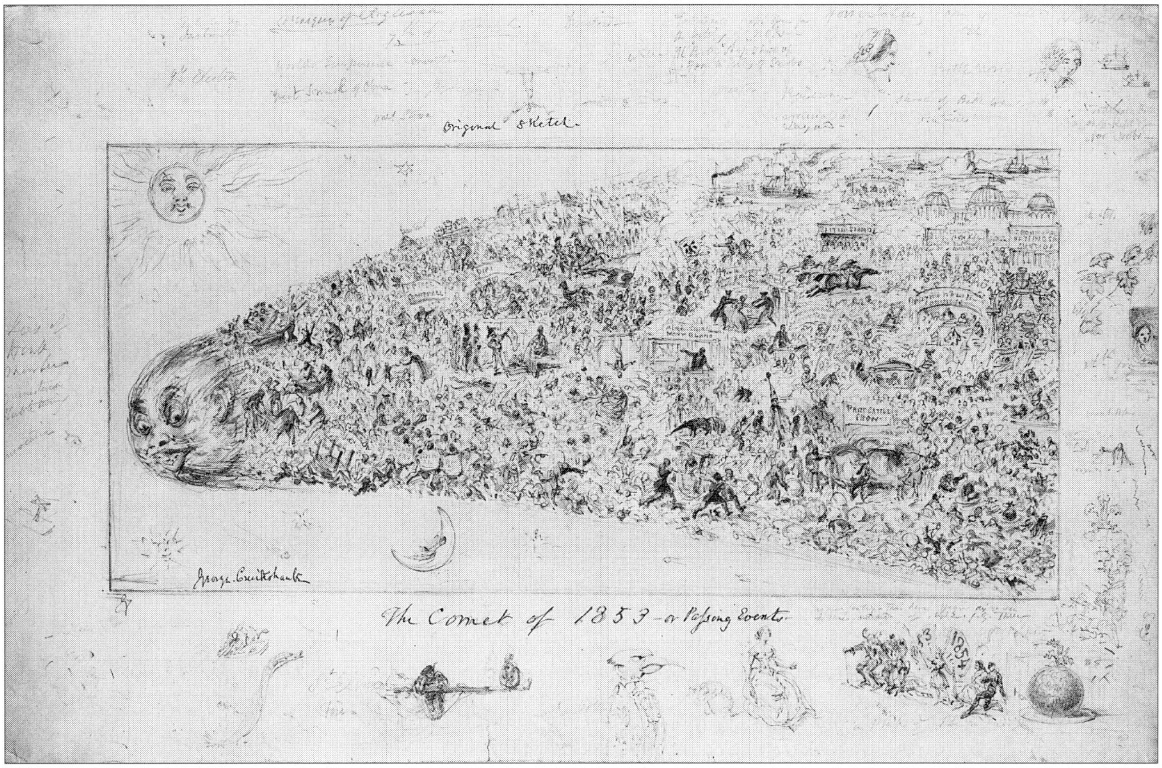

Figure 122. George Cruikshank, *The Comet of 1853—or Passing Events,* graphite and pen and ink with watercolor, 303 × 485 mm, The Victoria and Albert Museum, London.

whose tail circumscribes noteworthy and trivial events of that calendar year, drawn in ant-like scale. The drawing has, in addition to the droll comet device, other marginal doodles and notations about incidents for possible inclusion in the print. Moreover, the comet's head in the drawing has its tongue stuck out irreverently (as opposed to the grinning rictus in the print). In the print, an enthroned Queen Victoria and her consort Albert preside over the potpourri of contemporary scandals and countless Victorian curiosities, rendered with Lilliputian detail.[114] The work gives us a taste for the polyglot culture of the Victorians, as also demonstrated in the Great Exhibition of 1851 which showcased British accomplishments of the Industrial Revolution.

In the 1800s, between two and four comets are recorded per year, making them, while not always expected if their period was unknown, at least part of the rhythm of life. In 1857, seven comets were spotted, several of which were moderately bright. 1857 V (C/1857 Q1) seems to have been the brightest and naked-eye as well, while 1857 VII (6P/1857 X1) was observed by Maclear from the Cape Observatory. The latter is periodic comet 6P/d'Arrest, whose period is 6.38 years.[115] However, looking at the general record of comet observations from the 1840s and 1850s, it seems that British astronomers were not as active in the international observation and discovery of comets as previously.

Nevertheless, the public's awareness and fascination with comets did not abate, as suggested by alarmist pamphlets, whose covers reveal their sensationalist thrust. An example is one by a Dr. Young entitled "The Approaching Comet."[116] At the top of its cover is the quote: "'I trust the approaching Comet will not strike the Earth – but will not say it is impossible' – Professor Brewer." A second example of the genre, by "a Middle Bachelor," is entitled "The Comet and Its Consequences" [Fig. 123 color].[117] Its cover features a huge plummeting comet, colored in yellow and orange, with a star-like orange center, which in its technique and boldness reminds us of lurid sixteenth-century German broadsides. Other pamphlets of this nature were published in 1857.[118]

Moreover, *Punch, or The London Charivari* of 1857 contained several items that serve as barometers of contemporary sentiment. In the 7 March issue appears the poem "THE EXPECTED COMET (To Dr. Cumming)," which was meant to be recited to the air "Draw the Sword Scotland."

Hey! a Comet's coming, CUMMING, CUMMING,
Ho! a Comet's coming, expected very soon:
Unless folks are humming, humming, humming,
The Comet will be here on the Thirteenth day of June.
Prognostication
Spreads, consternation,
And with prostration,
Old women swoon,
Thinking of the Comet, coming, CUMMING
The Comet that is due on the Thirteenth day of June.

Because the Comet coming, CUMMING, CUMMING,
Because the Comet coming, astrologers declare, –
Silly people humming, humming, humming,
Silly people humming, – will blow us into air,
Fouling this planet:
Goodness! – how can it,

If we but scan it,
The spheres so untune.
By the Comet coming, CUMMING, CUMMING,
By the Comet coming and due this blessed June?

We know better, CUMMING, don't we CUMMING?
We are sure that any astrologer's a loon,
Or else a knave and humbug, humming, humming,
Who says the world is coming to its end so very soon,
Three years, if not more,
Lease it has got more.
May be a lot more,
Along with the moon,
Though a Comet's coming, CUMMING. CUMMING,
Though a comet's coming – possibly in June.

If the Comet's coming, CUMMING, CUMMING,
If the Comet's coming, ice will be a boon,
When the flies are humming on a sultry afternoon.
Hotter weather may prevail,
If it switch us with its tail,
How very like a whale,
Stung by a harpoon!
Let us hope the Comet, CUMMING, CUMMING
Won't come it quite so very strong as that in June.[119]

In June of that year a short article in the same satirical periodical, entitled "Effects of Comet Shock," narrates a long list of ridiculous follies heard by Mr. Punch after the Great Comet struck the earth.[120]

At mid-century these sensationalist offerings rather than David Scott's imaginative, romantic art represented one of the mainstream trends in popular culture's attitude toward comets. While Cruikshank's humor is related to them and reflects mid-century interests and social concerns, it marks the end to the whimsical satire that flourished earlier in the century. The other trend at mid-century was realism. It represented the pressing artistic muse, as witness what happened when an absolutely captivating comet appeared in 1858, Donati's Comet, and the photography of comets was first attempted.

Notes

1 Vsekhsvyatskii, 153-157.

2 Ms. Additional 34, 225, fol. 187, in the Department of Manuscripts, the British Library, London. The letter of 1820 addressed "To the Editor of Blackwood: Monthly Magazine," begins: "Various are the difficulties that oppose themselves to my comprehension of the Newtonian theory of comets. Some of these admit seemingly of a more natural solution than the Hevelian hypothesis (= the old Aristotelian idea, rectified and expanded by its adaptation into the Copernican system) that the substance of comets is meteoric and their curve of motion a parabola." Coleridge states he is against the idea that comets reflect sunlight and explains his own "philosophy," which he views as "caressing the heretical."

3 Vsekhsvyatskii, 161-164.

4 Ms. 15/149, fol. 43v, Royal Greenwich Observatory Archives, Cambridge University Library, in pen and brown ink measuring 70 × 80 mm (image). See T. Maclear, "Observations of Halley's Comet, made at the Royal Observatory, Cape of Good Hope, in the years 1835 and 1836," *Memoirs of the Royal Astronomical Society*, 10, 1838, 91-97; J.F.W. Herschel, *Results of Astronomical Observations Made During the Years 1834, 5, 6, 7, 8, at the Cape of Good Hope...*, London, 1847, 393-413. There is also a pen and ink drawing by Maclear in the archives of the South African Astronomical Observatory, dated 27 January 1836, in the account of his observations with Herschel through his 20-foot reflector. We thank Ethleen E. Lastovica for bringing it to our attention. Bennett, 1978, 37, lists other papers in the Royal Astronomical Society Library concerning the apparition of Halley's Comet in 1835-36 from the library and observatory of John Lee at Hartwell House, Buckinghamshire, including those of William Henry Smyth, the father of Piazzi Smyth. Bennett (*ibid.*, 56) lists John Herschel's sweeps, which after 476 individual ones take the form of a journal (1834-36), devoted mainly to observations of Halley's Comet, and also cites other material on this comet (*ibid.*, 59).

5 Pl. XV from Herschel's *Observations* (cited in n. 4), measuring 238 × 185 mm (image). It is inscribed: *J.F.W. Herschel del.t; London; Smith, Elder and C.o 65 Cornhill; Hullmandel & Walton, Lithographers.*

6 Pl. II from Maclear's article in *Memoirs* (cited in n. 4), a mezzotint measuring 192 × 310 mm (image), assisted by Piazzi Smyth. See also "Obituary: Thomas Maclear," *Monthly Notices of the Royal Astronomical Society*, 40, 1880, 202-204.

7 He became one of the most distinguished scientists and scientific celebrities, playing a role similar to that of Albert Einstein in the twentieth century. See J.H. Leech, et al., *John Herschel and Victorian Science*, exh. cat., Austin, 1966; Hoskin, 1990, 323.

8 Unnumbered in the Royal Astronomical Society Library, London.

9 He was eventually buried next to Newton in Westminster Abbey. See D.S. Evans, T.J. Deeming, B.H. Evans, and S. Goldfarb, eds., *Herschel at the Cape*, Austin and London, 1969; Buttmann; D.S. Evans, "John Herschel," *Dictionary of Scientific Biography*, ed. C.C. Gillispie, New York, 1974, 323-328; Hermann, 159-160.

10 Bell, 135.

11 *Ibid.*, 39. A model is in the Science Museum, London (inv. no. 1876-1013), and the original metal mirror (inv. no. 1914-370), the largest ever, is displayed nearby.

12 Hoskin, 1990, 331-344. See also *idem*, "The Telescope Wars," *Sky and Telescope*, 89:4, 1995, 26-28.

13 Westfall, 392. Newton had experimented c. 1682 with silvering the back of a glass mirror.

14 Ms. Add. 180 in the Royal Astronomical Society Library, London, a wood engraving in reddish brown and black ink measuring 334 × 258 mm. It was published by W. Marshall, London (cost two pence).

15 This association goes back to astrology. An early British example is *The Kalendar of Shepherds*, London, 1556 (published in Paris in 1493).

16 Whereabouts and measurements unknown. Listed as in the Norcross Collection, New York, in R.W. Lee, *A History of Valentines*, New York and London, 1952, 115-116. It was originally printed from a copperplate, but proved so popular that lithographic reprints were made and it was revived with the next appearance of the comet in 1910. Its poem reads:

> Cupid by the Comet's blaze
> Sets Lovers hearts on fire.
> Hymen leads them to the Church
> And makes their bliss entire:
> Venus Goddess of sweet Love,
> In beauty bright does shine.
> She with Cupid makes the Charm
> To bless each Valentine.

17 Inv. no. 1998-398 in the Science Museum, London, a carnelian intaglio in a silver mount measuring 10 × 10 × 20 mm.

18 Most are dated only approximately to the Victorian period c. 1830-70. They were used for pinning a *fichu* at the neck. One example with turquoise stones is in the Science Museum, London (inv. no. 1994-1081). However, a possible earlier example with topaz is in the Hull Grundy Collection, the British Museum (inv. no. H 718). Its label has the date c. 1815-20, whereas in the catalogue of the collection it is assigned to the late eighteenth century. See C. Gere, J. Rudoe, et al., *The Art of the Jeweller. A Catalogue of the Hull Grundy Gift to the British Museum*, 1, ed. H. Tait, London, 1984, 26, no. 114, who add that after the appearance of Halley's Comet in 1759, pins like these became popular. Judy Rudoe related verbally in January 1995, however, that subsequent to the book's publication, she and Charlotte Gere agreed that this pin seems later (c. 1815-20) and that it may not represent a comet at all. See also S. Bury, *Jewellery 1784-1910: The International Era*, 1, London, 1991, 157, 206, for a paste comet brooch in a private collection that she believes was inspired by the sightings of a comet by Sir James South at Kensington and John Herapath at Crawford Middlesex in 1831 [Vsekhsvyatskii, 159; Kronk, 1984, 33, C/1831 A1, formerly 1830 II] or by the reappearance of Halley's Comet in 1835-36). The Victoria and Albert Museum has two other comet pins, in paste (inv. no. M40-1977, dated c. 1820, and inv. no. 32C-1972, dated c. 1835). We thank Clare Phillips for bringing these to our attention. Comet jewelry remains desirable today with Chanel reissuing items in 1995, including for $385,000 Chanel's and Paul Iribe's platinum and diamond comet necklace (the original of c. 1932 is in the Bibliothèque Forney, Paris; see Olson, 1985b, fig. 107).

19 In a private collection, New York, in cut steel and brass measuring 114 mm.

20 Inv. no. A89-02 (a preliminary designation for the Harpur Collection of Halley memorabilia) in the National Maritime Museum, Greenwich, in brown, gray, and black watercolor wash with charcoal under- and over-

drawing, measuring 355 × 440 mm. We are grateful to Mr. Harpur for sending us the photograph. A visceral recording of just how topical comets were is found in a painting supposedly by John Orlando Parry, *London Street Scene with Posters*, 1835, in the Alfred Dunhill Collection, London, where on a wall near St. Paul's Cathedral, bills are posted; one is boldly lettered *COMET* (referring to a train). See D. Piper, *Artists' London*, New York, 1982, p. 90, for an illustration. His identification of the artist as John Orlando Parry, who was an actor and entertainer, may be mistaken. Thanks to Sally Gibson for this information. If the painting is by an artist named Parry, it may be by John Parry (1812-1865), a London painter and illustrator.

21 R.B. Beckett, *John Constable's Correspondence*, 3 (*Suffolk Record Society*, 7), Ipswich, 1965, 32; L. Parris, C. Shields, and I. Fleming-Williams, *John Constable: Further Documents and Correspondence* (*Suffolk Record Society*, 18), Ipswich, 1975, 190-192, for letters of John and Alfred Chalon to Constable. For Constable, see G. Reynolds, *The Later Paintings and Drawings of John Constable*, 2 vols., New Haven and London, 1984; L. Parris and I. Fleming-Williams, *Constable*, exh. cat., London, 1991. For the brothers, C.R. Leslie, *Memoirs of the Life of John Constable, R.A.*, ed. A. Shirley, London, 1937, who mentions Alfred on pp. xliv, 169, 236ff, 252, 280, 292, 296, 312, 324, and John on pp. xliv, 169, 271, 292, 296, 297, 301, 312, 324; Geneva, Musée Rath, *Deux Artistes Genevois en Angleterre. Alfred-Edouard Chalon (1781-1860) et Jean-Jacques dit John Chalon (1778-1854)*, exh. cat., Geneva, 1971.

22 Inv. no. 1987.185 in the Science Museum, London, a lithograph measuring 234 × 343 mm.

23 A.J. Finburg, *The Life of J.M.W. Turner, R.A*, Oxford, 1961, 325ff, entitles chapter XXX, on the years 1831-33, "A Meteor in Colouring." See also A. Wilton, *J.M.W. Turner. France Italy Germany Switzerland*, New York, 1982; *idem., Painting and Poetry. Turner's "Verse Book" and his Work of 1804-1812*, exh. cat., London, 1990.

24 Inv. no. B1977.14.7925 in the Yale Center for British Art, Paul Mellon Collection, New Haven, first state of an engraving measuring 436 × 295 mm (paper), 217 × 152 mm (plate impression), 90 × 83 mm (image). It was engraved by Robert Brandard and is Rawlinson no. 598 (W.G. Rawlinson, *The engraved work of J.M.W. Turner, R.A.*, 2 vols., London, 1913). Inv. no. B1977.14.13731 is the later state. See J. Piggott, *Turner's Vignettes*, exh. cat., London, 1993, 86, no. 31, who reproduces a preparatory watercolor in the Preston Hall Museum, Stockton Borough Council. M. Omer, *Die Landschaften der Bibel*, Vienna, 1985, 78-79, lists a preparatory watercolor as in the Mordechai Omer Collection, Israel. See also A. Wilton, *J.M.W. Turner. His Art and Life*, New York, 1979, 458, no. 1264. The meteor-like forms of the watercolor are not as pronounced as in the engraving, but Turner may have made corrections on the proofs, as was his habit, to intensify them. This print and the following one were part of *Milton's Poetical Works*, published in London by John Macrone in 1835. See also L. Lusk, "Milton and Art," *The Art Journal*, no vol. no., 1911-12, 131, for Turner's illustration of the Expulsion from Paradise, crowned by a glowing orb with a tail in which is drawn a sword with a twisted tail that symbolizes divine justice but also resembles a comet.

25 J. Gage, *J.M.W. Turner. "A Wonderful Range of Mind,"* New Haven and London, 1987, 222-224. See also Piggott, 60-62.

26 Inv. no. CCLXXX-87 in the Tate Gallery, London, Turner material (Rawlinson no. 360). See A.J. Finburg, *A Complete Inventory of the Drawings of the Turner Bequest*, 2, London, 1909, 895, no. 87. In a letter of 13 March 1995, Ian Warrell noted that "given Turner's very particular interest in all natural phenomenon [sic phenomena] (manifest in his studies of an eclipse and the Aurora Borealis), it is odd that there are apparently no sketches of the comets he could have seen during his lifetime." Turner drew eclipses in his "Eclipse Sketchbook" of 1804 (Finburg, 1909, 1, 224-225, TB LXXXV, pp. 1a-7) and the aurora borealis in 1830 (*ibid.*, 2, 826, TB CCLXIII-158).

27 Inv. no. B1977.14.7926 in the Yale Center for British Art, Paul Mellon Collection, New Haven, the first state of an engraving measuring 434 × 293 mm (paper), 214 × 150 mm (impression), 120 × 83 (image). It was engraved by Edward Goodall and is Rawlinson no. 599. Inv. no. B1977.14.13732 is the later state. Piggott, 87, no. 33, illustrates a preparatory watercolor in a private collection; Omer, 80-81, lists a preparatory watercolor in the Mordechai Omer Collection, Israel. See also Wilton, 1979, 450, no. 1265.

28 Book I, 44ff.

29 Piggott, 60. Also in Book II, 707-710, Milton wrote: "Incenc't with indignation *Satan* stood/Unterrified and like a Comet, burn'd... and from his horrid hair/Shakes Pestilence and Warr."

30 Gage, 1987, 222. Somerville's 610-page, highly technical and mathematical book was published in London by John Murray, Albemarle Street. On pp. 158-159 she discusses comets but does write about meteors. J. Gage, *Color in Turner. Poetry and Truth*, New York and Washington, DC, 1969, 120-121, states that this was Turner's textbook of astronomy. He believes that Somerville's rationalist approach harmonized with the artist's thinking.

31 In the Turner correspondence there is a letter to Turner from Chantrey on 24 June 1829 which refers to Somerville's husband, Dr. William Somerville, suggesting that the acquaintance had commenced earlier (Ms. Add. 50119, fol. 7, the British Library, London; see J. Gage, ed., *Collected Correspondence of J.M.W. Turner*, Oxford, 1980, 129; Wilton, 1987, 167). The Somervilles were also friends of Sir Walter Scott (Gage, 1980, 149).

32 M. Somerville, *Personal Recollections, From Early Life to Old Age, of Mary Somerville*, London, 1873, 269. She also notes her meeting with Giovanni Battista Donati in Florence and comments on comet 1843 I and Donati's Comet (*ibid.*, 255, 311).

33 Gage, 1980, xxv, n. 2. The Somerville Papers in the collection of Lady Alice Fairfax-Lucy are on deposit in the Bodleian Library, Oxford.

34 Gage, 1987, 226, writes: "The painter had long been familiar with the Newtonian notion of light and colour as material substances; and he will have been faced daily with the painter's paradox that light, the most evanescent and transparent of phenomena, could be represented only by white materials which were the most dense and opaque of pigments."

35 *Ibid.*, 222-224. Gage, 1969, 120, notes that in 1847 Turner drew Galileo's enclosed, concentric system into Solomon Hart's painting of Milton's visit to the astronomer, while in the 1820s he had proposed to introduce it prominently into the vignette depicting Galileo's villa. He believes that Turner's taste in astronomy, like that in art, was "classical."

36 In the context of Newton's Universal Law of Gravitation Somerville gives the equation for computing parabolic orbits, followed by a statement on

comets, which commences with a sentence that is not true today (Somerville, 1831, 158-159):

> The periodic time of a comet cannot be known from one appearance. Of more than a hundred comets, where orbits have been computed, the return of only three has been ascertained. A few have been calculated in very elliptical orbits, but in general it has been found, that the places of comets computed in parabolic orbits agree with observation; on that account it is usual to assume, that comets move in parabolic orbits.... It is probable that such bodies do exist in the infinite variety of creation, though their appearance is rare. Most of the comets that we have seen, however, are thought to move in extremely eccentric ellipses, returning to our system after very long intervals. Two hundred years have not elapsed since comets were observed with accuracy, a time which is probably greatly exceeded by the enormous period of the same bodies.

37 A. Wilton, *Turner and the Sublime*, exh. cat., London, 1980, 140, no. 52. Rawlinson recorded that the engraver's son informed him that when his father was engraving this plate, Turner wrote across the upper proof instructions to put innumerable figures in that location.

38 See G. Finley, *Landscapes of Memory. Turner as Illustrator to Scott*, Berkeley and Los Angeles, 1980, 198-199. Turner also illustrated J.G. Lockhart, *Memoirs of the Life of Sir Walter Scott*, Edinburgh and London, 1839, in which he used star symbolism to comment on the life of Scott.

39 Inv. no. 72.195 in the Indianapolis Museum of Art, in watercolor strengthened with pen and sepia ink measuring 174 × 148 mm, and inv. no. 1868-8-22-3584 in the British Museum, London, an engraving measuring 96 × 75 mm (image) and 210 × 150 mm (plate). The print is Rawlinson no. 538. See Wilton, 1979, 432, no. 1115. Piggott, 56-57, believes that these vignettes demonstrate Turner's own response to Napoleon, who is commonly seen as a fallen angel, although Byron compares him to Lucifer.

40 Ms. Add. 50118, fol. 24, in the British Library, London, quoted by Gage, 1980, 46. We thank John Walker for pointing out this reference.

41 *Ibid.*, 46, n. 3.

42 Inv. no. N00500 in the Tate Gallery, London, in oil on canvas measuring 147.3 × 238.8 cm. Wilton, 1979, 267, no. P138.

43 For the pen and brown ink and watercolor on a thin card vignette, see Christie's, sale cat., 30/VI/81, no. 100, where it is dated c. 1832.

44 See E. Shanes, *Turner's Human Landscape*, London, 1990, 114-118; C. Powell, *Turner's Rivers of Europe. The Rhine, Meuse, and Mosel*, exh. cat., London, 1991, 105-106, no. 12.

45 See I.B. Young, *A Concordance to the Poetry of Byron*, 1, Austin, Texas, 1965, 271; vol. 3, 935, for comets and meteors, respectively.

46 Inv. no. N00 529 in the Tate Gallery, London, in oil on canvas measuring 79 × 79 cm. See D.B. Brown, *Turner and Byron*, exh. cat., London, 1992, 119-20. For many of Turner's generation, Napoleon was first viewed as a shining reformer, so that his defeat was not celebrated because it restored the old reactionary regimes in many parts of Europe. This was also Byron's position. See also A.G.H. Bachrach, "The Field of Waterloo and beyond," *Turner Studies*, 1:2, 1981, 5-13.

47 Brown, 187-188.

48 See Olson, 1985b, 5-6, 39.

49 Inv. no. 1931-667 in the Science Museum, London.

50 Its engine is inv. no. 1862-53 in the Science Museum, London.

51 Inv. no. Eng. 365 in the National Gallery of Scotland, Edinburgh, an etching with aquatint measuring 244 × 342 mm. It is inscribed: *W. Turner de Lond. delin.t*; *Shipwreck and loss of the Comet Steam Vessel, near Greenock, on Friday 21st Oct.r 1825*; *C.T.*, the initials of the engraver. The stern of the sinking boat is inscribed *COMET of GLASGOW*. It sold for "one shilling 6' and for 2 s 6' colored." We thank Valerie Hunter for helping us to locate it. The best accounts of the incident in the British Library are: an anonymous four-page pamphlet, *The Loss of the Comet Steam Packet*, published in Leicester in 1825 by the Midland Society for Religious Tracts; *COMET, Steam Boat. Account of a most Melancholy and Dreadful Incident! Loss of the Comet Steam-Boat 70 persons drowned*, published in London by J. Calnach in 1825. For the artist, see A. Crookshank and the Knight of Glin, *The Watercolours of Ireland: Works on Paper, in Pencil, Pastel and Paint c. 1600-1914*, London, 1994, 147-148.

52 Inv. no. 1992-7253 in the Science Museum, London.

53 Inv. no. 1975-8022 in the Science Museum, London.

54 Inv. no. 1982-346 in the Science Museum, London.

55 In a private collection, New York, in brass measuring c. 24.5 cm.

56 In the Collection of H.M. the Queen, in oil on canvas measuring 142.9 × 218.4 cm. It is signed and dated *J. Martin. 1840*. See Lloyd, 194-195, no. 7; O. Millar, *The Victorian Pictures in the Collection of Her Majesty the Queen*, 1, Cambridge, 1992, 183, no. 489; as well as Balston, 147-148, 203; M.J. Campbell, 116-117, who reproduces the rare mezzotint with etching in the collection of Michael J. Campbell, measuring 263 × 356 mm (platemark), 191 × 283 mm (image). Balston, 203, adds that perhaps Martin painted a small picture of the subject at the time of the mezzotint and in 1838 exhibited a watercolor *The Assuagement of the Waters* at the New Water-colour Society. Martin's large mezzotint after the picture (395 × 656 mm) was published by S.H. Hawkins on 1 October 1842 and is inscribed to Prince Albert as owner of the original; Wees, 69-70, no. 55.

57 Quoted in M.J. Campbell, 117. There are also comets in Martin's various versions of *The Deluge*, but ones that look neither like the comets in Martin's *The Creation of Light* nor *The Eve of the Deluge* nor the comet of 1811. The extant painting of the subject, signed and dated 1834, is in the Yale Center for British Art, New Haven (inv. no. B1978.43.11, measuring 168.5 × 258.6 cm; see Sotheby's, London 22/ III, 1972, no. 28), while the mezzotint of 1828 after the lost painting of 1826 and the states of the 1831 mezzotint are discussed and reproduced by Wees, 33, no. 23; 42-45, nos. 30-32. See also L.R. Matteson, "John Martin's 'The Deluge,'" *Pantheon*, 39, 1981, 220-228, who provides evidence that Danby's *Opening of the Sixth Seal* [Fig. 91] and Martin's visit to Danby in 1826 influenced his 1826 and 1828 depictions of The Deluge. M.L. Pendered, *John Martin Painter*, London, 1923, 133, discusses the visit of the French paleontologist Baron Cuvier to Martin's studio. As reported by Ralph Thomas on 17 February 1834, Martin told him:

> Baron Cuvier came to see me when he was in London on account of my having, in my *Deluge*, made the event the consequence of sun, moon and comet in conjunction; the moon and comet drawing the water over the earth. This was Cuvier's opinion, he told me so, and he expressed himself highly pleased that I had entertained the same notion.

58 Vsekhsvyatskii, 164-166.

59 Velikovsky's theories, presented in his *Worlds in Collision* of 1950, have been entirely debunked. See C. Payne-Gaposchkin, "Worlds in Collision," *Popular Astronomy*, 58, 1950, 278-286; A. De Grazia, ed., *The Velikovsky Affair*, London, 1966; D. Goldsmith, ed., *Scientists confront Velikovsky*, Ithaca, 1977; H.H. Bauer, *Beyond Velikovsky: The history of a public controversy*, Urbana, IL, 1984.

60 In the British Library, London.

61 Vsekhsvyatskii, 167-168; B. Warner, "The Great Comet of 1843," *Monthly Notices of the Astronomical Society of South Africa*, 29, 1980, 69-75; Kronk, 1984, 35-36, who estimates that the magnitude was −7, over 100 times brighter than Sirius, the brightest star. *The Illustrated London News*, 2, 1843, 211-212, 240-241, 276, etc., also includes coverage. One of the engravings on p. 211 is entitled "GRAND METEORIC PHENOMENON – THE COMET," to demonstrate the public's lingering misunderstanding. See also B.G. Marsden, "The Sungrazing Comet Group. II," *The Astronomical Journal*, 98:4, 1989, 2316-2321.

62 J. Herschel, 1849, 366-370.

63 A.M. Clerke, *A Popular History of Astronomy During the Nineteenth Century*, London, 1902, 104.

64 Ms. Add. 183/1-4 in the Royal Astronomical Society Library, London. The sheets are from a sketchbook made during travels in the near-east and far-east. All four are mounted on horizontally lined paper. No. 1, a large foldout that plots the track of the comet across the sky; it is executed in pen and brown and black ink with white heightening on blue paper and measures 352 × 157 mm. No. 2 is in watercolor and pen and black ink with white heightening and measures 178 × 134 mm. No. 3 is described in the following note. No. 4 is executed in watercolor and white heightening with pen and red ink and measures 185 × 135 mm. His observations of comet 1843 I extend from 6 March to 30 March 1843.

65 Ms. Add. 183/3 in the Royal Astronomical Society Library, in pen and red and black ink with watercolor and white heightening measuring 186 × 135 mm. On the reverse and on the sheet on which it is mounted, it is inscribed: *March 7th 1843 Comet as observed from the Ship Justina/in Lat. 7.18 S. & Long. 67.2 E., about 7.15 p.m./and on the fifth from its first appearance. – / – R.S.=Ch.46m. 40s. S.D.-126.40.*

66 See D.J. Warner, "Smyth, Charles Piazzi," *Dictionary of Scientific Biography*, New York, 1974, 498-499; B. Warner, *Charles Piazzi Smyth, Astronomer-Artist; His Cape Years, 1835-1845*, Cape Town and Rotterdam, 1983; H.A. Brück and M.T. Brück, *The Peripatetic Astronomer: The Life of Charles Piazzi Smyth*, Bristol, 1988. See also "Obituary: Charles Piazzi Smyth," *The Observatory*, 23, 1900, 145-147; R. Copeland, "Obituary: Charles Piazzi Smyth," *Monthly Notices of the Royal Astronomical Society*, 61, 1901, 189-196; A.M. Clerke, "Charles Piazzi Smyth," *Compact Dictionary of National Biography*, 2, London, 1975, 2460.

67 W.H. Smyth, "Observations on Halley's Comet," *Memoirs of the Royal Astronomical Society*, 9, 1836, 229-246.

68 W.H. Smyth, *A Cycle of Celestial Objects*, 2 vols., London, 1844. For the elder Smyth, see K. Sugden, "An Eclectic Astronomer," *Sky and Telescope*, 63, 1982, 27-29. The author terms him "one of the last gentlemen of science," who made a contribution in several fields. His impact was particularly evident on the Reverend T.W. Webb, who in 1859 published a similar observers' guide (*Celestial Objects for Common Telescopes*) and who is discussed in chap. VII.

69 Maclear, 1838, 91-97.

70 Warner, 1980a, 69. See also Warner, 1983, 102-106, as well as T. Maclear, "Observations of the Comet of 1843, made at the Cape of Good Hope, re-examined and reduced," *Memoirs of the Royal Astronomical Society*, 20, 1851, 62-69; C.P. Smyth, "Remarks to Accompany the Foregoing Observations," *Memoirs of the Royal Astronomical Association*, 20, 1851, 68-69. C. Stott and D.W. Hughes, "Two Piazzi Smyth Comet Paintings," *Annals of Science*, 46, 1989, 170, note that lost drawings of comet 1843 I by Piazzi Smyth were sent to the Royal Astronomical Society. They consist of five naked-eye views dated March 3rd, 4th, 9th, 18th, and 25th, together with six telescopic views of its head on the nights of March 3rd, 4th, 6th, 9th, 22nd, and 31st – suggesting that his observations were actually of a longer duration than formerly thought.

71 Vol. 1, p. 95, RSE Piazzi Smyth archive, held on deposit at the Royal Observatory Edinburgh. Below the oval image the lithograph is inscribed: *C.P.S. del^{t}. H.A.S.sc5*

72 In the Don Africana Library, Durban Metropolitan Library, South Africa, in watercolor and mixed media (ink, body color, and scratching out) measuring 234 × 192 mm. See B. Warner, "Charles Piazzi Smyth at the Cape of Good Hope," *Sky and Telescope*, 59, 1980, 4-5; B.M. Spencer, *Don Africana Library. Select Acquisitions and Bibliographical Notes*, Cape Town, 1988, 15.

73 Vol. 4, p. 60, RSE Piazzi Smyth archive, held on deposit at the Royal Observatory Edinburgh, in watercolor and gouache measuring c. 124 × 164 mm.

74 Inv. nos. BHC4147, BHC4148 in the National Maritime Museum, Greenwich, both in oil on canvas, measuring 91 × 61 cm, and initialed and dated at the lower right: *C.P.S. 1843*. See Stott and Hughes, 165-172.

75 His observational notes are preserved in the Royal Greenwich Observatory Archives, Cambridge University Library. For the section on the Great Comet of 1843, see Warner, 1980a, 69-75.

76 Stott and Hughes, 168-169, note that there had only been one other account of a comet shining so brilliantly so near the sun – on 4 February 1106. They conclude (*ibid.*, 171), comparing data with Smyth's journals, where it seems he began observing the comet on 3 March, that the canvas was painted from reports made by someone else. They also reproduce as fig. 3 another drawing of comet 1843 I in March in the Durban Metropolitan Library, which is related to the one reproduced in the present study as fig. 112.

77 It was probably cited by Sir David Gill; see *ibid.*, 169, n. 18, who quote Gill in 1913 writing about Smyth and a twilight painting of this comet.

78 *Ibid.*, 170.

79 C.P. Smyth, "On Astronomical Drawing," *Memoirs of the Royal Astronomical Society*, 15, 1843, 71-83. According to Warner, 1983, 112, Piazzi Smyth first sent the paper on 30 November to the South African Literary and Scientific Institution. It was his father, to whom he sent a copy who, after neglecting it for a year, sent it to the Society. Somewhat to Piazzi's surprise and consternation the article was printed.

80 For example, Piazzi Smyth writes (*ibid.*, 81): "It is needless to enlarge on the rapidity with which astronomical drawings would increase in value, if they were rendered more accurate...."

81 Warner, 1983, 116.

82 C. Piazzi Smyth, "On photographic illustrations for books," *Transactions of the Scottish Society for the Arts*, 5, 1861, 87-92.

83 Vol. 26, unpaginated, RSE Piazzi Smyth archive, held on deposit at the Royal Observatory Edinburgh, in watercolor with pen and brown ink measuring approximately 124 × 164 mm (image).

84 The pertinent inscriptions are:
14 obs. Comet again in Transit Inst^r RA 19^h 47^m-D=N41° etc./
D3° southward per day/s[therefore with 3 dots] 3 days more will put it below h horizon.
15 Comet seen again Nucleus/very high=small star of 1 mg. but fog came on at Mer. Transit.

85 Vol. 26, unpaginated, RSE Piazzi Smyth archive, held on deposit at the Royal Observatory Edinburgh, in watercolor with pen and brown ink measuring approximately 124 × 164 mm (image).

86 The collection of Piazzi Smyth material on deposit at the Royal Observatory Edinburgh also includes a colored engraving of Encke's Comet as seen 12 September 1848 at the Hartwell Observatory in Buckinghamshire on p. 21 of vol. 1 of the sketchbooks. Below the plate mark Piazzi Smyth has written: *ENGRAVED BY BASIRE FROM THE DRAWING.* It consists of a telescopic circle of blue, framed by a black square, with stars in white and with a fuzzy, tailless comet. The image measures 175 × 175 mm, the paper 217 × 247 mm. It is followed on p. 22 by Piazzi Smyth's preparatory work in watercolor and bodycolor, inscribed: *ENCKE'S COMET, as seen in the Hartwell Telescope. Sept. 22, 1848.* Its slightly larger black square measures 200 mm. Piazzi Smyth continued his comet studies in the article "Some remarks on theories of cometary physics," *Transactions of the Royal Society of Edinburgh*, 20, 1853, 134-145.

87 E.C. Krupp, personal communication, 1995. See also *idem*, *In Search of Ancient Astronomers*, New York, 1978, 232-233; *idem*, "Recreating the Past: Powerful Pyramids, Lost Continents, and Ancient Astronauts," in *Science and the Paranormal*, eds. G.O. Abell and B. Singer, New York, 1981, 259-264.

88 For David Scott, see M. Campbell, *David Scott 1806-1849*, Edinburgh, 1990, as well as W.B. Scott, *Memoir of David Scott*, Edinburgh, 1850, in which the author has engraved some of his brother's drawings. See also J.M. Gray, *David Scott, R.S.A. and His Works. Catalogue of His Paintings, Engravings, and Designs*, Edinburgh and London, 1884; D.J. Black, *An Historical Account of the Lands and Mansion formerly called HERMITS AND TERMITS now known as the Coalyard House...*, Edinburgh, 1972; W. Hardie, *Scottish Painting 1837-1939*, London, 1976, 30-32; Halsby, 99-100. In addition, W.B. Scott, *Autobiographical Notes of the Life of William Bell Scott*, 1, New York, 1892, 214ff, discusses his brother David.

89 Inv. no. NG 1877 in the National Gallery of Scotland, Edinburgh; see M. Campbell, pl. 11. Scott, 1892, 1, 219, writes: "Here I may reiterate... that Samuel Brown .... and my brother David represent to me analogous powers in science and art."

90 Scott, 1850, 330.

91 *Compact Edition of the Dictionary of National Biography*, 1, 412-413, which mistakenly states that the ninth edition of Nichol's *Architecture of the Heavens*, has illustrations by David Scott. The ninth edition (which has no actual publication date although its preface is dated April 20, 1851) was published in London by Hippolyte Bailliere and associated presses in other cities in Europe and New York.

92 Demand for the book was so great that second and third editions were published in Edinburgh and London by N. Tait in 1838 and 1839. The first American edition was published in 1840 by H.A. Chapin in New York. They feature assorted numbers of astronomical diagrams, many of which are beautifully engraved with heavily inked black backgrounds. The preface to the 1840 edition states that the plates were chiefly taken from "graphic representations of the two Herschels," while later editions feature stunning engravings based mainly on the observations of nebulae by the Earl of Rosse. Nichol was an exponent of the nebular hypothesis that the solar system condensed out of a primordial nebula. Although a synthetic examination of all the editions of Nichol's book has not been tackled, volume 418 of the *National Union Catalogue*, 78-80, lists ten different editions in American libraries. The illustrations in each of these editions are reportedly different, although many diagrams are repeated. We have examined the editions of 1837, 1839, 1840, 1850, and 1851. Only the expanded edition of 1850 contains the engravings based on the works of David Scott, together with the scientific engravings, and is in a larger format; it is entitled *Architecture of the Heavens* (published in London by J.W. Parker in 1850 with no edition number).

93 Scott, 1892, 1, 268, remarks: "Besides the *Memoirs* I published several sets of designs – one set in a new edition of Professor Nichol's *Architecture of the Heavens*, his last and one of his most imaginative series of sketches." See Gray, 31-32. There are also twenty astronomical plates, enumerated on p. XV and followed on p. XVI with a list of the positions of nebulae, as well as an astronomical frontispiece of a nebula.

94 See W. Parsons, Earl of Rosse, *The Scientific Papers of William Parsons, Third Earl of Rosse, 1800-1867*, ed. Sir C. Parsons, London, 1926; H.S. Williams, *The Great Astronomers*, New York, 1930, 347-348; Abetti, 186-187; R. Berendzen, R. Hart, and D. Seeley, *Man Discovers the Galaxies*, New York, 1976, 13-15; Hermann, 159-160; Dreyer and Turner, 77, 107, 113-114; Hoskin, 1990, 331-344, who connects him with J.P. Nichol. The telescope's mirror is on display at the Science Museum, London, and the rest of the telescope is being restored as a Science Park on site in Ireland.

95 See *ibid.*, 341.

96 Inv. no. D 4709 in the National Gallery of Scotland, Edinburgh; see K. Andrews and J.R. Brotchie, *Catalogue of Scottish Drawings*, Edinburgh, 1960, 211-212. On the back of the cover is an Ex Libris featuring a representation of an open window with a crescent moon and stars, before which is displayed a book and palette. At the lower margin is a star with six points balancing on an unidentifiable object. It is inscribed: *WILLIAM+B+SCOTT+*, testifying to the fact that it came from his collection. The frontispiece is hand-lettered in orange and blue watercolor over graphite: *MAN/AND THE STARS./A SERIES OF EIGHTEEN DESIGNS/BY/ DAVID SCOTT./*[following in graphite only *SOME OF THEM FEEBLY*]*/EXECUTED 1848./FOR PROFESSOR NICHOL'S/ARCHITECTURE of the HEAVENS.*

97 The graphite drawings are of approximately the same dimensions. Each is numbered at the top of its album page in orange watercolor by William Bell Scott, and the subject inscribed above or more frequently below the image on the actual sheet. Some have other inscriptions that have been erased to suggest that Scott originally made notations on where he meant them to be inserted in Nichol's text. An * after the title indicates the drawing was engraved for the 1850 edition, sometimes with variations in

the titles, frequently accompanied by lines of poetry given in Scott's drawings, and listed on p. xx of Nichol's book. They are: 1) *A Preface*, consisting of three robed female figures on the zodiacal band; 2) *The Instructor** as plate VII; 3) *Nebulae. Struggling into Life** as plate X; 4) *Creation of a Star* (reproduced in M. Campbell, fig. 8); 5) *The Footprint of the Omnipotent** as plate VIII; 6) *The Comet*, which was cropped at lower left, where there are traces of an earlier inscription; 7) *The All-Sustaining Hand*; 8) *Worship of the Stars** as plate III; 9) *Emblem, the Astronomer*; 10) *Tycho Brahe and Kepler./Observation & Generization*, perhaps rethought as plate II with a representation of a man leaning on a telescope on top of a roof (Galileo?) and identified on p. xx as "The Astronomer;" 11) *Man raised by Thought** as plate I; 12) *VIRTUE*; 13) *Love and Confidence** as plate IV; 14) *Procession of Unknown Powers** as plate VI; 15) *Hearing the Eternal Melodies of the Spheres** as plate V; 16) *Age and Youth, expressive of desire*; 17) *Gaze into Eternity** modified as plate IX, the original drawing with a Dantesque conception that harmonizes with other near-death experiences; 18) *FINIS** as plate XI, the final plate designed by Scott for Nichol's book. The rare publication by Gray, reproduces near facsimiles of nos. 4, 5, 6, 7, 9, 14, 15. In the National Gallery of Scotland there is also a separate small drawing (inv. no. D 3855), which is signed: *David Scott*, and inscribed on the mat with the poetic lines that accompany Scott's plate X of "the Nebulae" in Nichol's book: *Pain is life, painful is death/ Passing into the higher life, the painfulist.* It is a variant of no. 3 used in the final engraving to suggest that David Scott may have executed others that are not in the album. In fact, inv. no. D 3855 is listed in Andrews and Brotchie, 208, together with three others related to designs for Nichol's book and/or related to its plates (inv. nos. D 3854 related to no. 13 above and pl. IV; D 3856 related to no. 7 above; and D 3857 related to no. 18 above and pl. IX). M. Campbell, 16, claims that the illustrations are "perhaps more mystical than astrological." We would like to characterize them as more philosophical than astronomical.

98 In the National Gallery of Scotland, Edinburgh, *The Comet* is in graphite with the possible addition of black chalk measuring 213 × 206 mm.

99 Gray, 32.

100 Inv. no. D 4710 in the National Gallery of Scotland, Edinburgh; see Andrews and Brotchie, 211-212; M. Campbell, 32. It contains 174 drawings and on the back of its cover is the same Ex Libris described in n. 96.

101 Fol. 17 in graphite and wash measuring 372 × 310 mm. It is inscribed: *Scandinavian Night—The day/of Ragnarock, when Loke shall/burst his chain & Surtur break/forth in flame. Odin fallen from his. throne. The arch of heaven (the rainbow)/ falls to pieces. The tree Zadasil/bends & breaks, & Hella, death or/ darkness, ascends triumphant over/ the Universe./D.S."*

102 See G.T.P., "Ragnarök," *Encyclopaedia Britannica*, 18, Chicago, 1972, 1103. Two sources for the event are the tenth-century *Völuspá* ("Sibyl's Prophecy") and the thirteenth-century *Prose Edda*. I. Donnelly, *Ragnarok: The Age of Fire and Gravel*, New York, 1887, discusses how Ragnarök was associated with the collision of a comet in the Eddas, and on pp. 141ff, notes that the Elder Edda relates that "The stars shall be hurled from heaven." See also New York, Pierpont Morgan Library, *Icelandic Sagas, Eddas, and Art*, exh. cat., New York, 1982, 81ff, for other English writers who were influenced by them (Thomas Gray, William Morris, W.H. Auden, Louis MacNeice, as well as Richard Wagner). To that list may now be added David Scott.

103 Scott, who was made a member of Scottish Academy in 1829, was much admired by his fellow artists but never achieved the public recognition he craved. See Hardie, 30-32. His work was technically and conceptually forward-looking and foreshadows the Victorian fairy painters and certain tendencies of the Symbolists.

104 Inv. no. 1978U1988-EH-1850 in the British Museum, London, a mezzotint measuring 280 × 424 mm (image). Below the pressmark at lower right it is inscribed in graphite: *From Matthew Cotes Wyatt/To his friend George Cruickshank.* It was published in London on 1 May 1850. The print came to the British Museum as part of the Cruikshank Collection, revealing that Cruikshank was interested in these celestial phenomena over and above his own employment of them in his own works of art (see below). For Wyatt, see "OBITUARY: Mr. Matthew Cotes Wyatt," *The Art Journal*, no vol. no., 1862, 86.

105 For seven accounts and Glaisher's appeal, see *The Illustrated London News*, 33, 16 February 1858, 118-120. Its two illustrations were *The Meteor Seen Near Hampstead*, where it looks very much like the Wyatt illustration, and *The Meteor, As Seen from Fulham Road*, where it appears to explode at a lower point. On p. 142 Glaisher announces that he is publishing the other accounts in the *Philosophical Magazine*. He also calculates that the body of the meteor ignited somewhere near the Bristol Channel, that it was seen within a 200 mile area, and that when it detonated it was heard 60 miles away. *The Illustrated London News* published reports and illustrations of meteors, as well as of comets and eclipses, together with portraits and obituaries of dignitaries, news of the empire, fashion, art, music, chess moves, etc. The relationship of the periodical to science is explored in K. Chew and A. Wilson, *Victorian Science and Engineering Portrayed in The Illustrated London News*, Phoenix Mill, Gloucestershire, 1993.

106 *The Illustrated London News*, 41, 8 April 1854, 321, based on a report written in a letter of 1 April 1854 to the Editor by Mackenzie Walcott of Belgravia.

107 Vsekhsyvatskii, 169-171; Kronk, 1984, 36-37.

108 Vsekhsyvatskii, 172-172; Kronk, 1984, 37-38.

109 Vsekhsyvatskii, 173-179; Kronk, 1984, 38-40.

110 Inv. no. 1988-1442 in the Science Museum, London, a lithograph measuring 194 × 139 mm.

111 Inv. no. TO1927 in the Tate Gallery London, in oil on canvas measuring 196.8 × 325.9 cm; see Feaver, 191-194. For the comet, see Vsekhsyvatskii, 188-191; Kronk, 1984, 43-44. *The Illustrated London News*, 23, 10 September 1853, 213-214, includes an article on the comet observed in Naples, illustrated by an engraving of it over the Bay of Naples with Vesuvius billowing smoke, as well as a reply about the English observations and calculations made at Greenwich. 1853 was also the year in which Honoré Daumier began in earnest to satirize the French bourgeoisie's gullible fascination with comets for the Parisian magazine *Charivari*; see Olson, 1985b, 90-93, who only mentions some of Daumier's 23 lithographic satires about comets that coincided with the apparition of very bright comets. Earlier, in 1843 Daumier had published one in *Le Charivari* (L Delteil, *Le Peintre-Graveur Illustré*, 22: *Honoré Daumier*, New York, 1969 [reprint of Paris, 1925-30 ed.], no. 1004). Olson, 1985b, 90-91, points out that its comet functions as an ironic metaphor of fame, poking fun at the opening of Victor Hugo's new play,

stating that the comet's tail was longer than its ticket lines. Then, in 1844 he published another (Delteil, 23, no. 1114). It was followed by three in 1853 (*ibid.*, 26, nos. 2428, 2373, 2374); ten in 1857 (*ibid.*, 27, nos. 2925-2934); eight in 1857 (*ibid.*, nos. 2935-2938, 2940-2942, 2944); two in 1858 (*ibid.*, nos. 2943, 3080). In some Daumier attacks the public's cometomania, while in others he satirizes amateur astronomers. Another French illustrator, Jean-Jacques-Isidore Gérard, called Grandville, began c. 1844 to employ comet and meteor imagery with a fantastically inventive facility. See Olson, 1985b, 101-103, for several of his pertinent illustrations.

112 The British Museum has five impressions, including one that may have been reworked and a first proof, measuring about 225 × 435 mm. The independent etching published by D. Bogue was circulated separately. The various impressions will be listed in numerical order of their accession numbers. (1) Inv. no. 1977U1308, which is inscribed, in graphite within the plate mark *First Proof* and signed: *George Cruikshank.* (2) Inv. no. 1977U1309, which is probably the first state, having a lightly etched inscription in larger letters than the proof of the final state: *Passing Events or [?] the Tail of the Comet of 1853.* (This early inscription was rubbed out and replaced). Below, the artist has inscribed two lines in graphite, demonstrating that he was experimenting with titles: *Passing Events - or the Comet of 1853*, below: *Passing Events or the Tail of the Comet of 1853.* Furthermore, this impression has no parallel lines on its background, and the sun and stars are only lightly etched. (3) Inv. no. 1977U1310 may be a recut version with slight differences in the background and stars, followed by (4) inv. nos. 1977U1311 and (5) 1977U1312. Reid, 1, 187, no. 2561, mentions the etching.

113 Inv. no. VAM 9448A in the Victoria and Albert Museum, London, in graphite and pen and ink with watercolor measuring 303 × 485 mm. It is inscribed in ink with the artist's earliest thought for the title: *The Comet of 1853 - or Passing Events.* It is signed in ink within the frame: *George Cruikshank.* Above the frame it is inscribed in ink: *Original Sketch.* See S. Lambert, *Reading Drawings, a selection from the Victoria and Albert Museum*, exh. cat., New York, 1984, 62; Olson, 1985a, 53.

114 For the list of anecdotes, many of them labelled in the print, see B. Jerrold, *The Life of George Cruikshank*, 2, London, 1882, 195-197, who adds: "It was his wont to open a serial with a *tour-de-force* of this description."

115 Vsekhsyvatskii, 196-200; Kronk, 1984, 47-48; Marsden and Williams, 18-19, 46-47, 99. For the period of 6P/d'Arrest, see Marsden and Williams, 76. An earlier discussion of periods appeared in B. Marsden, "Elements of Cometary Orbits," *Quarterly Journal of the Royal Astronomical Society*, 13, 1971, 428.

116 Inv. no. CR.C2.253 in the Crawford Collection, Library of the Royal Observatory Edinburgh. It was published in 1857 by G.J. Stevenson, London. Inside is Dr. Young's apocalyptic verse:

At the destined hour;
By the loud trumpet summ'd to the charge,
See all the formible sons of fire,
Eruptions, earthquakes, COMETS, lightnings, play
Their various engines: all at once disgorge
Their blazing magazines; and take by storm,
This poor terrestrial citadel of man.
DR. YOUNG

117 Inv. no. CR.C2. 243 in the Crawford Collection, Library of the Royal Observatory Edinburgh, a wood engraving measuring 214 × 140 mm. It was published in 1857 by Tallant and Allan, London.

118 For example, inv. no. CR.C2. 242-1857 in the Crawford Collection, Library of the Royal Observatory Edinburgh, an anonymous pamphlet entitled "Will the GREAT COMET now Rapidly Approaching Strike the Earth?" Its subtitle reads: "Being an Historical, Philosophical, and Prophetical Inquiry into the Probability of a Collision." Eleven thousand copies were printed by James Gilbert, London. In the British Library is another, *The Comet. Edited by an anonymous Tartar*, published by Hurst and Blackett, London. In the New York Public Library is James Bedford's *The Comet*, published by James Gilbert. Its only illustration is on the cover.

119 *Punch, or The London Charivari*, 32, 7 March, 1857, 94. Godfrey, 1984, 20-21, points out that the subsidiary title of this embodiment of Victorian humor – "*or the London Charivari*" – reveals the new ascendancy of the French artists Daumier and Charles Philipon, who published in the French journal *Le Charivari* and who had picked up where the English had left off in the art of caricature.

120 *Punch, or The London Charivari*, 32, 20 June 1857, 244.

Figure 116. Charles Piazzi Smyth, *Coggia's Comet 15 July 1874 at 10:30 p.m.*, 1874, watercolor with pen and brown ink, 124 × 164 mm (image), RSE Piazzi Smyth Archive, held on deposit at the Royal Observatory Edinburgh.

THE
COMET
AND ITS CONSEQUENCES
LONDON
TALLANT & ALLAN
PATERNOSTER ROW
PRICE 6D

Figure 123. *(Opposite)* Anonymous, *The Comet and Its Consequences*, cover of a pamphlet by "a Middle Bachelor," 1857, wood engraving, 214 × 140 mm, The Crawford Collection, Royal Observatory Edinburgh.

Figure 126. *(Above)* William Turner of Oxford, *Donati's Comet, Oxford, 7:30 p.m. 5 October 1858*, 1858, watercolor and body color over graphite, 258 × 367 mm, Yale Center for British Art, Paul Mellon Collection, New Haven.

Figure 127. *(Right)* William Turner of Oxford, *Donati's Comet on 5 October 1858*, 1858, in watercolor heightened with gum arabic measuring 252 × 360 mm, Private Collection, United Kingdom.

Figure 128. Anonymous, *Donati's Comet of 1858*, 1858, watercolor, 175 × 251 mm, Private Collection, New York.

Figure 129. J. Henry Griesbach, *Donati's Comet on 11 October 1858 (Ms. Add. 42/6)*, 1858, watercolor and gouache with brush and black ink, 385 ×484 mm (image), Royal Astronomical Society Library, London.

Figure 130. Samuel Palmer, *Donati's Comet of 1858 over Dartmoor*, 1858, watercolor, body color, and albumen, 311 × 699 mm, Private Collection, London.

Figure 131. Anonymous, *Victorian transparency of Donati's Comet over Balliol College and Trinity College, Oxford, near the Star Arcturus, 5 October 1858*, watercolor with pen and black ink, 197 × 280 mm, Maas Gallery, London.

Figure 132. C.S.C., *Donati's Comet from Markree Castle, Collooney, Ireland*, appears in *Observations of Donati's Comet, 1858, made at Markree Observatory, Ireland*, by Edward J. Cooper (Dublin, n.d.), chromo-lithograph, 178 × 245 mm, The Crawford Collection, Royal Observatory Edinburgh.

Figure 133. William Dyce, *Pegwell Bay: A Recollection of 5th October 1858*, 1859-60, oil on canvas, 63.5 × 88.9 cm, The Tate Gallery, London.

Figure 140. James Smetham, *Sir Bedivere Flinging Excalibur into the Lake*, 1871, oil on board, 110 × 305 mm, Satis House, Tuxedo Park.

Figure 153. Henry John Stock, *"And I saw a Star fall from Heaven," Revelation 9:1*, 1902, graphite and watercolor with touches of white heightening, 333 × 232 mm, whereabouts unknown, formerly Christie's, New York.

Figure 154. Henry John Stock, "*And the Moon became Blood...,*" *Revelation 6:12-13*, 1910, graphite and watercolor heightened with body color, 330 × 232 mm, whereabouts unknown, formerly Christie's, New York.

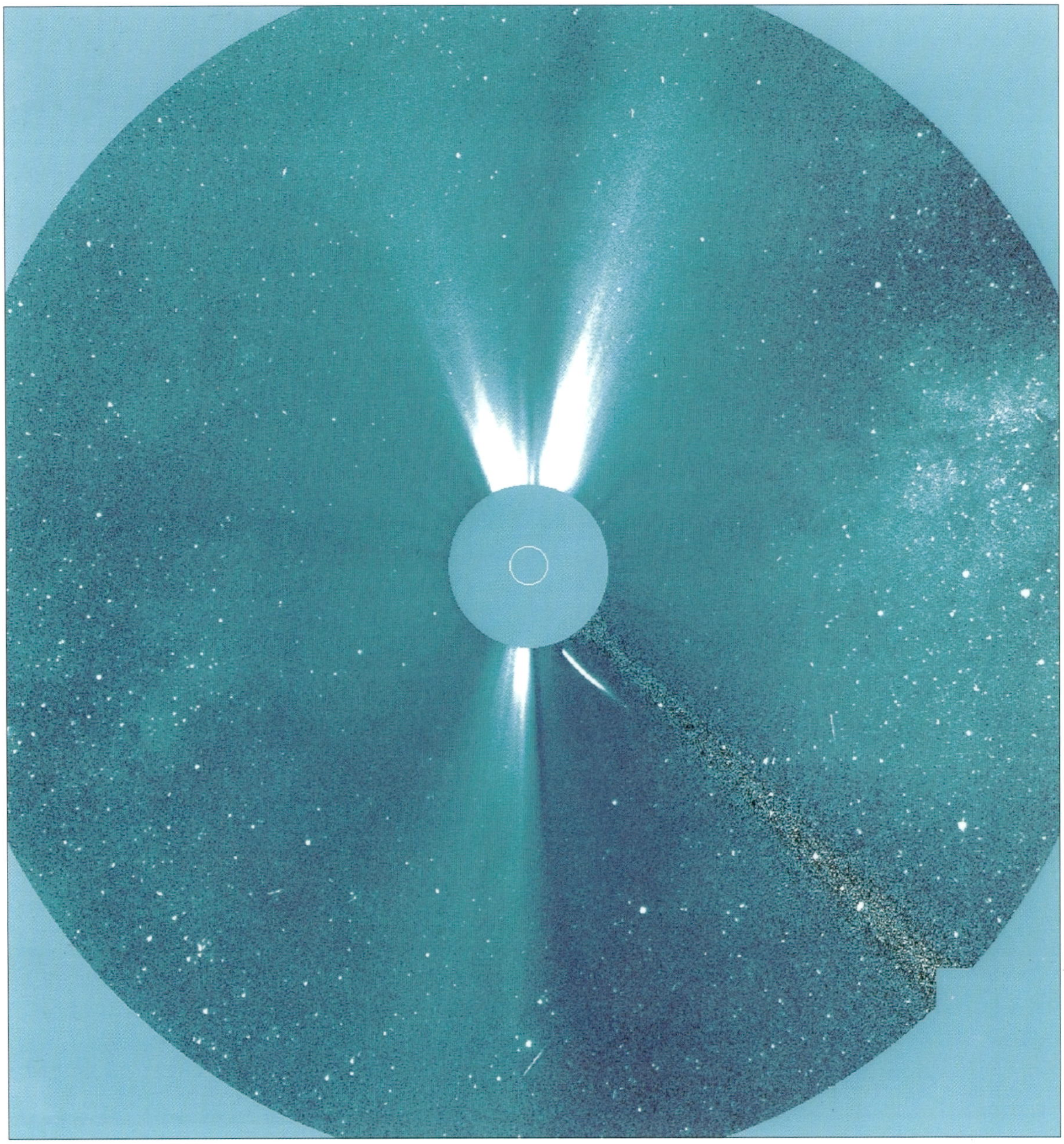

Figure 159. *(Above) A Comet Hits the Sun, as observed from the European Space Agency's Solar and Heliospheric Observatory (SOHO)*, 1996, coronagraph, Naval Research Laboratory.

Figure 161. *(Opposite)* Jay M. Pasachoff, *Comet C/1995 O1 (Hale-Bopp)*, photographed on 2 April 1997 from Williamstown, MA.

Figure 162. *(Overleaf)* Dennis di Cicco, *Comet C/1995 O1 (Hale-Bopp)*, photographed in April 1997 with a wide-field camera.

V

Donati's Comet, the Watershed

Donati's Comet (C/1858 L1, formerly 1858 VI) in its only recorded appearance in 1858 mesmerized astronomers and artists alike [Figs. 124-134]. It was one of eight comets that year and certainly one of the most beautiful ever recorded. Giovanni Battista Donati first sighted his famous namesake telescopically from Florence on 2 June, when it was ten times fainter than naked-eye visibility. By around 19 August it became naked-eye. (It remained so until 4 December and observable with a telescope until 4 March 1859, when it was last seen by Maclear at the Cape of Good Hope.) In late August its position allowed it both to rise before the sun and to set after it, causing some observers to believe that there were two comets. Calculations indicate that it is a periodic comet which returns to perihelion roughly every 2,000 years in its elliptical orbit, and some individuals believe that it may be identical with the comet recorded by Seneca in 146 B.C.[1] Donati's Comet was so large and bright that it stimulated not only more sophisticated drawings but also the first known attempts at comet photography (see chapter VI), although none recording its magnificent tail survives. This chapter will examine its apparition and the art recording this famous comet.

The fame of Donati's Comet derives in large part from its wonderfully broad, sweeping tail as well as from two less frequently seen thinner tails. Its curving dust tail grew to 60° on the 10th of October, while the two thin tails appeared after 1 October to the left of the dust tail. The thin tails have traditionally been identified as gas or ion tails, although it has been suggested that at least one of them was a synchrone, a special type of smaller dust tail.[2] Gas tails are usually more difficult to see than dust tails, because

Figure 124. Anonymous, *Donati's Comet Over St. Paul's Cathedral, 1858*, 1884, chromolithograph, 135 × 90 mm, Collection of Gerald H. Morris, London.

gas tails are blue and the human eye is less sensitive in that region of the spectrum than in the yellow, the typical color of dust tails. Thus, it comes as no surprise that most observers, including artists and astronomers, did not see or draw the thinner tails of Donati's Comet and concentrated on its spectacular dust tail.[3] The large broad tail and its two thinner companions appear in a chromolithograph of Donati's Comet over St. Paul's Cathedral in London [Fig. 124],[4] but this print is an illustration from a book published in 1884 in which its artist must have consulted one of the refined astronomical drawings or prints recording the thinner tails.[5] As we shall see from the discussion below, it is highly likely that the hazy atmosphere in Great Britain interfered with the detection of the two thin tails.

Tellingly, Comet Hale-Bopp (C/1995 O1) during its passage across the sky in 1997 resembled Donati's Comet in many ways [Figs. 161, 162, both color]. The greatest similarity was the sweeping dust tail, but the comet also exhibited a thin tail which showed clearly on long-exposure photographs for a few weeks near its perihelion. But this tail was usually neither visible to the naked eye nor to the eye aided by binoculars. It is, therefore, easy to understand how the straight gas tail of Donati's Comet could have been visible only on rare occasion.

In addition to its tails, other aspects of Donati's Comet were of interest. Many observers noted the fluctuations in the comet's head, especially a dark region behind the nucleus that appeared like a nucleus shadow. At one time, seven fluctuating luminous envelopes appeared. These changes were recorded zealously by astronomers in delicately shaded drawings (some were translated into various print media). The detailed information in these scientific studies is astounding and is still valuable to cometologists today. For example, in 1979, Whipple retroactively determined the rotation rate for Donati's Comet by deducing from them the intervals of dust emissions that formed its halos.[6] These emissions arose from an especially active area on the surface of the nucleus stimulated by its rotation into sunlight.

The American astronomer George P. Bond (1825-1865) produced some of the most famous studies of Donati's Comet. They were reproduced in a short preliminary publication in 1858 and in a major monographic work in 1862. The earlier one contained two

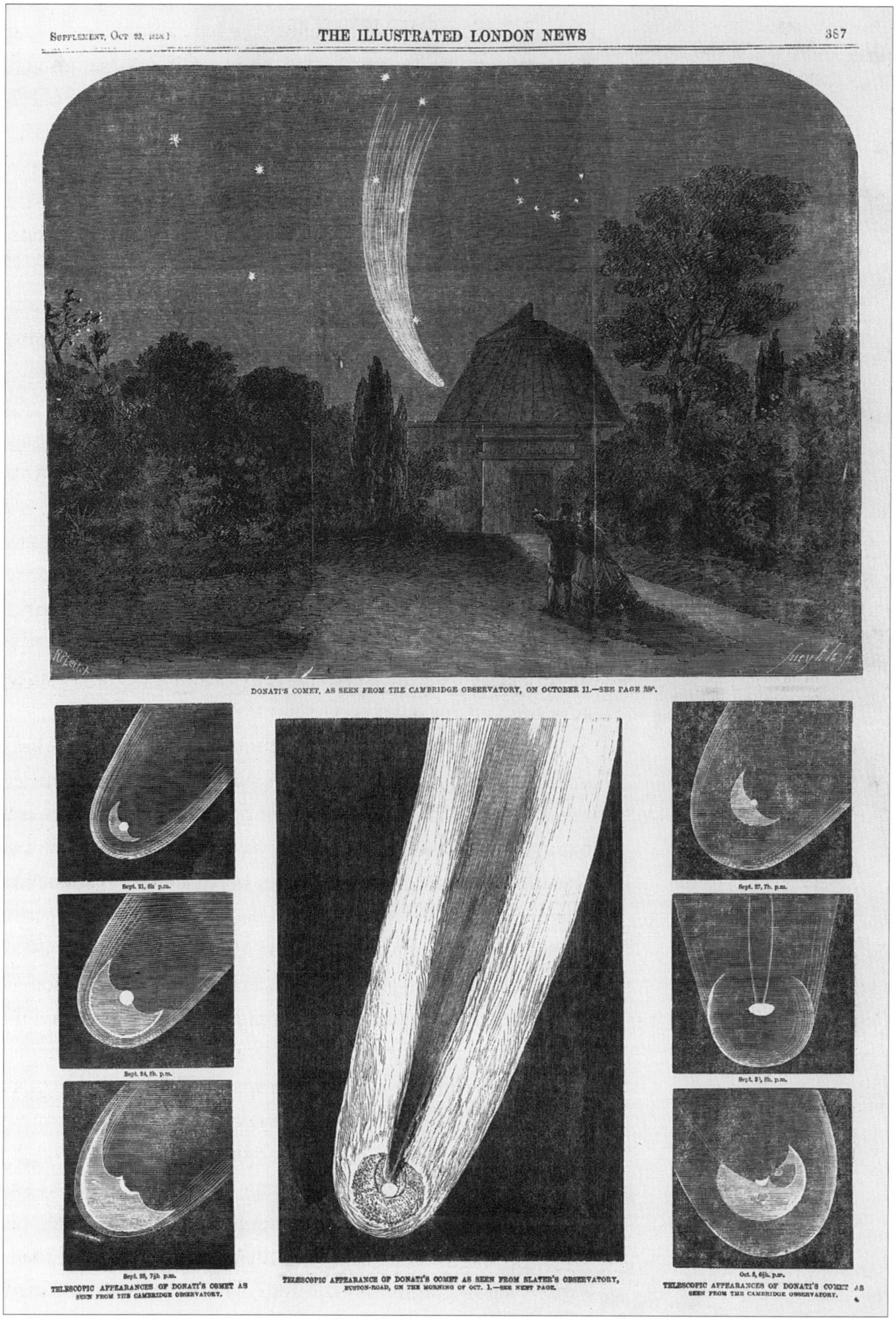

Supplement, Oct 23, 1858.] THE ILLUSTRATED LONDON NEWS 387

DONATI'S COMET, AS SEEN FROM THE CAMBRIDGE OBSERVATORY, ON OCTOBER 11.—SEE PAGE 38°.

Sept. 21, 8h p.m.

Sept. 24, 8h. p.m.

Sept. 28, 7½h. p.m.

TELESCOPIC APPEARANCES OF DONATI'S COMET AS SEEN FROM THE CAMBRIDGE OBSERVATORY.

TELESCOPIC APPEARANCE OF DONATI'S COMET AS SEEN FROM SLATER'S OBSERVATORY, EUSTON-ROAD, ON THE MORNING OF OCT. 1.—SEE NEXT PAGE.

Sept. 27, 7h. p.m.

Sept. 31, 8h. p.m.

Oct. 5, 6½h. p.m.

TELESCOPIC APPEARANCES OF DONATI'S COMET AS SEEN FROM THE CAMBRIDGE OBSERVATORY.

Figure 125. Anonymous, *Various Views of Donati's Comet* from *The Illustrated London News*, 23 October 1858, supplement, page 387, 1858, engravings, Williams College Library, Williamstown, MA.

plates, both close-ups of the comet's head, on 2 and 10 October, and 21 woodcuts.[7] Bond's later work, which collected observations from all over the world, included many detailed new engravings by James W. Watts, in addition to both large-field drawings of its tail and detailed drawings of its head.[8] Bond, who was director of the Harvard College Observatory (founded because of the public's interest in the comet of 1843 I) from 1859 to 1865 (succeeding his father, W.C. Bond, who was director from 1839 to 1859), himself found eleven comets, usually using a small comet-seeker. Shortly before he died, Bond was awarded the gold medal of the British Royal Astronomical Society, the first American to be so honored, largely for his work on Donati's Comet.[9] It is for this seminal, internationally important contribution and for his interaction with British astronomers in the early days of astronomical photography (chapter VI) that we have included him here.

British astronomers were avid followers of Donati's Comet and published a number of significant articles on their observations. The major one by Maclear in 1861 contains, true to his preferences, graphs and charts of positions and summarizes his observations between 11 October 1858 and 4 March 1859.[10] In the following volume of the *Memoirs* an extensive set of observing reports by amateur and professional observers appeared, together with eight plates wherein only one ion tail shows in the wide-angle views.[11]

On 5 October 1858, five days after perihelion, Donati's Comet passed in front of the very bright star Arcturus at 7:11 p.m. from Europe. With a magnitude of a little over 0, Arcturus, *alpha* in the constellation Boötes, is the second brightest star in the sky in England after Sirius (magnitude of – 1.45) and the fourth brightest in the whole visible sky if one includes the southern hemisphere. The comet appeared at sunset (c. 5:30 p.m.) and remained visible until it set (the head about 9:00 p.m. and the end of the tail hours later). It is estimated that at this time its tail (spanning some 40°) had grown, from a length of fourteen million miles in August, to fifty million miles. Later, on 10 October the tail reached 60°. The comet's progress was thoroughly covered by the press and illustrated. Among the handsomest are illustrations in *The Illustrated London News*, which frequently featured articles about comets, such as one of 23 October [Fig. 125].[12] The upper illustration on this page shows two people looking at the comet in front of Cambridge Observatory on 11 October as its tail sweeps through the constellations Serpens and Hercules with the circlet of stars forming Corona Borealis to its right.

Punch includes in its 1858 issues more light-hearted comet material. For example, a cartoon featuring a couple on a balcony gazing dreamily at the night sky; a nearby telescope blocked by a hat carries the caption: "SO FOND OF ASTRONOMY THAT THEY ARE ALWAYS ON THE BALCONY, LOOKING FOR THE COMET!"[13] On the same page is the article "COMET HUNT," which captures some of the enthusiasm as well as the ensuing skepticism about the public's infatuation with Donati's Comet.[14] Both *Punch* and *The Illustrated London News* were important journalistic sources that disseminated information about comets and meteors on a popular level until after the turn of the twentieth century.[15]

A number of British artists were among the dedicated watchers of Donati's Comet and several of them were aware of the articles on it in *The Illustrated London News*. Some of them concentrated their seemingly realistic renderings on its dashing tail without regard for the astronomical accuracy of the surrounding stars, while others represented the sky in a more documentary fashion.

The two arresting watercolors of Donati's Comet with nearly identical panoramic views of the countryside near Oxford by William Turner of Oxford (1789-1862) fall into the former category. The first is the freshest [Fig. 126 color],[16] while the second appears slightly sun-stained [Fig. 127 color].[17] In the first example, the artist with great assurance left the paper blank at the horizon to endow his view with breadth and to bestow on it the glowing effect of a sunset sky. This realistic attitude transfers to the textures he employed in the ground areas. Turner of Oxford also used slight touches of pink pigment in the comet's vaporous tail, a tonality remarked on by some observers, and also indicated the center of the comet's head with a pinpoint. However, he painted none of the constellations and stars visible in the sky in either work. Instead he rendered faint approximations of stellar forms to reveal that astronomical accuracy was not his aim. He did not even show Arcturus, which was visible in the comet's tail at the time this drawing was made (7:30 p.m. on 5 October) let alone the difficult-to-see thin tails to the left of its large magnificent tail, which may not have been visible in England. In the second version, the artist's point of view appears to have been a little higher, while the comet's tail is slightly longer and the arbitrary stars different. The existence of these two watercolors with nearly identical dimensions proves that Turner of Oxford was requested to at least duplicate his work.

A boldly executed watercolor by an unknown artist [Fig. 128 color] preserves the sweep of Donati's large dust tail around the same date, although again no stars or thin tails are depicted.[18] This fresh work, perhaps by an assured amateur, reveals the strength of the British amateur watercolor tradition. In this case, the omission of the stars was probably due to the artist's speed of execution, because the sky has been quickly brushed in and paper left blank where the comet's vaporous tail cuts through it.

We can judge just how generalized these watercolors are by comparing them with an exacting series of seven huge watercolors of Donati's Comet by J. Henry Griesbach [Fig. 129 color].[19] Nothing is known about Griesbach, other than the fact that these sheets all bear his name, together with the date of their execution and a sequential number. Griesbach's series represents the comet's varying profile against an evolving stellar backdrop on 11 and 25 September and 2, 5 (where the tail covers the still visible Arcturus), 9, 11, and 16 October. The seven works are astronomically correct and meticulously plotted, as confirmed by comparison with computerized reconstructions of the skies over England.[20] Because none of Griesbach's astronomically exacting works have either synchrones or ion tails, only the large scimitar-like dust tail, we can safely conclude that the thin tails were not visible to him in the hazy atmosphere of Britain. The example reproduced shows Donati's Comet on 11 October, when the comet was in the constellation Serpens. Its tail curves up through Serpens almost to Hercules, while Arcturus is at the far right. In several of the series, the artist has vividly conveyed the huge dust tail's effects by the addition of slightly viscous layers of gouache, applied with a dry brush, and has also indicated the comet's center of condensation successfully. Griesbach's attention to astronomical detail in the positions and relative brightness of the stars, coupled with the lack of a listing for him in either astronomical or art societies/sources, suggests that he may have been an amateur astronomer.

It is apparent that many English artists rendered Donati's Comet on 5 October, as though they had been notified that this date would be prime time viewing. Prior to that date *The Illustrated London News* had been keeping its readers abreast of developments about the comet via written and visual information. Then, on 2 October, it informed its readers in an unillustrated article that the comet was about to pass the star Arcturus on 5 October and that they could follow its path from the information provided.[21] One artist who may have taken this advice, Mary Lean, represented Donati's Comet on 5 October in a large lunette-shaped lithograph inscribed with its time (8:00), place (Birmingham), and position on that night.[22] It shows the star Arcturus just to the right of the comet's tail.

Samuel Palmer, who had included a more stylized comet in an earlier drawing [Fig. 90], rendered Donati's Comet on the same night in a wonderful watercolor [Fig. 130 color].[23] In this significant work, the artist, who frequently populated his evocative nocturnal scenes with stars and moons, captured Donati's Comet with an astronomical specificity previously unprecedented in his oeuvre. This watercolor reveals another more specific facet of the artist's general preoccupation with light, as evidenced in his writings of the time.[24] The subsequent appearance of Donati's Comet and its magnificent celestial display no doubt stimulated Palmer's already intense predisposition to light.

In Palmer's dramatic watercolor, the natural light of the comet at twilight contrasts with the artificial light in the cave at the right, where one can vaguely distinguish a hooded figure reading a book by candle or torchlight. Near the cave entrance, a woman points out the comet to three children – one of whom is supported on her right arm – who gaze in wonder at the apparition. At the far left in the middle distance, a shepherd with his flock and dog also witness the spectacular comet to add a pastoral note. This is Palmer's attempt at quiet narrative in a scene dominated by the celestial spectacle.

Most significantly, the watercolor can be dated with great precision because of the artist's unusual astronomical accuracy. Palmer depicted the comet after sunset on 5 October shortly before Donati's Comet passed in front of the very bright star Arcturus, seen just to the right of the comet's tail, about 7:00 p.m.

To the surprise of everyone at the time, neither dimmed the light of the other. Palmer also rendered with relative accuracy the stars *epsilon*, *delta*, and *eta* in Boötes close by, and further to the left, two stars of the constellation Corona Borealis. Moreover, he analyzed its visible structure, indicating in white the center of condensation and enlivening the surrounding coma with traces of orange and red pigment. Only a few stars are shown because of the largely cloudy sky (some stars peak through holes in the clouds). Like the other English artists who recorded this comet, Palmer has shown only the scimitar-like, curving dust tail and no thin tails, which reportedly lent a flamboyant effect to the comet's dramatically beautiful profile. Perhaps, as with Griesbach, these additional tails were impossible to see in England because of the haziness resulting from the humid English climate.[25]

It is estimated that at the time Donati's Comet passed in front of Arcturus, soon after Palmer rendered it, the comet's tail measured fifty million miles in length and spanned around 40°. While the comet's total magnitude on 2 October (of −1) was almost that of Sirius, the brightest star in the sky, by 5 October it was reduced by a factor of about four (to 0.5), making its brightness comparable to that of Arcturus, as Palmer has rendered it.[26]

Palmer almost certainly executed the watercolor or a preparatory sketch for it in the area of Dartmoor, as its title when first exhibited in 1859 indicates. The picturesque setting resembles this uninhabited region with its morasses, prehistoric remains, and isolated headlands called "tors," which rise in rugged grandeur, a land whose main resource was grazing. Further, we know that during 1858 Palmer visited Devon and Cornwall with his son Thomas More, who subsequently fell ill and died in 1861. From the gap in the artist's correspondence (from 17 September to 28 October 1858),[27] it seems that the trip fell between those dates, which marked the prime viewing time for Donati's Comet. Writing to Alexander Gilchrist, the biographer of Blake, in 1861 Palmer tellingly admits:

> I have never had a proper title to my Comet drawing – for want of knowing on what day of the month I saw the thing represented – viz: the passing of Arcturus *through* the comet. I left London so suddenly when I *did* leave as to have no time to get a sight of the Illustrated London News for that year, which I believe particularly

> describes the passing of Arcturus – I should therefore feel much obliged if by consulting any astronomical almanack or otherwise, you could, without giving yourself trouble *fill up* the day and date in the blanks I have left in the label behind the drawing.
> But if you have not the dates by you never mind – only *then*, have the goodness to alter the label – putting merely "The Comet of 1858 as seen from the heights of Dartmoor."[28]

This passage suggests several important points: that Palmer executed the watercolor, or at least a study for it, on his Devon trip, that he knew about the notices in *The Illustrated London News*, and that this watercolor was in the possession of Alexander Gilchrist (not listed in Lister's provenance) after its 1859 exhibition.

With this watercolor Palmer created his most exacting sky and at the same time a more visionary work in harmony with his earlier oeuvre. We can only speculate about why it is so astronomically precise. Moreover, he captured more than any artist what it was like to view this spectacular comet with a mixture of awe and dispassion.

The light qualities produced by Donati's Comet on the same night are also approximated in a transparency by an individual who depicted the comet over Balliol College Chapel and Trinity College in Oxford [Fig. 131 color].[29] The star Arcturus is above and slightly to the right of the Comet's head but inside the arcing dust tail, which means it was drawn slightly later on 5 October than the watercolor by Palmer. To increase the realism, the artist pricked the paper with a needle over 30 times in the positions of the stars to simulate, when backlit by a window or a strong artificial light, the pinpoint brightness and twinkling of actual stars. Both the comet's head and Arcturus (given a slightly red tinge) are also pierced, adding a scintillating effect. The circlet of stars composing Corona Borealis shows especially well to the top left of the comet's tail, as does the Big Dipper to its upper right. The work's accuracy raises the question, as it did for the Griesbachs and the Palmer, as to why the straight thin tails are not pictured, and endorses the theory that they were simply not visible from British sites because of the hazy atmosphere. The watercolor is backed by greenish-yellow paper, which adds to its glow when illuminated from behind (especially those areas painted with yellow pigment, such as the Gothic windows and horizon). Transparencies like this one are related to the Victorian fascination with dioramas and the trend toward realism.

A lovely chromolithograph in Edward J. Cooper's *Observations of Donati's Comet, 1858, made at the Markree Observatory, Ireland*, depicts Donati's Comet from that site in Collooney [Fig. 132 color].[30] Below the image at the left is the inscription that identifies the artist only by initials: *Drawn by C.S.C.* Interestingly, the title page of the small book is inscribed: *Professor C. Piazzi Smyth*, indicating that it came from his collection. Edward J. Cooper of Markree Castle possessed one of the largest refracting telescopes in the world and published a pamphlet with his observations of Donati's Comet. According to an anonymous article, Cooper made "well-executed drawings...which may be advantageously compared with the delineations of the same sort made at other observatories."[31] In the illustrations, the asterism of the Big Dipper is directly above the comet's tail, indicating that the work was drawn around 30 September. Except for a star embedded midway in its tail, these are the only stars in the sky. Since this print shows Donati's Comet at an earlier hour with brighter skies than most other nocturnal views, the comet does not stand out starkly against the sky.

Another well-known depiction of Donati's Comet before twilight is *Pegwell Bay* [Fig. 133 color] by William Dyce (1806-1864).[32] The painting's precise title is *Pegwell Bay: A Recollection of 5th October 1858*, which indicates both a realistic attitude and a retrospective, contemplative point of view. It depicts the chalk cliffs and shell-strewn beach of Pegwell Bay, near Ramsgate in Kent, east of London, around sunset. Since comets are brightest when closest to both the sun and the earth, the brightest with the longest tails are usually visible just after sunset and just before sunrise. However, Dyce's sky is so light that it bleaches out the light of the comet that bleeds into the sky. It also appears that Dyce may have presented the comet smaller than it appeared in nature, but then again, this might be due to the blending of its light into the sky, illustrating the difficulty of capturing direct light from a bright object with mere paint. Dyce's subtler treatment of the comet does not allow the sensational apparition to steal the show.

The seemingly photographic realism of the painting not only reflects Dyce's Pre-Raphaelitism,[33] but also demonstrates his familiarity with photography. He had been a friend of David Octavius Hill since the 1830s when both were working in Edinburgh, at the very time when Hill was involved with calotype equipment. It was precisely at the time of Donati's Comet when photography began entering into the popular mainstream.[34] Pointon believes, however, that Dyce's way of looking at landscape was not necessarily modified by using photographs, but rather that Dyce was affected by his knowledge of how landscape looked through a camera lens.[35] Willsdon disputes her view and believes, based on Dyce's correspondence with Hill, that c. 1846, when photography was becoming a fully fledged alternative to the pen and brush, Dyce was experimenting with the new medium and working from photographic studies. In addition, she thinks that the general impossibility of obtaining a fusion of sky and landscape, a problem in photography of the 1850s and 1860s solved only by a combination of two or more negatives, is reflected in *Pegwell Bay*.[36] However, as we shall see, the photography of comets had not progressed far enough to obtain a photograph as detailed as Dyce's comet (Fig. 134 and chapter VI).

Upon closer analysis, Dyce's intense painting is more than just an objective rendering of Donati's apparition over the coastline of Kent, where the artist's family was on holiday in the autumn of 1858. It is a complex allegory of the passing of time and people's vulnerability in the universe, and perhaps a Victorian confrontation between the exciting scientific discoveries and old superstitions. In the context of Dyce's painting, Donati's Comet symbolizes the longevity of the forces of the universe and nature, which dwarf humans and eventually wear them down, like the background cliffs of the Kent coastline by the tide. Moreover, the mood of impending doom is heightened by the sense here that Donati's Comet will become increasingly awesome as night descends. As noted, there was also a widespread fear at the time, supported by some astronomers, that the comet would actually hit the earth.

Furthermore, we know that Dyce produced an exacting watercolor study of the same area when he was on holiday with his family in 1857 to prove that he worked up the canvas from a watercolor executed the year prior to the appearance of Donati's Comet.[37] Pointon believes that the main difference between the watercolor and the later canvas is one of mood. In the later work, the composition has been deliberately subjected to the imagination and is not true to the Pre-Raphaelite insistence on visual accuracy.[38] In fact, the painting, which was exhibited at the Royal Academy in 1860, may not have been painted until 1859.[39]

The people Dyce included wandering on the shore with the tide out can be identified. Because the male figure in the middle ground at the far right of the canvas carries artist's materials, Pointon believes that he represents the artist in this very autobiographical work.[40] Dyce, like the scientist, is an observer of nature and the comet, while everyone else in the scene ignores it. The inclusion of identifiable individuals personalizes and intensifies the universal statement made in the canvas.

In the final analysis, Pointon believes that the painting is a result of artistic manipulation and that its photographic realism is a cultivated illusion.[41] She argues that the artist chose to include a date for the event – unlike Constable and other landscape painters who desired to convey the meteorological accuracy of their observations – because he wanted to stress the day on which Donati's Comet appeared most brilliant – almost in a journalistic fashion. She cites the 2 October *Illustrated London News* announcement that the comet was about to pass Arcturus on 5 October.[42] Further, she believes it is significant that Dyce represented aspects of both astronomy and geology (in the fossil-embedded cliffs), citing Tennyson's description of astronomy and geology as "terrible muses." Both, with the vastness of their telescopic and microscopic investigations, respectively, represent unrelenting processes of nature and add to the ominous tone in *Pegwell Bay*, that is missing from the frenetic holiday fun other Victorian painters celebrate in sea-side scenes.[43] Rather, the figures, who regard neither the natural spectacles nor each other, are vulnerable, as highlighted by the slow, unrelenting geological erosion of the cliffs into pebbles and by the apparition of Donati's Comet.

It is not surprising that Dyce was attracted to Donati's Comet, for he was no stranger to scientific matters. The son of a Fellow of the Royal Society of Edinburgh and a lecturer in Medicine at the Marischal College in Aberdeen, he was brought up in an atmosphere of scientific inquiry and had studied Natural Philosophy in Aberdeen. In fact, he was an amateur scientist himself and had even written a prize-winning essay on "Electro-Magnetism." After moving to Edinburgh in the early 1830s, his friends were academics and men of science.[44] In fact, his nephew James Clerk Maxwell became one of the greatest scientists of all times, ranking with Einstein and Newton.[45] In London, Dyce's circle included Sir Charles Eastlake, President of the Royal Academy and first President of the Photographic Society (later the Royal Photographic Society). Dyce was particularly interested in geology and must have been drawn to Pegwell Bay for its chalk cliffs rich in fossils. It has been suggested that the inclusion of the comet in this context has less to do with any symbolism than with the general importance and implications of science in Victorian society.[46]

Donati's Comet of 1858, because of both its brightness and the beauty of its long dust tail, inspired artists and astronomers to produce excellent work. The representations by Palmer and Dyce, for example, are balanced on the scientific side by the studies of British astronomers. From his observations, Bond determined that the comet's nucleus was very small, less than 500 miles (800 kilometers) across.[47] (We know from spacecraft measurements that the nucleus of Halley's Comet, presumably similar, is only about 10 miles across, that of Comet Hale-Bopp around 25 miles; thus Bond's estimates while generous were useful for the time.) He also found, by carefully drawing the head of Donati's Comet, that it rotated every four to seven days. Even highly technological, twentieth-century observations of Halley's Comet did not succeed in determining the comet's rotation period any more accurately.

Scientific advances often occur at a confluence of new instrumentation and a prepared mind, as with the apparition of Donati's Comet. Significantly, it was the first comet ever photographed, though its tailless image is difficult to discern [Fig. 134]. Donati's Comet was bright enough that scientific measurements could be prepared and carried out. For example, polarization measurements were made; since reflections polarize light, these showed that the light from the tail was sunlight being reflected toward earth.[48] Unfortunately, spectroscopic methods were not quite advanced enough to use at this date, and it took Donati himself, working in Florence, until Comet Tempel in 1864 (C/1864 N1, formerly 1864 II) to obtain useful spectra.[49]

Donati's Comet provided mid-nineteenth-century astronomers and artists with a sight so beautiful that many individuals chose to render it in various media. Its lengthy apparition was pivotal and marshaled an extraordinarily objective investigatory spirit in both artists and astronomers, an attitude rejected in art during the last decades of the century which reacted against realism and the advances of photography (chapter VI).

Notes

1 Vsekhsyvatskii, 200-206; Kronk, 1984, 48-49. It passed perihelion on 30 September, and its elliptical orbit is inclined 117° to the plane of the solar system; calculations show that it will return in 1950 years. We have duplicated the comet's positions using computer programs – Carina Software, Voyager II, San Leandro, CA, 1993; Maris Multimedia, RedShift, London, 1994, CD-ROM – and thank Brian Marsden for his calculations of its path.

2 For straight yellow dust features known as synchrones, see L. Wilkening, *Comets*, Tucson, AZ, 1982, 750, who relies on a particular mechanical theory to define a synchrone as "the locus of particles that leave the nucleus at nearly the same time but have a full range of initial velocities and are subject to a range of repulsive forces. . . ." For a reproduction of the straight dust tail and synchrones of Comet Mrkos (C/1957 P1, formerly 1957 V), see J.E. Bortle, "How to Observe Comets," *Sky and Telescope*, 61, 1981, 212. Vsekhsyvatskii, 204-205, notes that beginning 1 October Arthur Auwers and August (Friedrich August Theodor) Winnecke observed a straight thin tail tangent to the nucleus from the convex side of the large tail; Kronk, 1984, 49, notes that on 16 September the two straight tails were first noticed by Winnecke (Pulkova Observatory, St. Petersburg).

3 This is true even in the twentieth century when Brian Marsden did not see the synchrones of Comet Mrkos with the naked eye, although they showed up in photographs.

4 In the Gerald H. Morris Collection, London, a chromolithograph measuring 135 × 90 mm; it is the frontispiece of A. Giberne, *Sun, Moon, and Stars: A Book for Beginners*, published in London by Seeley and Co., 1884.

5 There were a few by astronomers that record the thin tails. For example, in 1859 Otto Struve and Winnecke published a series of prints of Donati's Comet, at least two of which show an ion tail in addition to the large curving dust tail; see O. Struve and A. Winnecke, "Pulkowaer Beobachtungen des Grossen Cometen von 1858," *Mémoires de L'Académie Impériale des Sciences de St.-Pétersbourg*, 2:1, 1859, 1-69. See also references to Bond's studies in nn. 8 and 9. Another illustrative chromolithograph (fig. 124 in Amédée Guillemin, *The World of Comets*, trans. and ed. J. Glaisher, London, 1877, opp. p. 220), usually published as anonymous, shows Donati's Comet over the Conciergerie in Paris on 5 October 1858 with the same three tails. (We thank Donald K. Yeomans for helping with this identification.) This edition was issued by Sampson, Low, Marston, Searle and Rivington, although it was first published by Librairie Hachette as *Les Comètes*, Paris, 1875, with illustrations by the engravers Maximillian Honoré François Rapine (1840-1925), Philip Benoist (1813-?), and Emile Coriolan Hippolyte Guillemin (1841-1907), the author's son (in this edition, Donati's Comet over the Conciergerie is the frontispiece, labelled plate IX, and plate VII, a double-view of Donati's Comet with one and two thin tails, is labelled as after Bond). Both the views were probably informed by Bond's observations because in an earlier book by the author (*The Heavens, An Illustrated Handbook of Popular Astronomy*, ed. J.N. Lockyer, London, 1867, published by Richard Bentley), the frontispiece with Donati's over Harvard College Observatory has the caption: *Donati's Comet. From the observations of G.P. Bond, at Cambridge, United States, October 4 1858*. It was executed by E. Guillemin, one of the illustrators of the later book, and Hubert Clerget (1818-1899), the lithographer and painter. The stars, including the constellation Corona Borealis, in both chromolithographs are shown as they appeared from Europe with Arcturus in the middle of the comet's tail at 7:11 p.m. (W.R. Dawes, "Observations of Donati's Comet made at Haddenham, Bucks," *Monthly Notices of the Royal Astronomical Society*, 19, 1859, 88-91.) Computer calculations by Brian Marsden confirm this observation.

6 F.L. Whipple, "The Rotation Period of Comet Donati," *Nature*, 273, 1978, 134-145.

7 G.P. Bond, "An Account of Donati's Comet of 1858," Cambridge, MA, 1858 (An Account of the Comet of Donati. 1858," *Mathematical Monthly.*, 1:2, 1858, 61-67; "An Account of Comet Donati. 1858. Part II," *Mathematical Monthly*, 1:3, 1859, 88-116).

8 G.P. Bond, *Account of the Great Comet of 1858* (as *Annals of the Astronomical Observatory of Harvard College*, 3), Cambridge, MA, 1862. Bond's book, billed on the title page as having 51 plates, has one plate in the introduction plus 51 plates in the text.

9 Warren De la Rue, President of the Royal Astronomical Society, wrote about Bond's work: "...the splendid graphic illustrations...are a remarkable feature of this altogether remarkable production. The pains taken to ensure a truthful representation of the Comet of 1858, both in its eye- and telescopic-features, few are better able than myself to appreciate." (B.Z. Jones and J.G. Boyd, *The Harvard College Observatory: The First Four Directorships, 1839-1919*, Cambridge, MA, 1971, 35, quoting from "Addresses on Presenting the Gold Medal of the Society to Professor G.P. Bond," *Monthly Notices of the Royal Astronomical Society*, 25, 1865, 125-137.) De la Rue, who himself attempted unsuccessfully to photograph the comet, made drawings of it; see W. De la Rue, "Observations on Donati's Comet. 6: Observations by Mr. Warren De la Rue, *Memoirs of the Royal Astronomical Society*, 30, 1862, 78-84. Henry Norris Russell, perhaps the greatest astrophysicist of the first half of the twentieth century, wrote, "Bond's beautiful drawings of Donati's comet of 1858 are, and probably will remain, the classic example of what can be accomplished by the combination of a keen eye and a skilful hand." (H.N. Russell, "America's Role in the Development of Astronomy," *Proceedings of the American Philosophical Society*, 91, 1946, 1-16, quoted in Jones and Boyd, 461. For Bond, see also S.I. Bailey, *The History and Work of Harvard Observatory, 1839 to 1927*, New York and London, 1931, 94-95.

10 Maclear, 1861, 59-83.

11 They included those of W. Lassell at Liverpool; the Reverend W.R. Dawes; the Reverend T.W. Webb; J.L. Challis and James Breen at Cambridge; the Astronomer Royal (G.B. Airy), the Reverend R. Main, and W.H.W. Christy at Greenwich; and Warren De la Rue, President of the Royal Astronomical Society. See W. Lassell et al., 58-84, which mentions that all of these reports were read separately but published collectively.

12 *The Illustrated London News*, 33, 23 October 1858, supplement, p. 387, includes a view in the upper section inscribed: *R.P. Leitch*, Richard Principal Leitch, a painter active 1844-1860; and *Smyth ft.*, as well as seven vignettes after drawings of telescopic views of the head. The six lower lateral ones were made at Cambridge Observatory between 21 September and 5 October, while the lower central illustration depicts a telescopic view from Slater's Observatory, Euston Road, on 1 October. The 28 August issue (p. 205) features a report on Donati's Comet, while

the half-page cover illustration of the 25 September 1858 issue depicts a group of people observing it from Greenwich Park on 17 September.

13 *Punch, or The London Charivari*, 35, 2 October 1858, 140.

14 *Ibid.*

> The poor Comet has been leading a sad life of it of late. People have been fairly hunting him to death. There have been nightly meets in London at the corners of the streets, and in fact the country generally has been a field in chase of him. Packs of people everywhere have joined in the pursuit.... No fox's brush has ever been more hunted than his tail, although nobody of course has any hopes of getting hold of it.... To leave this vulpine metaphor, All England has in fact been star-struck for the nonce, and amateur astronomers have been plentiful as partridges.... No sooner does the Comet show himself in public, than there is a general shout raised of "Here he is again!"... Like all great notorieties his goings on are watched...and there is a daily record of them in the paper. An astreal *Court Circular* is published every morning, in which the Comet's movements are continually reported.... Comet-hunters who have telescopes, almost take a glass too much in their eagerness to spy him; and those who are not blest with even spectacles to help them, are nightly in the habit of straining their nude optics (propriety forbids we should say their naked eyes).... *Punch*, for one can sympathise with the Comet. *Punch* having risen as a "bright particular star," of course well knows what it is to be pointed at as such.

15 In the same issue (*ibid.*, 136) is a comment that reveals the public's lack of scientific understanding about comets:

> The appearance of the Comet, which is now starring it for a hundred series of nights has given rise to much speculation, which, as usual, has resulted in failure. Nobody has as yet offered a satisfactory conjecture in solution of the question what the Comet is? Some astronomers say that they can see through it, and the stars on the other side of it; but for all that they have not been able to penetrate its mystery. Two theories of the constitution of Comets which have not yet been published, shall now be and may deserve to be as much as some others. Perhaps, as Comets are both luminous and transparent; and ghosts likewise, according to some observers, both shine themselves and are also diaphanous; Comets are the ghosts of departed planets. Natural as well as supernatural substances, however, both reflect and transmit the rays of light: and your Comet generally has a tail. Analogies subsist throughout nature. May not a Comet be a planet in the first stage of its development, a celestial body corresponding to a terrestrial tadpole?"

The writer's idea, that comets are both luminous and transparent, is correct. Their material is so spread out that starlight passes through a comet's tail unaided, while enough dust is present to scatter sunlight, making the dust tail visible. There were many accounts of the comet, including a pamphlet in the New York Public Library with only a cover illustration, *The Great Comet of 1858*, published by Bert Hardwick in London (sold for sixpence).

16 Inv. no. B1975.4.1767 in the Yale Center for British Art, Paul Mellon Collection, New Haven, in watercolor and body color over graphite measuring 258 × 367 mm. It is signed in pen and brownish ink: *W. Turner/ Oxford*. In the sky to the right of the comet's tail there is slight damage, perhaps due to the artist's reworking. See J. Bayard, *Works of Splendor and Imagination: The Exhibition Watercolor, 1770-1870*, exh. cat., New Haven, 1981, 69-70, no. 65.

17 In a private collection, United Kingdom, in watercolor heightened with gum arabic measuring 252 × 360 mm. It is signed *W. Turner/Oxford* and dated *at a half past seven o/c pm OCT 5th 1858*, according to Sotheby's, London, sale cat., 18/III/82, no. 114. The same watercolor was subsequently at Agnew's, London, where its inscription was recorded as: *William Turner of Oxford, Donati's Comet on 5 October 1858*. We thank Andrew Wyld for the information and photograph.

18 In a private collection, New York, in watercolor measuring 175 × 251 mm.

19 Ms. Add. 42/1-7 in the Royal Astronomical Society Library, London. Each is rendered in watercolor and gouache with brush and black ink, and no. 6, reproduced here, measures 463 × 574 mm (paper) and 385 × 484 mm (image). All are signed in pen and black ink below the image at right: *Henry Griesbach* (or a close variant) and have similar dimensions. See Bennett, 1978, 50, no. 42.

20 Computer plotting – on Maris Multimedia, RedShift, using an ephemeris calculated by Brian Marsden – agrees with Griesbach's watercolors. Bond, 1858, 24, writing on Donati's Comet states:

> We must add a few words on the appearance presented by the tail between 6th and 10th October. At the date first named, one of the supplementary rays, Fig. 19, attained a distance of 55°, or fifty millions of miles from the nucleus, somewhat exceeding that of the principal tail, and in a direction, as usual, nearly in a line from the sun. Others less perfectly developed could be discerned near a point where the curvature of the main stream was pretty suddenly changed. On the 8th , Fig. 20, five or six traverse bands could be distinguished in the tail half a degree or less in breadth, with clear, well-defined outlines, and perfectly resembling auroral streamers, excepting that they kept their position permanently, that is without motion sensible to the eye, they diverged from a point between the sun and the nucleus.

One sees some of the same phenomena, except for the "supplementary rays" or ion tails, in Griesbach's series.

21 *The Illustrated London News*, 33, 2 October 1858, 309. An earlier article with a chart in the 25 September issue on p. 276, had advised: "The course of the comet after Sept. 29 may be traced...that it will be visible till the middle of October (at least) and with increasing brilliancy."

22 Inv. no. EH 1858-inv. no. 1916-11-10-17 in the British Museum, London, a lithograph measuring 264 × 400 mm. Its letterpress inscription reads: *DRAWN FROM NATURE BY MARY LEAN. T. UNDERWOOD.LITH./ THE GREAT COMET OF 1858,/AS IT APPEARED AT BIRMINGHAM ABOUT 8 OCLOCK P.M. ON THE 5TH OF OCTOBER./JUST AFTER THE TAIL HAD PASSED BEFORE THE/STAR ARCTURUS.*

23 In a private collection, London, in watercolor, body color, and albumen measuring 311 × 699 mm. It is signed: *S. PALMER*. It was exhibited at the Royal Society of Painters in Water Colours in 1859 as "The Comet of 1858 as Seen from the Skirts of Dartmoor" and again in 1881 at the Fine Art Society. See R.J.M. Olson, "A water-colour by Samuel Palmer of Donati's Comet," *The Burlington Magazine*, 132, 1990, 795-796. In a letter

of 31/5/96 Raymond Lister stated that he believes that it is "too richly conceived, its setting too contrived to have been painted on the spot; that it was probably based on sketches made during the comet's appearance and then ripened, as was Palmer's practice, at leisure."

24 Earlier, on 18 July of the very same year, Palmer wrote: "In these days, there is an *immense power of light* searching, vast depths of gloom, and woods are *very deep*. Do not forget; all is full of colour." See Lister, 1974a, 1, 545-548.

25 Since a German watercolor of the comet (Olson, 1985b, fig. 92) also does not show any tails, and because German artists tend to be amongst the exacting, the thin tails simply may have not been visible to the naked eye.

26 Vsekhsyvatskii, 202-206; Kronk, 1984, 48-49.

27 Lister, 1974a, 1, 545-548.

28 *Ibid.*, 2, 603.

29 In 1995 at the Maas Gallery, London, in watercolor with pen and black ink measuring 197 × 280 mm. It is inscribed in pen and brown ink: *Transparency.; Donati's Comet with the Star Arcturus./over Baliol Chapel and Trinity Coll. Oxford/Octbr. 5th 1858.*

30 Inv. no. CR.C2.265 in the Crawford Collection, Library of the Royal Observatory Edinburgh, a chromolithograph measuring c. 178 × 245 mm (image). It is inscribed: *Lith.d by Maclure Macdonald & Macgregor, 67 Walbrook London; MARKREE CASTLE, COLLOONEY, IRELAND.* Before this inscription someone has written in pen and brown ink *from*.

31 Anon., "Observations of Donati's Comet made at the Observatory of Edward J. Cooper, Esq., with the large Equatorial," *Monthly Notices of the Royal Astronomical Society*, no. 35, *Memoirs of the Royal Astronomical Society*, 29, 1859-60, n.p.

32 Inv. no. 1407 in the Tate Gallery, London, in oil on canvas measuring 63.5 × 88.9 cm. For Dyce, see M. Pointon, *William Dyce 1806-1864*, Oxford, 1979, as well as *idem*, "The representation of time in painting: A study of William Dyce's *Pegwell Bay:* A Recollection of October 5th, 1858," *Art History*, 1, 1978, 99-103. Dyce's papers – unindexed typescripts of an unpublished autobiography (including transcribed letters) by the artist's son – are in the Aberdeen Art Gallery.

33 See A. Staley, *The Pre-Raphaelite Landscape,* Oxford, 1973, for Pre-Raphaelite attitudes and techniques regarding the painting of landscape.

34 For example, *The Art Journal* for the years 1856-58 is full of advice to photographers and in 1856 a chair of photography was established at King's College, London.

35 Pointon, 1979, 170-171, maintains that the foreground figures are painted in such a manner and with such detail as one might expect to see at close proximity and yet we view them and the scene across a vacant space. She compares this effect with early cameras that had a positive (biconvex) lens of crown glass cemented to a negative (biconcave) lens of flint glass. They also had a long focal length which meant that, when looking through the view-finder, everything was distanced from the viewer. The popular stereoscope picture device had a similar effect.

36 C. Willsdon, "Dyce 'in camera': new evidence of his working methods," *The Burlington Magazine*, 132, 1990, 760-765.

37 In a private collection measuring 247 × 349 mm and reproduced in Pointon, 1978, as fig. 68; *idem.*, 1979, pl. 33.

38 *Ibid.*, 174.

39 *Ibid.*, 169.

40 *Ibid.* From right to the left, the women are: the painter's wife, Jane, and her two sisters, Grace and Isabella. The boy is one of the artist's sons.

41 It has been criticized and praised for this realism, as in by M. Brion, *Art of the Romantic Era*, New York, 1966, 98.

42 *The Illustrated London News*, 33, 2 October 1858, 309.

43 Pointon, 1979, 173. She also notes that Dyce's painting reflects the uniformitarian view, established by the geologist Charles Lyell in his book *Principles of Geology* (1830-38), that the greatest geological changes were not the result necessarily of stupendous events but the result of processes of time that were still at work. (This view has been recently modified by the realization that asteroids and comets hit the earth – and Jupiter, viz. Comet Shoemaker-Levy 9 – with catastrophic consequences.) Pointon ends her discussion with a comparison to Matthew Arnold's "Dover Beach" with its "grating roar of pebbles" and its "eternal note of sadness."

44 *Idem*, 1978, 100, who points out that after becoming Professor of Fine Arts at Kings College in 1844, Dyce maintained his scientific contacts via friends at the newly established Chemical College.

45 In the 1850s Maxwell found equations that unify electricity and magnetism, which were thought of as separate forces. One of the most dramatic moments in nineteenth-century physics occurred when Maxwell combined the laws of electricity and magnetism to show that light is an electromagnetic wave. His work led to the discovery of radio waves and so to television and radio. J.G. Crowther, *Men of Science*, New York, 1936, 261; J. Tolstoy, *James Clerk Maxwell: A Biography*, Chicago, 1981.

46 It has also been allied to the artist's High-Church religious beliefs, and the religious controversies stirred by the new discoveries about time and space in geology and astronomy. L. Parris, ed., *The Pre-Raphaelites*, London, 1994, 182-183, no. 106.

47 Bond, 1858, 29.

48 Yeomans, 213.

49 *Ibid.*, 215. The first polarization measurements were made by Dominique F.J. Arago in 1819 (*ibid.*, 213); Donati's drawing of the comet's spectrum is on *ibid.*, 217.

VI

The Origin of Comet (and Meteor) Photography

Photography has been linked with astronomy since its inception.[1] While Louis Jacques Mandé Daguerre (1781-1851) in France is often credited with its invention in 1839,[2] William Henry Fox Talbot (1800-1877) in England developed it concurrently.[3]

The ability to make permanent photographic images had an earlier base in British science. In 1802 Sir Humphrey Davy and Thomas Wedgwood independently rediscovered that silver chloride turns black upon exposure to light, when they used paper wet in a solution of silver to make an impermanent image of an object in contact with it. The subsequent stages in the development of photography led to lengthy battles for priority between Nicéphore Niépce (1765-1833)/Daguerre and Fox Talbot. Niépce and Daguerre, who collaborated starting in 1829, used a "positive" process, while Fox Talbot used a "negative" process. (Those terms were suggested by John Herschel, who was involved with many of the early chemical aspects of photography and produced the first photograph on glass, an image of his 40-foot telescope, in 1839.) While the daguerreotype name is still commonly known, Fox Talbot's process, known as "talbotyping" or "calotyping," is actually more the ancestor of today's positive/negative photography. Both techniques contributed to the development of astronomical photography.

Although the new techniques of Daguerre and Talbot were difficult to use, and exposures needed to be long, it did not take long for people to turn their cameras toward the sky. Indeed, the French Academy of Sciences had asked Daguerre to photograph the moon even before François Arago described the daguerreotype process to a joint meeting of the French Academy of Science and the French Academy of Fine Arts in 1839.[4] Nevertheless, even at this early date Arago reported Daguerre's attempt and predicted that lunar photography would soon be possible.[5] The moon was chosen as the first object to photograph because after the sun, which would burn film, it is the brightest object in the sky. It would be some time yet before comets would be successfully photographed because they are thousands and usually millions of times fainter. Needless to say, several astronomers quickly became engaged with the new methods. For example, in South Africa, Piazzi Smyth soon started experiments in 1843.

A daguerreotype is a photograph made directly onto a light-sensitive surface. Thus it is a unique work, just like a photographic negative or a drawing. The light-sensitive surface is silver halide, made by exposing a highly polished silver-plated copper plate to vapors of iodine and bromine and then promptly to light. This procedure makes a latent image, which must be brought out by holding it over hot mercury vapor. Then, a sodium thiosulfate treatment (hyposulphite of soda) removes the silver halide, leaving the image behind.[6] (This "fixing" solution had been found twenty years earlier by John Herschel.) Finally, the plate is gilded in a gold-chloride solution. (In these days of environmental sensitivity, the risks that earlier photographers took are obvious.) Because the daguerreotype is a positive image, it is viewed in reflection by shining light onto it to produce a dark, metallic image that is difficult to see. To copy a daguerreotype, another daguerreotype or a photograph must be taken of it. (Negatives were not in use until later in the century.)

It is little known that the moon is about the same brightness as a daylight scene on earth. After all, it is only a quarter million miles from the earth, while the sun is about 93 million miles away from both. Thus about the same strength of sunlight falls upon both the earth and moon, even though the moon appears dark gray, reflecting only about ten per cent of the light that hits it, not very different from foliage on earth. The moon appears bright to us on earth only because we see it against the dark sky. The result is that the same exposures are necessary to photograph the moon as an outdoor scene on a sunny day, making the moon a likely subject to photograph. Since daguerreotypes were thousands of times slower than photographs of today, exposures had to be several minutes long.

Comets are more difficult subjects to photograph than the moon. Even though they are extended objects, like the sun and the moon, they are much fainter. Calculations of exposure time for such extended objects are very different from those for stars. Stars are point objects, and the difference is significant. For a star, one needs a large collecting area in the telescope mirror or lens, and all the light from the star is deposited onto one point on the film. (Because stars appear as points, their images are not magnified by telescopes.) But for a comet or another extended object, it matters not only how much light is collected each second but also how far across the plate or film that light is spread out. The measure of the combination is the focal ratio, or f-stop, of the telescope or camera. (A telescope is but a giant camera.) The focal ratio is the ratio of the focal length of the lens or mirror – the distance from it at which the image of a faraway object forms – to the diameter of the lens. When the ratio is high, a longer exposure is necessary to expose properly an extended object than for a low focal ratio. The longer the focal length, the farther the image is from the lens, and the more the light spreads out. Since the same amount of light from, say, the moon or a comet would be more concentrated on the film if the focal length were shorter, this concentrated image would have a shorter exposure time necessary to record the object.[7] This effect played a decisive role in the quality of the earliest comet photographs.

Earlier large telescope lenses did not have surfaces so curved as to have relatively short focal lengths and thus relatively fast focal ratios. Thus the Great Refractor of the Harvard College Observatory, the largest telescope in the world at mid-century, had a lens with a 15-inch diameter and a focal length of 22½ feet, which gave it a focal ratio of f/18. It had a near twin at the Pulkova Observatory. This high focal ratio meant that exposures would have to be relatively long compared to those made with portrait lenses in ordinary cameras, which might have a focal ratio of f/4. The long exposures have one advantage for extended objects only: the images are bigger. Therefore, comets and celestial objects other than stars have larger images than on photographs taken with shorter focal lengths.

Early Astronomical Photography

The first lunar and solar photographs are described by Barger, who in 1985-86 participated in a team that used old techniques to make new daguerreotypes of the moon.[8] The first lunar images were probably made by John William Draper (1811-1882) of New York on 23 March 1840, using lenses into which a lunar image was reflected by a tracking mirror that kept the image motionless during 30-minute exposures. He used a variety of lenses, whose focal lengths gave him images of the moon that ranged in size from 1/6 inch to one inch. All his images are thought lost; the best were reportedly burned in a fire that destroyed the New York Lyceum in 1866, although the possibility exists that one was saved and has been found.[9] In spite of Draper's daguerreotype, Barger reports that a general belief persisted in the scientific community that the light from the moon was unable to produce photographic effects.[10] Images of the sun were the next to be taken.[11]

Because of their fast lenses, some of the first astronomical daguerreotypes were taken with ordinary portrait cameras, if any camera at that time can be called "ordinary." The earliest generally accepted extant astronomical photograph of the moon was taken with such a camera by Samuel Dwight Humphrey at Canandaigua in upstate New York on 1 September 1849 (in the collection of the Harvard College Observatory). Humphrey, who was a prominent daguerreotypist, reported his success in *The Daguerreian Journal*, which he edited, and sent one to his friend Jared Sparks, the president of Harvard College.[12]

Before mid-century, the newly completed Great Refractor of the Harvard College Observatory, planned for visual observation, had already been turned toward the moon for photography.[13] In 1847, John Adams Whipple, a Boston daguerreotypist who had experimented with a smaller telescope and decided that he needed a larger one with a tracking drive, approached William Cranch Bond (1788-1859), the Director of the observatory, whose telescope had been in use only a few months, to collaborate. Though exposures would be long, the Harvard telescope was mounted so that it tracked the objects in the sky as they rose and moved across the celestial vault.

According to Bond's diary entry for 23 October 1847, Whipple and Bond attempted to make daguerreotypes of the sun and moon with the Great Refractor, inadvertently setting Bond's coat sleeve on fire when he put it at the focus. Consequently, they switched to a smaller telescope, and over the next few years they used both telescopes.[14] Whipple and his partner, William B. Jones, subsequently collaborated with both William Bond and his son George P. Bond (1825-1865).[15] The Whipple/Bond team did not succeed in making an astronomical daguerreotype until they photographed the star Vega with a 100-second exposure and then the star Castor on 16-17 July 1850. During that process they spotted a comet ("Comet has a fine star nucleus"), but did not try to photograph it before they took off the micrometer equipment to change to the daguerreotyping equipment.[16] Then, it took many more months before they obtained a lunar daguerreotype that satisfied them on 14 March 1851.[17] Whipple had made a daguerreotype enlargement of one of these, which won a prize at the Great Exhibition at the Crystal Palace in London (1851). Unfortunately, even though it was only a copy, it was destroyed in a fire at an exhibition in Glasgow.[18]

The British astronomer and chemist Warren De la Rue (1815-1890), who has been called the "Father of Astronomical Photography,"[19] was inspired by the Bond/Whipple daguerreotype shown at the Crystal Palace to employ a newer process.[20] He used a "collodion," guncotton dissolved in alcohol and ether, or "wet-plate" process in which the glass backing is coated by a wet, light-sensitive material that must be used before it dries. The process gives a negative image, from which positive prints can be made. While De la Rue had attempted to photograph a solar eclipse in 1851 using a daguerreotype camera, he went on to organize cooperative efforts between astronomers and photographers, and set up large-scale international cooperation for eclipse photography.[21] He obtained collodion lunar photographs in 1853, but then abandoned the work until 1857, when he finished building his own observatory with its improved tracking of astronomical objects. At the meeting of the British Association at Aberdeen in 1859, De la Rue presented an exhaustive report entitled "On the State of Celestial Photography in England." He noted that while attention had been directed to celestial photography in England for eight years and some observers had made occasional experiments, there had been no systematic pursuit of this branch of astronomy in

England, except by himself at his observatory.[22] Nevertheless, work on lunar and solar eclipse photography continued in England as well as in continental Europe and the United States.

Back at Harvard in 1857, after the photographic hiatus of five years and the installation of a better clock drive in the telescope to improve its tracking precision, photographic work resumed using the collodion technique. The scene was thus set for the arrival of the extremely bright comet in 1858.[23]

Comet Photography

By the time Donati's Comet appeared in 1858, photography had advanced greatly and was widespread. George Bond attempted to photograph it on 28 September at Harvard [Fig. 134], but was supposedly scooped, the night before, by an English commercial photographer named Usherwood.[24]

In the only published reference to Mr. Usherwood from the time, he is characterized in 1859 as "an artist residing on Walton Common" [now known as Walton Heath or Walton Downs, common land close to Walton-on-the-Hill in Surrey], who succeeded in obtaining, in seven seconds, a good negative of the comet with a portrait lens of short focus. According to a contemporary report: "The camera was stationary, hence the image is somewhat imperfect; nevertheless it bears enlargement of four times tolerably well. Mr. Usherwood's residence is situated about 700 feet above sea-level, and it is possible that his success is in some degree attributable to this circumstance, but it is chiefly due to the large area of the portrait lens and the relative shortness of its focal distance."[25]

William Usherwood (1821-1916) was established commercially with a store front in Reigate, Surrey, about 20 miles south of London.[26] His obituary, which includes a photograph of him, also describes his career as a painter.[27] In the local post-office directory, he was listed as a portrait painter in 1862, as a portrait painter and photographer in 1867, and as a photographer in 1874. This progression demonstrates the gradual ascendancy of photography and, in this case, its integral connection to other visual arts.

Additional newly uncovered correspondence in the Harvard University Archives informs us about what is arguably the first comet photograph. The well-known British amateur astronomer

Richard C. Carrington wrote to George Bond on 26 May 1859: "Herewith I send a copy of Usherwood's photograph of Donati's Comet, same size as the original on plate, of which, he has obliged me with 4 or 5 copies for distribution."[28] No mention is made of the date of the photograph, and indeed Bond responded by requesting the date and, curiously, whether the copy was the same size as the original. In Bond's reply, he asks for information from Usherwood that can refer only to this photograph, although Usherwood's name is nowhere mentioned. Carrington was replying to a letter from Bond dated 4 April 1859, in which Bond had briefly mentioned his own photograph: "We obtained a photograph of the Comet of Sept. 28th. On Oct. 5th it was visible in the daytime."[29] Though this Usherwood photograph was once at Harvard, no one has been able to locate it or a reference to its current existence.[30]

Responding to Carrington on 11 June 1859, Bond notes that he is:

> much interested in the photograph which you had the kindness to send. . . . Here is a very singular fact. The camera lens, with its short focus, affords a strong image of the *nebulosity* of the tail at a point where the *intensity* of the light was probably a thousand times less than that of the nucleus. And this, too, in seven seconds; whereas, with an object glass of probably eight or ten times the area, we barely obtained an impression of the *nucleus* itself in 360 seconds on the following day, September 28.

Very interesting is Bond's concluding postscript:

> I take the liberty of enclosing two notes with the request that you will forward them. Mr. Usherwood's address I do not know. I have written to him for particulars about the photograph.[31]

In addition, the Harvard University Archives contains not only Bond's letter, but also Usherwood's reply, which includes new information about the photograph and the camera that took it. The form of address used by Bond, "— Usherwood Esq.," reveals that he did not know Usherwood's first name.[32] Most importantly, Usherwood is unsure of the date of his photograph; perhaps it did not even precede Bond's. Bond's letter to Carrington cites Usherwood's priority by one day, although none of the correspondence we have encountered gives a reason for this precedence (perhaps something was written on the photograph itself), and

Figure 134. George P. Bond, *Donati's Comet, 28 September 1858, plate III*, 1858, full collodion plate and detail *(right)*, Harvard College Observatory, Cambridge, MA.

Usherwood's tentative statement of the date, months after the fact, should make everyone question the chronology.

At Harvard, George Bond reported in his 1862 *magnum opus* on Donati's Comet that an "attempt was made to photograph the image of the Comet in the focus of the Great Refractor, but only the nucleus and a little nebulosity 15" in diameter acted on the plate in an exposure of six minutes."[33] The record book from 28 September 1858 shows that Plate III [Fig. 134] is listed with an exposure of six minutes, and the logbook contains the note next to the exposure time, "Think this took." Bond also sketched the comet in the log-book as seen in a "Comet Seeker." He reported that the sky was "not quite clear" and that he was observing in "strong twilight."[34]

On the successful collodion plate with the comet image, the few darkened grains that represent the comet nucleus are visible

with a magnifying glass. Only a tiny black spot appears to the unaided eye. On closer inspection, the image in figure 134 shows the hazy nebulosity of the comet's head but not its tail. (The glass plate reveals the uneven edges of the emulsion caused when it was added wet to the plate.)

In summary, this plate would be the first comet photograph had it not been for Usherwood's reputed success a day earlier. At the time, the Royal Astronomical Society knew only of Usherwood's photograph, but not that of Bond.[35] In the absence of Usherwood's original, however, Bond's plate is the oldest extant comet photograph, and thus occupies a seminal point in cometology. Further, given the uncertainty of the date of Usherwood's image, Bond's plate may even be the first comet photograph ever taken.

Though De la Rue, in England, was very interested in photography, his attempt to make his own photograph of Donati's Comet failed.[36] De la Rue, who served as President of the Royal Astronomical Society, was given that Society's Gold Medal for his work on astronomical photography.

Dry Plate Photography

According to the astronomer Edward Emerson Barnard, astronomical images were made with wet plates through 1871, while dry plates were used only from 1876 onward and dependable five or six years later.[37] With the use of (silver bromide) dry plates as a recording medium, comet photography was further established during the bright comets of 1881 and 1882.[38] When Comet Tebbutt appeared in 1881 (C/1881 K1, formerly 1881 III/1881b), Pierre Jules César Janssen (1824-1907) in France and Henry Draper (1837-1882) in the United States took the photographs that are generally cited as the first comet photographs, since those of Usherwood and Bond had been forgotten.[39] Janssen's image of 30 June was a 30-minute exposure, showing a tail of 2.5°, while Draper's image was 2 hours 42 minutes long, showing a tail of 10°. On the night of 24 June 1881, Draper took a 17-minute exposure of the head of comet 1881 III Tebbutt through breaks in the clouds. In the days immediately following he took successful photographs of its spectrum employing a two-prism slit spectrograph that he had been using for stellar spectroscopy. Spectroscopic techniques had sufficiently advanced that its spectrum was photographed by a dozen observers.[40] While direct photography gives only the appearance of a comet, spectroscopy allows analysis of the molecular and chemical constituents of a comet, and thus represents a major scientific advance. The British astronomer A. Ainslie Common also photographed this comet on 24 June.[41] At his observatory at Upper Tulse Hill in London, so did the British amateur astronomer William Huggins (1824-1910), who took a spectrum with a one-hour exposure. This photograph and Huggins's drawing of it show not only the continuous radiation but also three molecular bands, which he suggested (and which has been verified) came from carbon, possibly combined with hydrogen.[42]

The next year, another very bright comet, which could be seen naked-eye during the day, appeared in the southern hemisphere and was called the "Great September Comet" (C/1882 R1, formerly 1882 II).[43] When it moved into the night sky, Sir David Gill (1843-1914), director of the Royal Observatory at the Cape of Good Hope, used a portrait lens to make a series of photographs of it [Fig. 135].[44] This magnificent comet, which never had a spectacular tail in the northern hemisphere, delighted astronomers for the next four or five months "with the splendor of its display."[45] Further, the Great September Comet was bright enough to have its spectrum photographed.[46]

As Gill wrote in his lecture presented to the Royal Institution, "It was, in fact the introduction of the gelatin dry-plate process in 1876 which really paved the way for the rapid development of celestial photography."[47] Gill's photographs, by showing many more stars than expected, precipitated the dependence of astronomy on photography. But Gill's photographs also led to the greatest debacle in the history of astronomy: the *Carte du Ciel*.[48] This international project, organized by the director of the Paris Observatory in 1886 with eighteen participating observatories, aimed to make a photographic map of sky. It was to consume the major resources of many of the world's observatories for decades, and thus kept them from starting newer programs. It is commonly thought that because American observatories did not participate that they went on to novel projects, such as astrophysically related

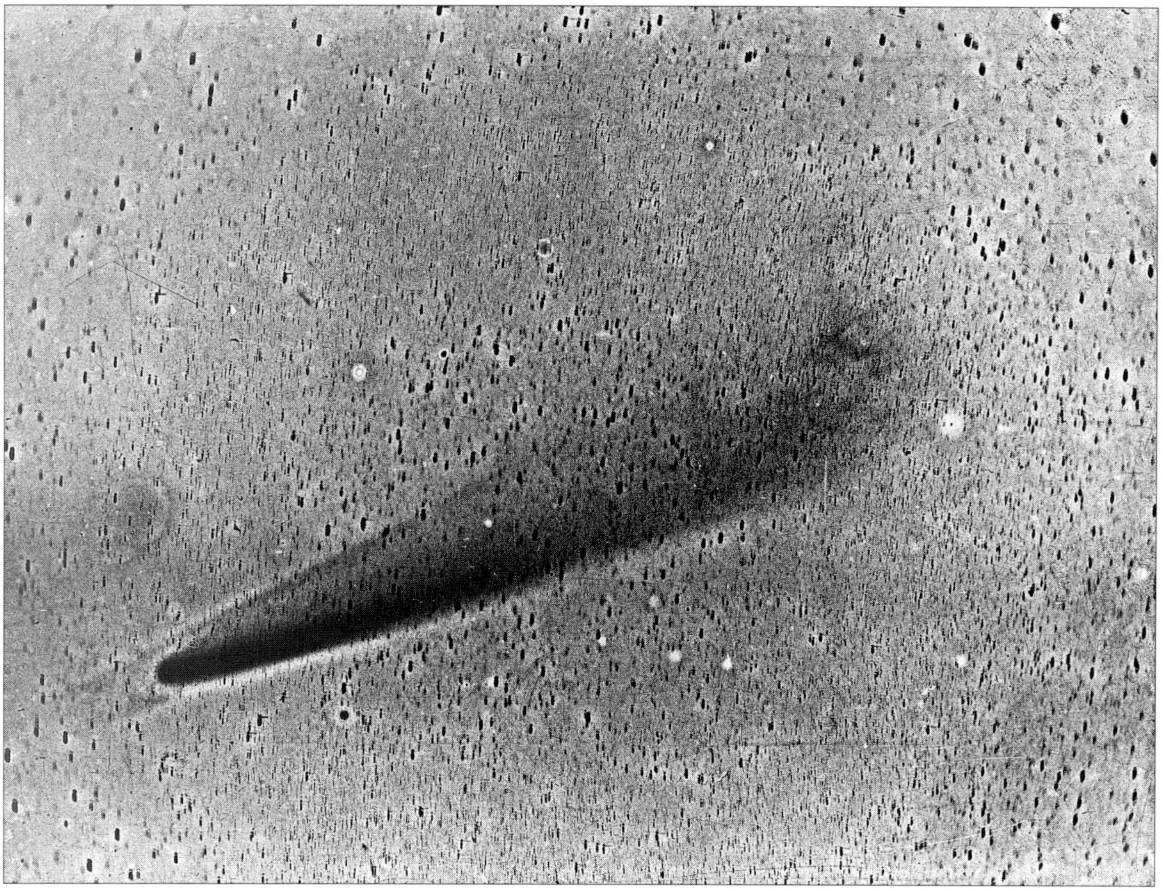

Figure 135. David Gill, *The Great September Comet of 1882 (1882 II), Enhancement (© ROE) by Photolabs of the Royal Observatory Edinburgh, of an original image on glass held by the Royal Astronomical Society*, Royal Observatory Edinburgh.

researches, which we now think more valuable.[49] The *Carte du Ciel* project may even have contributed to the minimization of comet research in England and the other countries participating in it.

After 1881, when photographs began to record the structure of comet tails, the technique grew increasingly valuable. Though the Great September Comet of 1882 provided important photographs, Swift's Comet of 1892 (C/1892 E1, formerly 1892 I) was a more interesting astronomical object.[50] Though not prominent to the naked eye, its tail underwent rapid transformations that were recorded on film. Sometimes it had a dozen narrow streamers and at others only two broad ones. These changes, and those of Brooks's Comet of 1893 (C/1893 U1, formerly 1893 VI), led to a realization of the interaction between the head of the comet and the changes seen in the tail. Photographs of the latter showed "great masses of cometary matter drifting away into space, probably to become meteor swarms."[51] Even views through telescopes did not show on this occasion what photography revealed very clearly.

Today, most comets are discovered photographically, but only a few were so spotted in the nineteenth century. In 1882 an exciting discovery was made: a comet near the sun during the total eclipse. It was visible to the naked eye when the sky darkened and was recorded on the dry plates that had just come into use. The comet was seen only on that single occasion.[52] In addition to this eclipse comet, others were discovered photographically at the Lick Observatory in California on 12 October 1892 (D/1892 T1, formerly 1892 V, Comet Barnard 3) and in July 1898 (C/1898 L1, formerly 1898 VII, Comet Coddington-Pauly), and at the Yale University Observatory in 1898 (C/1898 V1, formerly 1898 VIII, Comet Chase).[53] Another comet was observed photographically in Paris in 1893.[54] There have also been some false "comets" detected in photographs that later proved to be other objects.[55]

Earlier Meteor Photography

Photography also paid off for the studies of meteors. The first known was of an Andromedid meteor photographed at Prague in 1885. In 1887, a meteor trail showed up accidentally on a photographic plate exposed on a Harvard expedition to Colorado Springs.[56] Meteors were intentionally photographed after 1895, as cameras were set up to take wide-field views of the sky.[57] Since meteor trails appear accidentally and frequently, astronomers soon began, with notable exceptions, to view meteor trails on their wide-field images as contaminants. Today, the ability of photography to detect meteors plagues wide-field astronomical photography.

In 1898 a world-wide organization of observers was put together to study the expected Leonid storm of 1889-99, given the prior spectacle in 1833 and the lesser but still spectacular 1866-67 storms. By that time, photography had come into its stride. Though valuable data were obtained in the pre-peak of 1898 (when Harvard astronomers recorded 32 on 14 November), the hoped-for high peak did not materialize. However, some brilliant showers were seen occasionally in the following years at various places.[58]

Astronomical Photography and Aesthetics

A recent study by Pang raises fascinating issues about the technical problems involved with practices in astronomical photography and aesthetics. In this article he reacts to the writer in the *Edinburgh Review*, who predicted in 1888 that photography would bring about "a new birth of knowledge regarding the structures of the universe" in which "stars should henceforth register themselves," a phrase which suggests photographic neutrality and, as the writer claimed, an absence of aesthetics.[59] Through a study of archival documents in the Lick Observatory from the 1880s to the 1920s, Pang reveals how astronomers talked to one another about these issues and how they looked at photographs. He suggests that the reliability of early astronomical photographs depended on the successful use of aesthetic standards as guides in their production and evaluation, and the employment of experienced craftsmen to correct errors of mechanical processes. His research also demonstrates how difficult it was for printers to reproduce astronomical images and how necessary was human intervention to compensate for mechanized printing processes. For example, he examines the technical problems of halftone printing and photoengraving that kept these prints from being copies of photographs, as well as the need for management and intervention from engravers. The imperative, he believes, was to use engravers' skills and aesthetics but to make them self-effacing.[60] He points out that even the actual printing could not be entrusted to machines alone, as in the case of Barnard's experience in printing photographs of Comet Morehouse (for the December 1908 *Astrophysical Journal*). While the proofs of Barnard's photographs were excellent, the prints themselves were disappointing due to the steam-press used.[61] Slowly standards were set for the improvement of images (removing dust or background contamination) versus an unacceptable alteration of the images. Even today, it is proved impossible to reproduce satisfactorily the detail in the finest sun-grazing comet images taken with the Solar and Heliospheric Observatory [Fig. 162].

The camera's power to gather light gave it the ability to resolve and find objects that were beyond the powers of the human eye. Eventually, astronomical photographs recorded a wide variety of objects, including the spiral "nebulae" discovered earlier by Lord

Rosse. Pang believes that the development of astrophotography represents the larger eclipse of human observations by mechanical technology and a general distrust of human judgment.[62] Nevertheless, photography has greatly enhanced astronomy's frontiers and abilities. Today's astrophysics and our ability to study fainter objects farther out than our solar system have developed out of nineteenth-century photographs of such solar-system objects as the sun, the moon, and comets.[63]

Notes

1 The earliest relevant work is that of Leonardo da Vinci in the late fifteenth century and his conception of the *camera obscura*. Jean-Baptiste Porta added a lens in the seventeenth century, bringing the *camera obscura* into the optical age. The permanent recording of images began with Georg Fabricius in Germany, who discovered in 1566 that silver chloride turned black when exposed to light. Heinrich Schulze in Germany made images in 1727 by placing a stencil over a plate treated with a chalk containing silver grains. N. Rosenblum, *A World History of Photography*, New York, 1989, 15-18, 35-37.

2 *Ibid.*, 24-31, 37.

3 D. Hoffleit, *Some Firsts in Astronomical Photography*, Cambridge, MA, 1950, 8, quoting R. Wolf, *Handbuch der Astronomie*, Zurich, 1890.

4 F. Arago, "Le Daguerréotype," *Compte Rendu des Séances de l'Académie des Sciences*, 9, 1839, 250-267; E.E. Barnard, "The Development of Photography in Astronomy," *Popular Astronomy*, 6:8, 1898, 425-455; D. Norman, "The Development of Astronomical Photography," *Osiris* 5, 1938, 560-594; Hoffleit, 1950, 9-10; G. de Vaucouleurs, *Astronomical Photography: From the Daguerreotype to the Electron Camera*, trans. R. Wright, London, 1961; J. Lankford, "The impact of photography on astronomy," in *General History of Astronomy*, 4: *Astrophysics and Twentieth-Century Astronomy to 1950*, ed. O. Gingerich, Cambridge and New York, 1984, 16-39.

5 Norman, 560, reports that a faint image without detail had been obtained.

6 Hoffleit, 1950, 10.

7 This tradeoff is known as "the law of reciprocity," with focal ratio and shutter speed compensating for each other. For the same focal length and same brightness of an object, a short exposure at a certain focal ratio gives the same image as a longer exposure for a shorter focal ratio. Actually, this relation holds for only certain ranges, and the long exposures often necessary for astronomical photography of faint objects brings astronomers to a region of so-called "reciprocity failure." The effect has disappeared for electronic imaging, such as the new charge-coupled devices (CCD's) that have now taken over astronomical photography, replacing film for almost all purposes.

8 J.K. Herman and M.S. Barger, "The Moon on a Silver Plate," *Astronomy Magazine*, 15:10, 1987, 98-103; M.S. Barger, "The Moon, 6 August 1851," *1989 Annual Report, The Christian A. Johnson Memorial Gallery*, Middlebury, VT, undated, unnumbered pp. 7-16.

9 D. Trombino and J. Pazmino inform us (private communication, 1996) that the daguerreotype of the moon referred to by Barger and described in D.F. Trombino, "Dr. John William Draper," *Journal of the British Astronomical Association*, 90, 1980, 565-571, matches the materials and markings found on a set of faded daguerreotypes labeled as taken by W.C. Draper in the New York University Archives. It also convinces them that the daguerreotype Pazmino purchased in a Greenwich Village bookstore in the 1970s is the lost one, the best of Draper's lunar daguerreotypes from 1840, presumably the one exhibited at the New York Lyceum and thought destroyed. Dennis di Cicco (private communication, 1996) holds that the matter is still in doubt, since the quality of the image does not match Draper's notes.

10 Barger, unnumbered p. 8.

11 The first photographs of the sun were the daguerreotypes and bromide paper images of the solar eclipse of 8 July 1842 by G.A. Majocchi of Milan (F. Arago, *Popular Astronomy*, 1, trans. W.H. Smyth and R. Grant, London, 1855, pl. XII). The first non-eclipse daguerreotype of the sun was taken at the Observatoire de Paris on 2 April 1845 by Léon Foucault (1819-1868) and Hippolyte Fizeau (1819-1896); it is in the Musée des Arts et Métiers, Paris. The Foucault/Fizeau photograph, which was printed by Arago, shows two large groups of sunspots. Nevertheless, because of the challenge involved, photography of the surface of the sun did not resume until 1857. Hoffleit, 1950, 22.

12 Barger, unnumbered p. 9. It is a composite image containing nine exposures on its several-centimeter-diameter field. Each is only a millimeter or so in diameter, as results from a portrait lens whose focal length is about four inches (10 cm). The nine exposure times noted on a paper attached to its mat range from ½ second to two minutes. Unlike Draper's setup, Humphrey's lens did not track the moon, so the longer exposure images are elongated and blurred.

13 *Ibid.*

14 Jones and Boyd, 71-81.

15 There were many problems sharing the telescope. The changeover of equipment to mount the daguerreotype materials was not easy or quick, and afterwards left the ordinary equipment slightly out of its proper calibration. A major technical problem was that the tracking rate of the telescope was not uniform enough for photography. Further, daguerreotypes had to be made at times that coincided with the intervals at which the nonuniform tracking rate corresponded with the rate of lunar motion across the sky. Moreover, the sensitivity of the daguerreotype to ultraviolet and blue light meant that it must be exposed at a different focus from the yellow and green light to which the eye is most sensitive.

16 Barger, unnumbered pp. 9-10.

17 Hoffleit, 1950, 25.

18 Barger, unnumbered pp. 11-12.

19 Hoffleit, 1950, 28; Barger, unnumbered p. 13.

20 Anonymous, "How Mr. Warren De La Rue Photographed the Moon," *The British Journal of Photography*, 25, 1868, 256-257, 270-271, 279-280. For additional information, see E.B.K. [E.B. Knobel], "Obituary: Warren De la Rue," *Monthly Notices of the Royal Astronomical Society*, 50, 1889-90, 155-164; H. Rothermel, "Images of the Sun: Warren De la Rue, George Biddell Airy and celestial photography," *The British Journal of the History of Science*, 26, 1993, 137-169.

21 W. De la Rue, "The Bakerian Lecture. – On the Total Solar Eclipse of July 18, 1860, observed at Rivabellosa, near Miranda de Ebro, in Spain," *Philosophical Transactions*, 52, 1862, 333-416.

22 E.B.K., 160. In 1860 De la Rue was the first British astronomer to photograph an eclipse. See also A. S.-K. Pang, "Victorian Observing Practices, Printing Technology, and Representations of the Solar Corona, (1): The 1860s and 1870s," *Journal for the History of Astronomy*, 25, 1994, 253; *idem*, "Victorian Observing Practices, Printing Technology, and Representations of the Solar Corona, (2): The Age of Photomechanical Reproductions," *Journal for the History of Astronomy*, 26, 1995, 63-75.

23 Photography was in a fledgling state with the Great Comet of 1843, but there is no record of any attempt to photograph it.

24 E.H. Grove-Hills, "The Decade 1850-1860," in *History of the Royal Astronomical Society 1820-1920*, eds. J.L.E. Dreyer and H.H. Turner, London, 1923 (reprint 1987), 113; Norman, 575. See also J.M. Pasachoff and R.J.M. Olson, "Early Comet Photography: Usherwood and Bond for Donati," *Abstract: Bulletin American Astronomical Society*, 27:4, 1995, 1331; J.M. Pasachoff, R.J.M. Olson, and M.L. Hazen, "The Earliest Comet Photographs: Usherwood, Bond, and Donati 1858," *Journal for the History of Astronomy*, 27, 1996, 129-145.

25 Anonymous, "Report of the Council," *Monthly Notices of the Royal Astronomical Society*, 19, 1859, 138-139. See also De Vaucouleurs, 41.

26 See J. M. Pasachoff, "Not so common," *New Scientist*, 147, 1993, 51; Pasachoff, Olson, and Hazen, 129-145. We thank Jeffrey Hill, who provided the photograph of Usherwood's establishment (*ibid.*, fig. 4) from J. Janaway, *Surrey–A Photographic Record*, Newbury, Berkshire, 1984. We are also grateful to Pamela Roberts, for trying to find references to Usherwood in photographic societies; Peter Hingley, for checking that Usherwood had not been a Fellow of the Royal Astronomical Society; and others, such as Brian Overell. Usherwood appears neither in the *International Guide to Nineteenth Century Photographers*, Boston, 1988, nor in *Nineteenth Century Photography: An Annotated Bibliography, 1839-1879*, Boston, 1990.

27 Anonymous, "Death of Mr. W. Usherwood," *Dorking and Leatherhead Advertiser*, November, 1916.

28 Harvard University Archives, UAV 630.6, Carrington folder.

29 *Ibid.*

30 No one at either the Harvard College Observatory or the Harvard University Archives knows of the print, which apparently was already lost by 1897. A staff member identified as E.S.H. wrote in "Notices from the Lick Observatory, Photographs of Donati's Comet in September 1858," *Publication of the Astronomical Society of the Pacific*, 9, 1897, 89: "Mr. Carrington sends another photograph of the comet to Bond, taken on September 27th by ??? [sic] with an exposure of 7 seconds, using a camera lens.... Carrington's enclosed photograph is not now to be found, I believe. The photograph was unknown to Dr. De la [sic] Rue, apparently (see *Monthly Notices*, R.A.S., Vol. XIX, p. 353). Both these photographs – the first ever made of comets – have remained unnoticed by all the historians of astronomical photography, up to this time, so far as I know."

31 Letter in Harvard University Archives, UAV 630.6, Carrington folder, 2-4, which includes the postscript omitted by E.S. Holden, *Memorials of William Cranch Bond and of his son George Phillips Bond*, San Francisco and New York, 1897, 167. Norman, 576, calculates that not only was Usherwood's lens faster but also he must have had a more sensitive collodion plate.

32 Harvard University Archives, UAV, 630.6, Usherwood folder. For the text of Bond's letter, see Pasachoff, Olson, and Hazen, 136. Usherwood's reply reads:

> I rec^{d} your letter of Mr Carrington & in reply I beg to state that the Plate sent to you by Mr C is a copy the same size with the original negative my camera is for plates nine inch square Lens three & a quarter inch. a Portrait Lens twelve inch focal length Time of exposure from seven to nine seconds The Development by Gallic acid and acetate of Lead. then the Collodion was prepared by my self. I think it was on the 27 of Sep^{t} last I did the negative

33 Bond, 1862, 210.

34 Book H.28, box C0004 in the Harvard Deposit Library. See Pasachoff, Olson, and Hazen, 130, 141, n. 6, fig. 1.

35 Anonymous, "Report of the Council," 139, "So far as the Council has been informed this is the only instance of a photograph of Donati's comet having been obtained."

36 The anonymous writer reports (*ibid.*, 138): "... [Mr. De la Rue] informs the Council that he made one attempt to obtain a photograph of Donati's comet, but that he did not obtain any trace of an image in 60 seconds...." See also De la Rue, 1862, 84; Anonymous, 1868, 256-257, 270-271, 279-281.

37 Barnard, 427.

38 De Vaucouleurs, 41.

39 De Vaucouleurs, pl. 8, fig. 1, reproduces Janssen's 30 June 1881 photograph of Comet 1881 III made with a dry plate. See also Barnard, 437-439; Norman, 587; Hoffleit, 1950, 38; De Vaucouleurs, 41-44; Yeomans, 205-206. See also H. Draper, "Photographs of Comet b 1881," *The Observatory*, 4, 1881, 239; D. Gill, "On Photographs of the Great comet (b) 1882," *Monthly Notices of the Royal Astronomical Society*, 43, 1882-83, 53-54, whose photographs are described and not reproduced.

40 Norman, 587, citing articles in *The Observatory* of 1881 (see n. 39). See also A.A. Common, "Photography as applied to Comet b 1881," *The Observatory*, 4, 1881, 232-233; *idem*, "Note on Photographs of the spectrum of Comet b 1881," *The Observatory*, 4, 1881, 252-253.

41 R.A.S. President E.J. Stone ("Address delivered by the President on presenting the Gold Medal of the Society to Mr. Common," *Monthly Notices of the Royal Astronomical Society*, 44, 1883-84, 221-222), noted: "Thus availing himself of every increase of the sensitiveness of the prepared photographic plates and continually improving the control of the driving-clock, Mr. Common was able, on 1881 June 24, to obtain a photograph of the Comet *b*, which is probably the earliest successful photograph of any comet (although a photograph of the comet was obtained on the same night by Dr. Draper)...."

42 Pasachoff, Olson, and Hazen, 144, n. 41. Huggins described his spectroscopic work in a Friday Evening Discourse at the Royal Institution in 1882. W. Huggins, "On Comets," in *Astronomy: The Royal Institution*, 1, ed. B. Lovell, New York, Barking, Essex, and Amsterdam, 1970, 191-202; Yeomans, 206-207.

43 R.L.J. Ellery, "Observations of the Great Comet (b), 1882 made at the Melbourne Observatory," *Monthly Notices of the Royal Astronomical Society*, 43, 1882-83, 29.

44 The original Gill plate is in the Archives of the Royal Astronomical Society, London. The important series of photographs was made by Gill and E.H. Allis, a local photographer. The one reproduced was made on the morning of 7 November using the 6-inch equatorial (Gill plate C4 no. 12, a 100-minute exposure). High-contrasting and unsharp-masking techniques used in this reproduction, made at the Photolabs of the Royal Observatory Edinburgh, have brought out the maximum yield of information from the original.

Gill, 1882-83, 53, lists three photographs:

No. 1. 1882 Oct. 19^{d} 15^{h} 45^{m}	Cape M.T.,	Exposure 30^{m}.
No. 2. 1882 Oct. 20^{d} 15^{h} 30^{m}	"	" 1^{h}.
No. 3. 1882 Oct. 21^{d} 15^{h} 40^{m}	"	" 40^{m}.

A following note lists additional photographs:

No. 4. 1882 Nov. 7^d 14^h 42^m Cape M.T., Exposure 1^h 40^m.

No. 5. 1882 Nov. 13^d 14^h 0^m " " 1^h 20^m.

No. 6. 1882 Nov. 14^d 14^h 15^m " " 2^h 20^m.

See G. Forbes, *David Gill: Man and Astronomer*, London, 1916, 135; Lankford, 1984, 25; D. Dewhirst and M. Hoskin, "The Memory of Starlight: The Rise of Astrophysics," in *Cambridge Illustrated History of Astronomy*, ed. M. Hoskin, New York and Cambridge, 1996, 256-243.

45 Barnard, 438.

46 Yeomans, 208.

47 D. Gill, "The Applications of Photography," *The Observatory*, 10, 1887, 269-270.

48 Barnard, 438-439. Norman, 588, notes that: "Gill's plates of the Comet of 1882... showed such faint stars that they added enormous impetus to the idea of a photographic chart of the sky."

49 Lankford, 1984, 29-32.

50 Barnard, 445.

51 *Ibid.*, 446. This idea has stood the test of time.

52 *Ibid.*, 430-431. Because the comet was observed only once, no orbit could be computed and it was not assigned a number.

53 See Kronk, 1984, 222, for Comet Barnard 3. For Comet Coddington-Pauly, Barnard, 452, says "July," while Kronk, 1984, 91-92, records "June 10." Brian Marsden (private communication) concludes that "Coddington-Pauly clearly discovered in June, not July." For Comet Chase, see Kronk, 1984, 92, as well as W.L. Elkin, "Observations of the 1898 Leonid meteors and discovery of a comet (Chase, Nov. 14) at the Yale Observatory," *Astronomical Journal*, 19, 1898, 151-152; W.L. Elkin, "Report of Dr. Elkin," *Report for the Year 1898-99 of the Observatory of Yale University*, New Haven, 1899, 4-5, discovered on plates taken 12 and 14 November 1898 to record the Leonid meteor shower.

54 The two oldest comet photographs that are classified and catalogued in the collection of l'Observatoire de Paris at Meudon, France, were taken by E.M. Antoniadi on 19 July 1893 (Comet Rordame-Quénisset – C/1893 N1, formerly 1893 II) and by Antoine (Antonin) Benoit on 25 October 1902 (Comet Perrine – C/1902 R1, formerly 1902 III). We thank Zadig Mouradian for checking.

55 See Barnard, 432ff, who refers to a drawing (by J.M. Schaeberle which would today be recognized as a coronal mass ejection rather than a comet) in "*Astronomy and Astro-Physics*, p. 307, Vol. XIII," and 435; E.W. Cliver, "Was the Eclipse Comet of 1893 a Disconnected Coronal Mass Ejection?," *Solar Physics*, 122, 1989, 319-333.

56 See L. Weinek, "Der grosse Sternschnuppenfall am 27 Nov. 1885," *Astronomische Beobachtungen an der K.K. Sternwarte zu Prag in den Jahren 1885, 1886 und 1887*, Prague, 1890, 48; W.J. Fischer and M. Olmsted, "Catalogue of the Harvard Observatory Meteor Photographs," *Annals Harvard College Observatory*, 87, 1931, 233-254; D. Hoffleit, "The Thousandth Fireball-Portrait, a Harvard Antique," *The Telescope*, 8, 1941, 6-11; Hoffleit, 1950, 38.

57 For example, the Science Museum, London, has a photo (Neg. no. 609/37) of an exploding Andromedid meteor, credited to Butler, 1895.

58 Bailey, 96-7; Yeomans, 200-201.

59 A.S.-K. Pang, "'Stars should henceforth register themselves': astrophotography at the early Lick Observatory," *The British Journal for the History of Science*, 30, 1997, 2.

60 *Ibid.*, 1-50.

61 *Ibid.*, 19.

62 *Ibid.*, 3.

63 Norman, 591.

VII

The Triumph of the Imagination

After the stupendous pyrotechnics of Donati's Comet in 1858, the second part of the Victorian age experienced sweeping changes in comet/meteor astronomy as well as in art. Astronomy underwent an increased specialization and explored the potentials promised by photography. Partially in reaction to the camera's ability to capture reality and as an integral part of complex cultural-social imperatives, including blatant historical stylisms, artists, beginning with the Pre-Raphaelites, turned to the territory of the imagination and the resonance of old symbols, among them comets and meteors.

While the previous chapter discussed how the technology of photography changed astronomy, it did not consider the impact photography had on the appearance and content of images in the visual arts. Nor did it grapple with all the important issues of how photography affected scientific and popular illustration, specifically the illustrations of comets. Scientific illustrators were often persons of scientific distinction and artistic skill and their ranks included professional artists, book illustrators, architects, military and naval officers, and engineers, a number of whom were also Fellows of the Royal Astronomical Society.[1] A case in point is Piazzi Smyth, who was conscious of many of these issues at an early stage. As noted, he addressed them in an essay on astronomical drawing in 1843, the same year he began experimenting with photography. Piazzi Smyth believed that the prime objective of astronomical drawing was accuracy, to record "a faithful imitation," for which he urged astronomers to develop a correctness of the eye, a facility of the hand, and an appreciation of the subject.[2]

Even though an especially large number of bright comets appeared in the latter part of the nineteenth century, which were observed by important individuals,[3] the attention of many of the best British astronomers was turned to more distant celestial objects, with Sir John Herschel and Lord Rosse mapping nebulae and clusters in the Milky Way and beyond. Nevertheless, British observations of comets continued on a regular basis in the *Monthly Notices of the Royal Astronomical Society*.

Chapter VII will trace, again in a roughly chronological manner, the developments in art and astronomy from the time following Donati's Comet to the early years of the twentieth century. The discussion will highlight significant individuals in both fields and the connections between them as well as the issues that concerned them.

Tebbutt, His Comets, and the State of British Astronomy

John Tebbutt (1834-1916), the distinguished amateur astronomer, helped to establish Australia as a significant astronomical site for the British Empire in the southern hemisphere (along with South Africa). His fame, which rests largely on his comet work, grew to the point where he was honored by having a crater on the moon named after him, and his portrait was on the $100 Australian currency note issued in 1984. Tebbutt had the talent, and the luck, to discover two of the greatest comets of all time. The first appeared in May 1861.[4] C/1861 J1, formerly 1861 II, a bright naked-eye object with a long tail was so bright between 29 June and 1 July 1861 that it cast shadows at night and was thoroughly covered as a newsworthy object in *The Illustrated London News*.[5] Tebbutt's Comet arrested the attention of De la Rue, who tried, without success, to photograph it (as he had Donati's) but recorded it only in drawings made with the aid of a telescope.[6] Tebbutt had correctly predicted that the earth would pass through its tail on 30 June when the comet's appearance was accompanied by a spectacular meteor show in Europe, a day after it became visible in England. Soon, the Great Comet of 1861 was so bright that it could be seen in the daytime.[7] Then, in 1881, Tebbutt discovered the Great Comet C/1881 K1, formerly 1881 III. It was bright enough to be observed with a spectroscope and became the first for which a spectrum was obtained for an entire comet (previously 14 comets had been observed spectroscopically), as well as the first comet whose tail was successfully photographed, aside from the lost photograph by Usherwood mentioned in chapter VI. Tebbutt even founded a society of comet searchers (the "Corps of Amateur Comet Seekers"), and he himself recorded 700 comet positions between 1880 and 1900, over twice as many as were made by Australia's state observatories.[8]

Some Pre-Raphaelite Comets and Meteors

During the period of Tebbutt's activity, artists associated with the ever-expanding circle of the Pre-Raphaelite Brotherhood sometimes depicted comets and meteors. (The PRB was founded in 1848 by seven artists who admired the simplicity of Italian Renaissance painting before Raphael and wanted to reform British art with those models in mind.) The group's literary bias and interest in symbolic content as well as their passion for observed accuracy in landscapes makes it seem odd that so few of these images are known, although several Pre-Raphaelite artists were enthusiastic about depicting generic stars.

Before turning to the Pre-Raphaelite artists who represented comets and meteors, let us mention one of the painters associated with the group, Henry Holiday (1839-1927), who was also an amateur astronomer.[9] He was clearly interested in both artistic and scientific matters. Together with Alexander Herschel, son of the astronomer (John), he was a founding member of the Philomanthic Society in the early 1850s, whose object was to read papers on chiefly scientific as well as on artistic and literary subjects.[10]

Holiday was famous for his astronomical draftsmanship and his sketch of himself drawing the solar corona during an eclipse through a telescope on 12 December 1871, in his autobiography, preserves his method.[11] Holiday observes the corona with his left eye while drawing with his right eye, his hand on a vertical drafting board. At the time, he was on the Royal Astronomical Society's solar eclipse expedition, observing from the roof of his bungalow in Poodoocottah, Ceylon, as a guest of the Maharajah. He had been invited on the expedition by Sir Norman Lockyer, who knew his drawings of lunar craters and sunspots as well as his unique

Figure 136. Dante Gabriel Rossetti, *Paolo and Francesca*, 1855, watercolor, 248 × 445 mm, The Tate Gallery, London.

drafting technique. Holiday's opinion is even quoted in a report of this expedition.[12]

Holiday, who had many acquaintances in the scientific and astronomical community – such as William Spottiswoode, President of the Royal Society, and Sir John and Lady Herschel – was also a friend and associate of many of the Pre-Raphaelite artists discussed below.[13] His passion for Urania grew steadily and in 1865 he began drawing the moon with a telescope.[14] Unfortunately, we do not know whether Holiday observed any of the spectacular comets of his time, but it is highly likely that he did, and perhaps drawings by him of these comets will eventually surface.

One of the founding members of the Pre-Raphaelite Brotherhood, Dante Gabriel Rossetti (1828-1882), poet and painter, used much stellar and cosmological symbolism in his literary-oriented art.[15] In this same vein and dependent on trecento literature are three watercolors concerning the star-crossed lovers Francesca da Rimini and her brother-in-law, Paolo Malatesta.[16] Dante Alighieri narrates their tale in Canto V of his *Inferno* (*Divina Commedia*), where the adulterous lovers, who were both murdered by her husband Giancotto Malatesta, are condemned to eternal damnation.

Figure 137. Anonymous, *Meteor Shower of 1866 over Greenwich*, 1884, chromolithograph, 90 × 135 mm, Collection of Gerald H. Morris, London.

Rossetti heightens the drama of their illicit love via symbolic forms that lend a layered meaning to the work. For example, the cosmic nature of the theme and the resulting fate of the two in the circles of Hell is foreshadowed by the round window behind the embracing lovers.

Rossetti's earliest version from 1855 is tripartite in design [Fig. 136].[17] It depicts from left to right: the seated lovers kissing, Dante and his guide Virgil despairing of the lovers' fate, and Paolo and Francesca intertwined in an eternal embrace to symbolize their souls writhing together on the wind in the eternal darkness of the second circle of Hell, where the sins of the flesh are punished. Thus, the left and right sections contrast earthly and spiritual love. The flames in the background on the right are not in Dante's text, where Hell is described as a void having black air and a hurricane-like wind. Moreover, the background through which the lovers twist at an unstable diagonal in endless pathos is depicted by other artists as dark. Nevertheless, Surtees interprets Rossetti's

Figure 138. Sir Edward Coley Burne-Jones, *Sir Galahad*, 1858, pen and black ink on vellum, 156 × 192 mm, Fogg Art Museum, Cambridge, MA.

flames as the sulfurous ones of Hell.[18] Instead, we believe that they represent a sophisticated image possibly conflating flames with a meteor shower (symbolic of doom). In fact, Rossetti's flames resemble recorded history's most prodigious meteor shower, the Leonids, which reaches its peak in November every 33 years [see Fig. 137].[19] Those of 13 November 1833, which Rossetti could have seen as a child, featured hundreds of shooting stars simultaneously pouring from the heavens for several hours. While flames represent blessed souls in Dante's *Paradiso*, in Rossetti's depictions of Paolo and Francesca they must allude to the inverse, symbolizing suffering individuals, who, like Paolo and Francesca, are plummeting to their destruction like meteor showers. Ironically, meteors have been associated with goddess of love Aphrodite/ Venus (n. 98), adding to the layered, learned symbolism of Rossetti's work. This complex and wonderful conflation would only have been possible for an artist like Rossetti, who had a thorough knowledge of Dante.[20]

Rossetti's colleague Edward Coley Burne-Jones (1833-1898), a later member of the Pre-Raphaelite group, drew another "medieval" scene with a comet [Fig. 138].[21] He placed it in the central right section of the intricately delineated sheet, where it decorates the chalice held by Sir Galahad. In order to understand this small but significant comet image, we must first discuss its context. The drawing, which dates from 1858, was made at Little Holland House in Kensington, residence of Mrs. Thoby Prinsep, hostess of an intellectual, bohemian salon, who, together with her son Valentine, invited Burne-Jones to convalesce there from a persistent respiratory infection. Included among the frequenters of Mrs. Prinsep's salon were the artist George Frederick Watts, who was a permanent house guest, and the writers Robert Browning and Alfred Lord Tennyson, who in July 1858 was in residence there.[22] During that month, Burne-Jones and Tennyson enjoyed an enthusiastic friendship, and their favorite topic of conversation was the Arthurian legend. Therefore, it is not surprising that the artist's drawing under consideration derives from Tennyson's poem *Sir Galahad* (written 1834, published 1842).[23]

Burne-Jones's drawing of Sir Galahad with the comet chalice is eclectic in both its literary and visual sources.[24] Waters and Harrison believe that while the drawing contains elements influenced by various sources, the knight's self-absorption derives from Tennyson, where he is portrayed as virginal.[25] And Mancoff allies the drawing with Tennyson's description of Galahad as a contemplative youth, charged with innocent conviction. The temptations he disdains are depicted in the background via the earthly pleasures of amorous flirting and making music (love).[26] However, Burne-Jones's perfect knight, while possibly informed by more than one literary source, is the artist's highly individual conception. In his early college days, the artist had already revealed his own deep sentiment about Galahad in a letter (dated May 1853): "Remember, I have set my heart on our founding a

Brotherhood. Learn Sir Galahad by heart. He is to be the patron of our Order."[27]

The comet on the chalice held by Galahad is also Burne-Jones's original invention. According to legend, Galahad was one of the pure knights able to fulfill the quest for the Holy Grail, the cup of Christ's Last Supper used by Joseph of Arimathea to collect Christ's blood at the Deposition.[28] Galahad was, in fact, the hero of the quest for the Grail, physically beautiful, gifted in his prowess, pure, and morally virtuous.[29] Burne-Jones decorated the Grail with at least one star emitting three rays on the side facing the viewer (there may be another on the opposite side from which five rays seem to emanate). The comet-star signals the celestial and divine nature of the mystical Grail, which held the blood of the "King of Kings," whose birth was foretold by a comet in both medieval and Renaissance literary and visual traditions.[30] Although Burne-Jones drew other stars decorating the knight's saddle, the conjunction of a comet with the Grail itself creates a highly pointed image.[31] Because this comet-star is not present in the literary sources associated with the drawing, it must have originated in the artist's own knowledge about comets' associations with kings as well as the Christmas/Magi comet-star tradition. Further, this significant, albeit minor, motif could have been encouraged by contact with Tennyson, who used cometary symbolism explicitly in some of his writings. For example, in his later drama of doom, *Harold* (1876), wherein Tennyson echoes old stereotypical notions about comets:

> Lo! there once more – this is the seventh night
> You grimly glaring, treble-brandished scourge of England...
> It glares in heaven, it flares upon the Thames....
> Three rods of blood-red fire up yonder mean
> The doom of England and the wrath of Heaven.[32]

Most importantly, we should not ignore the fact that in 1858-59, the very year of the Galahad drawing, no one could have missed the appearance of Donati's Comet. Its apparition probably functioned as the catalyst for the inclusion of this diminutive comet.[33]

We know, in fact, that the artist was keenly interested in astronomy. In the *Memorials*, his wife, Georgina Burne-Jones, remarks: "Humboldt's Cosmos and Views of Nature he read, and followed with imperative interest the vast schemes of which they deal.

Astronomy had a great fascination for him – almost a terrible one."[34] Further, in 1896, Burne-Jones commented, according to his wife:

> "So I've had to take to my astronomy again." What he read concerning this science was distilled to his friends in less technical language than that of the books in which he found it. For instance: "... Do you remember what fusses there used to be about a collision with a comet? How angry astronomers used to be with each other: they took sides about it as though it were politics.... They have a fine time of it; they can hand on their processes so that the science goes steadily forward for hundreds of years as though it were being done by one man, but we poor artists are not nearly so lucky, we can hand on next to nothing." Rashly enough, he kept astronomical books by his bedside, and often turned to them when unable to sleep.[35]

Despite Burne-Jones's interest in astronomy, the artist – an illustrator as well as a painter of idealistic, languorous, and decorative works, which are occasionally peppered with stars – placed the only comet(s) that we have identified in his oeuvre in this drawing.

Frederic James Shields (1833-1911), another artist/illustrator associated with the Pre-Raphaelite circle, employed comets in two of his preparatory watercolor designs for editions of Daniel Defoe's *History of the Plague of London* [Fig. 139].[36] Defoe's book, originally published in 1722, is an historical work written as an eyewitness account of the Great Plague of London in 1664-65. It includes descriptions of the menacing comets of 1664 (C/1664 W1) and 1665 (C/1665 F1). The scene with the comet is one of the most climactic. It represents Solomon Eagle, a religious fanatic who saw the plague as a sign of divine judgment, denouncing the impenitent as the flaming comet of 1665 hangs overhead.[37] Shield's depiction of this subject is considered one of the finest of his designs for Defoe's work.[38]

Defoe mentioned the comet depicted by Shields at the beginning of his book to set the proper apocalyptic tone: "In the first Place, a blazing Star or Comet appear'd for several Months before the Plague." Then he graphically relates an account of that plague which hit London in 1665, the last of a series of virulent outbreaks which England suffered since the infamous Black Death had swept Europe in the mid-fourteenth century. At the time, semi-naked

Figure 139. Frederic Shields, *Study for "Solomon Eagle Denouncing the Impenitent,"* 1863, graphite and grisaille watercolor, 708 × 480 mm, Manchester City Art Galleries.

and bearing a pan of burning charcoal on his head, Solomon Eagle went about the streets calling people to repent. In both Defoe's and Shield's works, the comet functions as a portent to the people and a sign of God's wrath to Solomon Eagle, an anachronistic attitude in both the eighteenth and nineteenth centuries, but a prevalent one in 1665. Because the artist has positioned the comet intersecting the church's cross, he has reinforced the idea that it signals God's judgment. Shields's adaptation of the comet from Defoe's text dovetails nicely with the ideals of the Pre-Raphaelites who, in their evocative works, aimed for authenticity by returning to the styles and motifs of the past.

More than one early edition of Defoe's classic included Shields's illustrations, one of which may even predate the edition cited by Mills (n. 38). This undated edition in the New York Public Library has an opening section entitled "The Comet."[39] It features the illustration of Solomon Eagle juxtaposed with the following revelatory passage:

> In the first place, a blazing star or comet appeared for several months before the Plague, as there did the year after, a little before the great Fire...that these two comets passed directly over the city...that it was plain they imported something peculiar to the city alone; that the comet before the pestilence was of a faint, dull, languid color, and its motion very heavy, solemn, and slow; but that the comet before the Fire was bright and sparkling, or, as others said flaming, and its motion swift and furious; and that, accordingly, one foretold a heavy judgment, slow but severe, terrible, and frightful, as the Plague was; but the other foretold a stroke, sudden, swift, and fiery, like the conflagration.... I saw both these stars, and I must confess, had so much of the common notion of such things in my head, that I was apt to look upon them as forerunners and warnings of God's judgments; and especially, when after the Plague had followed the first, I yet saw another of the like kind, I could not but say, God had not yet sufficiently scourged the city. But I could not at the same time carry these things to the height that others do, knowing too that natural causes are assigned by the astronomers for such things; and that their motions, and even their evolutions, are calculated, or pretended to be calculated: so that they cannot be so perfectly called the forerunners or foretellers, much less the procurers, of...pestilence, war, fire....these things had a more than ordinary influence upon the minds of the common people, and they had, almost universally, melancholy apprehensions of some dreadful calamity and judgment coming upon the city; and this principally from the sight of this comet.... books frightened them terribly; such as *Lilly's Almanack*, *Gadbury's Allogical Predictions*; some were so enthusiastically bold, as to run about the streets with their oral predictions, pretending they were sent to preach to the city; and one in particular, who, like Jonah to Ninevah, cried in the streets: "Yet forty days, and London shall be destroyed." Another

> ran about naked, except a pair of drawers about his waist, crying day and night. . . .[40]

In this undated edition the comet of the illustration has its head above its tail as in the smaller sheet in Manchester, opposite from that in figure 139, suggesting that it was derived from the smaller drawing.[41] The New York Public Library owns another edition dated 1863[42] in which this same Solomon Eagle illustration is placed later in the text, where Defoe narrates the specific episode of Solomon Eagle.[43] In both printed illustrations, Solomon Eagle is facing left with his hands extended in that direction, reversed from figure 139.[44]

Another artist associated with the Pre-Raphaelites, James Smetham (1821-1889), depicted a comet or a shooting star in the small painting *Sir Bedivere Flinging Excalibur into the Lake* (1871) [Fig. 140 color].[45] Smetham, like Burne-Jones and other Pre-Raphaelites, was interested in the Arthurian legends. Already in a letter of January 1864, Smetham wrote of his great admiration of Tennyson.[46] Moreover, Smetham, a poet as well as a painter, was intensely religious and subject to prolonged periods of depression. He believed that his age had gone awry and hence became involved with subjects of knights and chivalrous conduct.[47] Smetham continued to create works in this vein, like *Sir Bedivere Flinging Excalibur*, where Bedivere is the sole companion to Arthur at his death.[48] It is only in this small painting, however, which includes a comet-like form, that Smetham's invented image rises to the level of the Arthurian legend that inspired it. This work illustrates the climactic moment when loyal Sir Bedivere, after twice failing to obey the dying Arthur's wishes to throw Excalibur into the lake, finally acts. Casteras believes that the artist identified strongly with Bedivere and derived the idea for the painting from Acton Bell's poem "Sir Bedivere with the load o Prince Arthur on his Shoulder."[49] However, while Casteras notes the probable comet in the sky, she misses its highly significant symbolism (outlined previously by Olson): comets signaled both the birth and death of kings, rulers, and important people.[50] In her interpretation, Casteras also ignores the snow-capped mountains, which since prehistoric times have been identified with the domain of kings. Knowing that Smetham made one of the most effective artistic reevaluations of Blake, who, as we have seen, employed comets and meteors prodigiously, it comes as no surprise that he used a symbolic comet metaphor in this context.

Smetham's source for *Sir Bedivere Flinging Excalibur* was not Bell's poem, but rather Tennyson's *The Passing of Arthur* (dated variously and eventually included in the *Idylls of the King*, published in 1869 just before Smetham's painting):

> Then quickly rose Sir Bedivere, and ran,
> And, leaping down the ridge lightly, plunged
> Among the bulrush beds, and clutch'd the sword,
> And strongly wheel'd and threw it. The great brand
> Made lightnings in the splendour of the moon,
> And flashing round and round, and whirl'd in an arch,
> Shot like a streamer of the northern morn [which Tennyson
> could have meant as a reference to an aurora borealis,
> a meteor, or a comet]...
> So flash'd and fell the brand Excalibur
> But ere he dipt the surface rose an arm...
> And caught him by the hilt and brandished him....

Smetham seems to have converted the poet's rather vague simile evoking a comet, an aurora borealis, or meteor (which could be used interchangeably in the eighteenth century) into a comet. His depiction, however, is not detailed enough to ascertain definitively the celestial body's identity from its physical characteristics. Because of his painting's subject, however, Smetham most likely intended it to be a comet, indicating that he was no stranger to the historical/literary tradition wherein comets signal the death of kings. While his comet image certainly depended on traditional iconographic associations, it may also have been at least partly engendered by the rash of comet apparitions in the nineteenth century. Moreover, the sweeping arc of the comet against the starry backdrop echoes the trajectory of Excalibur, lending a dynamic unity to the painting's extremely horizontal format.[51]

The awareness of comets in poetic and artistic circles can be further demonstrated in a poem from 1864 by Gerard Manley Hopkins (1844-1889).

> —I am like a slip of comet,
> Scarce worth discovery, in some corner seen
> Bridging the slender difference of two stars,
> Come out of space, or suddenly engender'd

By heady elements, for no man knows:
But when she sights the sun she grows and sizes
And spins her skirts out, while her central star
Shakes its cocooning mists; and so she comes
To fields of light, millions of travelling rays
Pierce her; she hangs upon the flame-cased sun,
And sucks the light as full as Gideon's fleece:
But then her father calls her; she falls off,
And as she dwindles sheds her smock of gold
Amidst the sistering planets, till she comes
To simple Saturn, last and solitary;
And then goes out into the cavernous dark.
So I go out: my little sweet is done:
I have drawn heat from this contagious sun:
To not ungentle death now forth I run.

David Levy, an avid comet-hunter himself with 21 comet discoveries to his credit (8 visually plus 13 photographically as part of a joint project with Carolyn and Eugene Shoemaker), has allied this poem to the naked-eye Tempel's Comet (C/1864 N1, formerly 1864 II), which appeared the year of Hopkins's poem.[52]

Another Pre-Raphaelite associate, George Frederick Watts (1817-1904), may have represented a fireball in *The Denunciation of Cain* (1868-72) [Fig. 141].[53] The canvas features a broad central shaft of glowing light that despite its resemblance to a fireball is not a clear-cut candidate. The form may, in fact, only represent an intense light symbolic of God's intervention, a physical embodiment of the branding of Cain after he had slain Abel. Less convincingly, the fire has been interpreted as the pillar of flame from Abel's sacrifice that the Lord found more acceptable than Cain's, thus provoking his fratricide. However, no such flame is described in the fourth chapter of Genesis. This kind of loosely evocative, unstructured symbolic association is characteristic of Watts's oeuvre. *The Denunciation of Cain* testifies to his exposure to Michelangelo's fresco *The Conversion of St. Paul*, in the Paoline Chapel in the Vatican with its divine shaft of light. However, the differences between the two works suggest that Watts's more concentrated shaft, blazing through the darkness, may have been in part, at least, influenced by some celestial phenomenon. And, it is not inconceivable that he envisioned the branding like the

Figure 141. George Frederick Watts, *The Denunciation of Cain*, c. 1872, oil on canvas, 147 × 66 cm, Watts Gallery, Compton, Guildford.

rapid descent of a fireball, which would have harmonzied with his ambitious plan for his *magnum opus*, *The House of Life*, a frescoed ceiling with a cosmological/astronomical focus.[54]

Coggia's Comet and Its Depictions

Not all comet illustrations in late nineteenth-century Britain involved imaginative, literary sources. Since the photography of comets was not viable until the 1880s, astronomers recorded their observations for themselves and for publication in various media, just as amateur astronomers continue to draw comets to this day. This was the case when Jerome Eugene Coggia on 17 April 1874 sighted from Marseilles one of the brightest nineteenth-century comets, which hence was called "Coggia's Comet" (C/1874 H1, formerly 1874 III). At various times it exhibited multiple structures in its head. A bright jet – sometimes oscillating and fan-shaped and other times spiral but always dynamic – spurted from its head, and a black line was observed bisecting its tail. By the middle of June it was naked-eye and moved quickly to the south to become invisible from European observatories beginning 17 July. It was so bright that its spectrum could be examined with a spectroscope and carbon compounds identified. Thus, scientists obtained from it some of their first important clues about the chemical composition of comets and carbon compounds – which are so important for the existence of life – in space. Astronomical drawings attest to the fact that the comet's head was surrounded by many envelopes. Its changes were described by a multitude of observers, including John Russell Hind.[55] As noted in chapter IV, Piazzi Smyth observed Coggia's Comet and drew it [Figs. 115, 116 both color]. Following this famous comet, Coggia discovered yet another (C/1874 H2, formerly 1874 IV) that same year, a year in which altogether six comets were spotted.

Three previously unpublished black and white studies of Coggia's Comet by George H. With (1827-1904) detail its head and changing configurations [Fig. 142].[56] Each, made with the aid of telescopic magnification, is carefully annotated in its lower margin as to the date, the diameter of the telescope's mirror and magnification used, and the time. The *versi* have inscriptions on how to best view the representations. The first is dated 5 July,

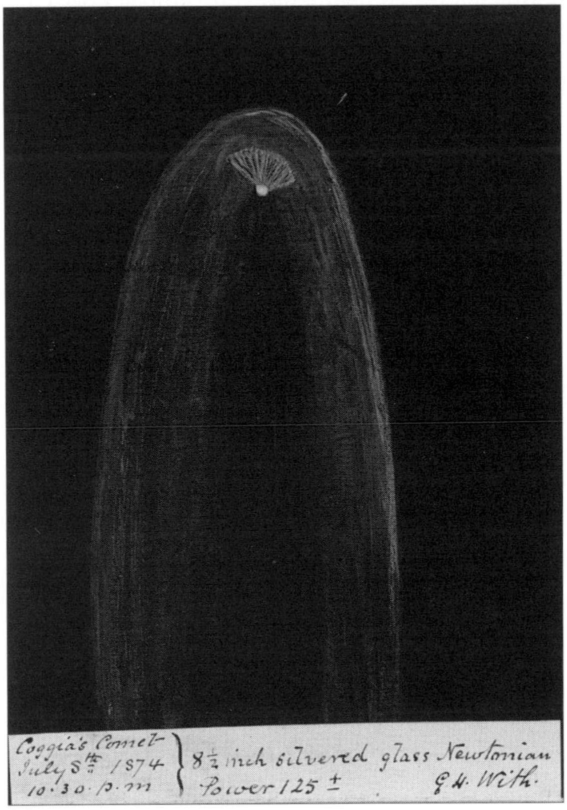

when the eyepiece used was 125 power.[57] The second, brighter image, from 11 July, shows a bright condensation curving out from the nucleus.[58] The third image, from 14 July with a higher magnification, has material flowing out of its nucleus that resembles the fan-shaped form in the first of the three only brighter.[59] As the studies of Halley's Comet in 1986 have demonstrated, jets of gas and dust are expelled from fissures in a comet nucleus's black crust when they are on the side illuminated by the sun. This outflow could lead to the shapes seen in these studies.[60]

With, master of the Blue Coat School in Hereford, was a great friend of the Reverend Thomas William Webb, another influential amateur astronomer who will be discussed shortly. In addition to his pedagogical duties, With made mirrors and telescopes. (They were excellent because Webb used one for observations that formed the basis of his famous *Celestial Objects for Common Telescopes* of 1859, which set the style for amateur astronomy and is still in print.) With's careful renderings of Coggia's Comet seem crude

Figure 142. G.H. With, *Three Drawings of Coggia's Comet of 1874 (Ms. Add. 161/1-3)*, white wash on paper prepared with black ink, each c. 162 × 114 mm, Royal Astronomical Society Library, London.

next to the delicately shaded plates of Bond, which they resemble; the difference may be partly due to the media.[61] Importantly, they reveal that With had kept current with the developments of astronomical illustration and also dabbled in the visual arts.

The Comets of the 1880s

In the 1880s several remarkable comets appeared (see Appendix II). Among them was C/1880 C1, formerly 1880 I, called "Large Southern Comet" because it was first spotted and observed in Argentina. Its 40° tail was reported to have extended from below the horizon.[62] After perihelion in mid-June, it was visible in the northern hemisphere. In July it became naked-eye, and brilliant jets issued from its nucleus. It had at least two tails and was so bright that its spectrum could be studied and, for the first time, photographed by Janssen and Draper with excellent results revealing its structure (chapter VI).[63]

Although in 1881, eight comets are recorded, it was the surprise appearance the following year of the "Great September Comet" (C/1882 R1, formerly 1882 II) that marks a watershed in cometology. Bright as Venus, this comet was spotted in the southern sky by sailors on board an Italian ship on the first day of the month, and was independently discovered by several astronomers. Beginning on 16 September, Tebbutt in Australia saw the naked-eye comet in the daytime with a 20° tail. He observed an oval head with slight condensation and followed it with the naked eye until 13 February. Observers at the Cape saw it disappear instantaneously when it went in front of the solar disk (transited). This exceptional celestial phenomenon remained visible at night to the naked eye south of the equator until the beginning of March 1883. Its spectrum was studied in a novel fashion, and the sodium discovered in it contributed an additional element to the knowledge of comets. A member of the Kreutz sungrazers like the comet of 1843, it was most spectacular in the southern hemisphere, so that while British astronomers studied it no British artists depicted its profile.[64]

As discussed in chapter VI, the Great September Comet of 1882 was photographed by Gill, the British astronomer at the Royal Observatory at the Cape of Good Hope who is generally credited with demonstrating the usefulness of astronomical photography [Fig. 135].[65] Despite Gill's seminal photographs, many people correctly thought that drawings showed more detail. Indeed, only in the late 1980s did electronic imaging allow sufficiently short exposures to surpass visual observations for planetary and cometary detail. Tellingly, Gill's article on his observations, which yields interesting material on the evolution of this comet, is illustrated not only with six photographs taken between 12 October and 14 November but also with four plates (including on plate I fifteen studies of alterations in the nucleus). These are based on the drawings of William H. Finlay, Chief Assistant at the Cape Observatory, and William L. Elkin. The illustrations demonstrate that while photographic images convey a great deal of immediacy for general views, astronomers still relied on detailed drawings made through a telescope.[66] Nevertheless, by the end of the 1890s scientific photographs of comets became quite common. While it was the English and American amateurs who had initially led the field in astronomical photography, most professionals recognized the medium's potential when coupled with the use of a telescope, once telescope drives and film sensitivities were adequate.

Thomas Webb and Celestial Objects

Among the most dedicated of the large number of astronomers – professionals and amateurs alike – who observed the comet of 1882 was With's friend the Reverend Thomas Webb (1807-1885), who lived at Hardwick, Herefordshire. Webb began observing the moon precociously when he was twelve years old. As was common, he made his own mirrors for telescopes although he also used those of With. He was the author of several books, most notably *Celestial Objects for the Common Telescope* (which went through four editions in his lifetime starting in 1859 and has been reprinted in this century without being modernized). Perhaps no other book so completely and authentically describes the realm of the amateur visual observer, and his influence is demonstrated today by the active Webb Society of enthusiasts honoring his style of observing. When Webb wrote *Celestial Objects*, the gap between amateur astronomers and professionals was much narrower than today, and visual methods reigned supreme in a more simplified and less specialized age. While much of the book is now outdated (it is basically concerned with nineteenth-century descriptive astronomy), it remains a valuable and readable characterization of the era.[67] Moreover, observing columns in astronomical periodicals of today are direct descendants of Webb's book.

Webb observed and drew a number of important comets. In fact, his forte was his descriptions of astronomical objects, including comets. His obituary in the *Monthly Notices* noted his tendency to make his drawings of astronomical phenomena small, so as not to be tempted to fill in unrealistic detail, except for comet diagrams in which he used a larger scale because of the "delicate differences of brightness."[68] Webb's observations of the very bright Tebbutt's Comet in 1861 (C/1861 J1, formerly 1861 II), published in volume 22 of the *Monthly Notices*, were especially singled out for praise. In this paper, Webb carefully describes the day-to-day changes in the comet and provides an outline sketch of its envelopes on 10 June. Each of these arcs is scientifically labelled,

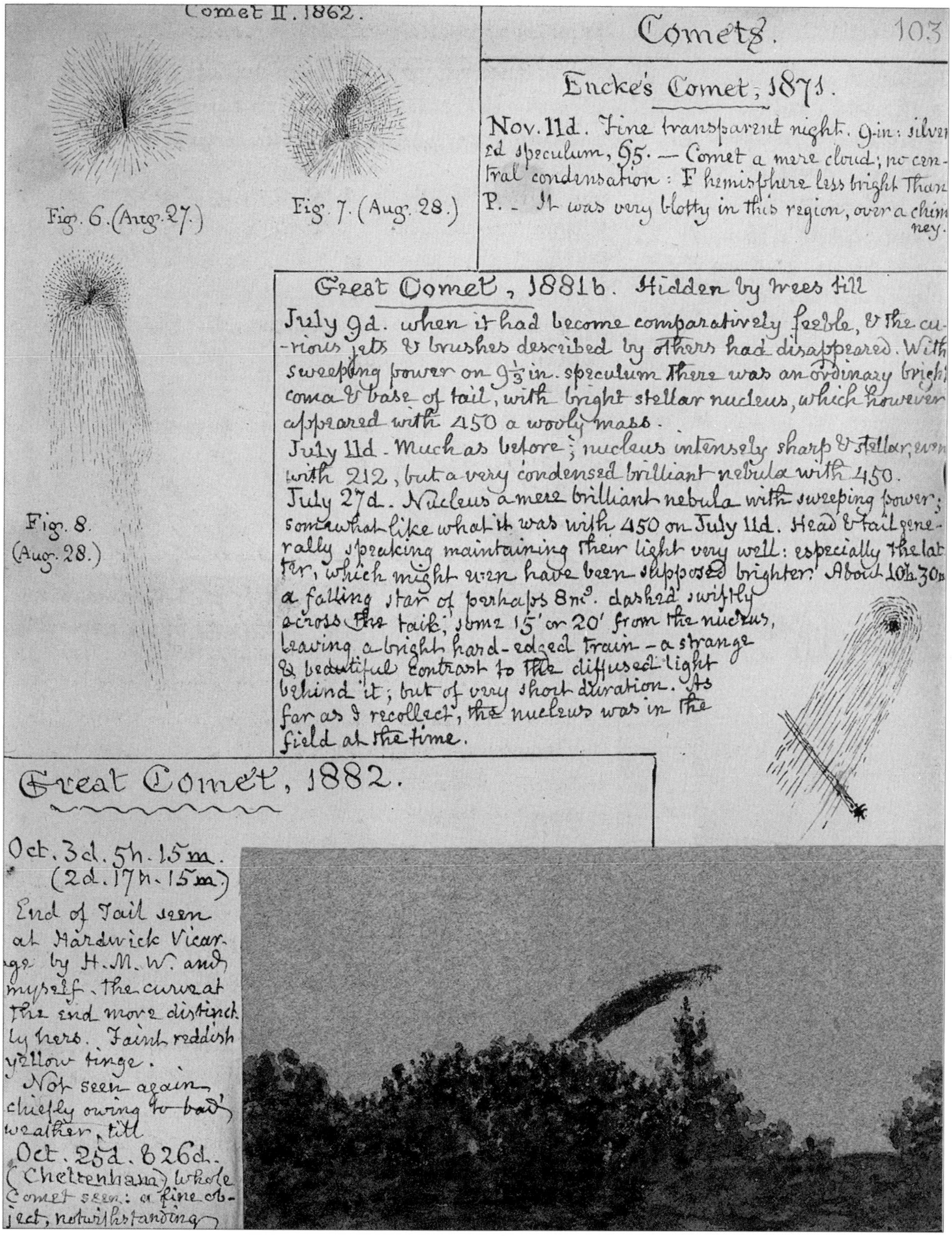
Comet II. 1862.

Comets. 103

Encke's Comet, 1871.

Nov. 11d. Fine transparent night. 9-in: silvered speculum, 95. — Comet a mere cloud; no central condensation: F hemisphere less bright than P. — It was very blotty in this region, over a chimney.

Fig. 6. (Aug. 27.)

Fig. 7. (Aug. 28.)

Great Comet, 1881b. Hidden by trees till

July 9d. when it had become comparatively feeble, & the curious jets & brushes described by others had disappeared. With sweeping power on 9⅓ in. speculum there was an ordinary bright coma & base of tail, with bright stellar nucleus, which however appeared with 450 a wooly mass.

July 11d. Much as before; nucleus intensely sharp & stellar, even with 212, but a very condensed brilliant nebula with 450.

July 27d. Nucleus a mere brilliant nebula with sweeping power; somewhat like what it was with 450 on July 11d. Head & tail generally speaking maintaining their light very well: especially the latter, which might even have been supposed brighter. About 10h.30m a falling star of perhaps 8m. dashed swiftly across the tail, some 15' or 20' from the nucleus, leaving a bright hard-edged train – a strange & beautiful contrast to the diffused light behind it; but of very short duration. As far as I recollect, the nucleus was in the field at the time.

Fig. 8. (Aug. 28.)

Great Comet, 1882.

Oct. 3d. 5h. 15m. (2d. 17h. 15m.) End of Tail seen at Hardwick Vicarage by H. M. W. and myself. The curve at the end more distinctly here. Faint reddish yellow tinge.

Not seen again chiefly owing to bad weather, till Oct. 25d. & 26d. (Cheltenham) Whole Comet seen: a fine object, notwithstanding

Figure 143. Thomas William Webb and Henrietta Moore Webb, *Astronomical Observations of various comets, including the Great Comet of 1881 II and the Great Comet of 1882 (RAS Archive Webb 5A, folio 103r)*, pen and brown ink on white paper and small insert pasted on at lower right in watercolor and body color with oxidized white heightening on blue paper, 196 × 155 mm (page) and 62 × 116 mm (blue insert at lower right), Royal Astronomical Society Library, London.

making his illustrations very different from the dramatically shaded renderings with which astronomers had recorded Donati's Comet.[69]

Webb's handwritten observing book, dating from 1834 to 1872, has a number of comet sketches, including those on folio 103 with observations of four different comets [Fig. 143].[70] They are labelled *Comet II. 1862.*; *Encke's Comet, 1871*; *Great Comet, 1881b*; *Great Comet, 1882.* Two of these, one represented on three different days, are drawn in pen and brown ink inside a rectangular outline in brown ink, while Encke's Comet is not illustrated. At the lower right, the Great Comet of 1882 is singularly depicted in a different medium on a separate tiny sheet of blue paper pasted over the outlined border. It appears to have been added after the fact, perhaps masking a drawing in pen and ink beneath. Moreover, the care with which the illustrations are inserted into the accompanying neatly laid out text implies that they, like the exacting, analytical, and calligraphic notes, were copied from separate more spontaneous sketches.

Next to the drawing of the "Great Comet, 1881b, Hidden by trees till July 9d" (C/1881 K1, formerly 1881 III), Webb notes that on 27 July a falling star of "perhaps 8m [8th magnitude, four times fainter than naked-eye visibility] darted across the tail; some 15' or 20' [about the radius of the moon] from the nucleus, leaving a bright hard-edged train – a strange & beautiful contrast to the diffused light behind it; but of very short duration." He illustrated this celestial crisscross, a legitimate conjunction of the two types of celestial cousins we have been examining, in an adjacent thumbnail pen and ink sketch. It is a lucky observation to have seen through a telescope, made all more probable only by Webb's extensive observing schedule. In fact, it is the type of occurrence one expects to see on a photographic plate's long exposure, testifying both to Webb's ability and sophistication.

Webb was certainly excited about the appearance of the Great Comet of 1882, because he emphatically underlined "Great Comet" with a snaky line. Its adjacent illustration on the piece of blue paper is executed in a different medium – watercolor and body color with white heightening applied in a blatesque technique that looks almost as though a sponge had been used – and style from the other more linear drawings. The white heightening has oxidized black, a fate that is reversible by conservation, which is especially noticeable in the comet's tail, visible above the stand of trees. This small vignette depicts the comet on 3 October at 5:15 p.m., and its note in a rather meticulous hand states:

> End of Tail seen at Hardwick Vicarage by H.M.W. and myself. The curve at the end more distinctly here. Faint reddish yellow tinge. Not seen again chiefly owing to bad weather, till Oct. 25d. & 26d.. (Cheltenham) whole comet seen: a fine object, notwithstanding.

This annotation supports what the stylistic contrast suggests, that Mrs. Webb [the aforementioned H.M.W.], née Henrietta Montague (died 1884), who was known as a watercolorist, executed the small watercolor. That explanation would account for the difference in technique and for the immediacy of the small impressionistic view. Unfortunately, while Henrietta M. Webb is known to have been an artist, none of her watercolors have been identified, rendering a stylistic comparison impossible. Henrietta Webb may also have transcribed the observing book, because, like Caroline Herschel, she seems to have been an amanuensis.[71] This page of Webb's observing book, therefore, may preserve an important collaboration between the husband-and-wife team as well as revealing another intersection of art and astronomy, two fields in which visual acuity is paramount.

The Comets of John Brett, Artist and Astronomer

An artist with strong links to both art and astronomy – perhaps of all those surveyed in this study with the strongest commitment to both – was the painter John Brett (1831-1902). Although Brett was first a precocious astronomer, he is best known as a painter attached to the later Pre-Raphaelites as well as a protégé for a brief time of the critic and artist John Ruskin. Prior to the age of fourteen, when he bought his first telescope, Brett made his own from pieces he collected in second-hand shops. In June 1871 he was elected a Fellow of the Royal Astronomical Society (he submitted at least eleven papers to the *Monthly Notices*), after which he was twice elected to its Council (1876, 1877), and from 1881 he was an Associate Member of the Royal Academy. In his observations, Brett progressed from a 3-inch reflector to an 11-inch Browning reflector in the observatory of his house, "Daisyfield,"

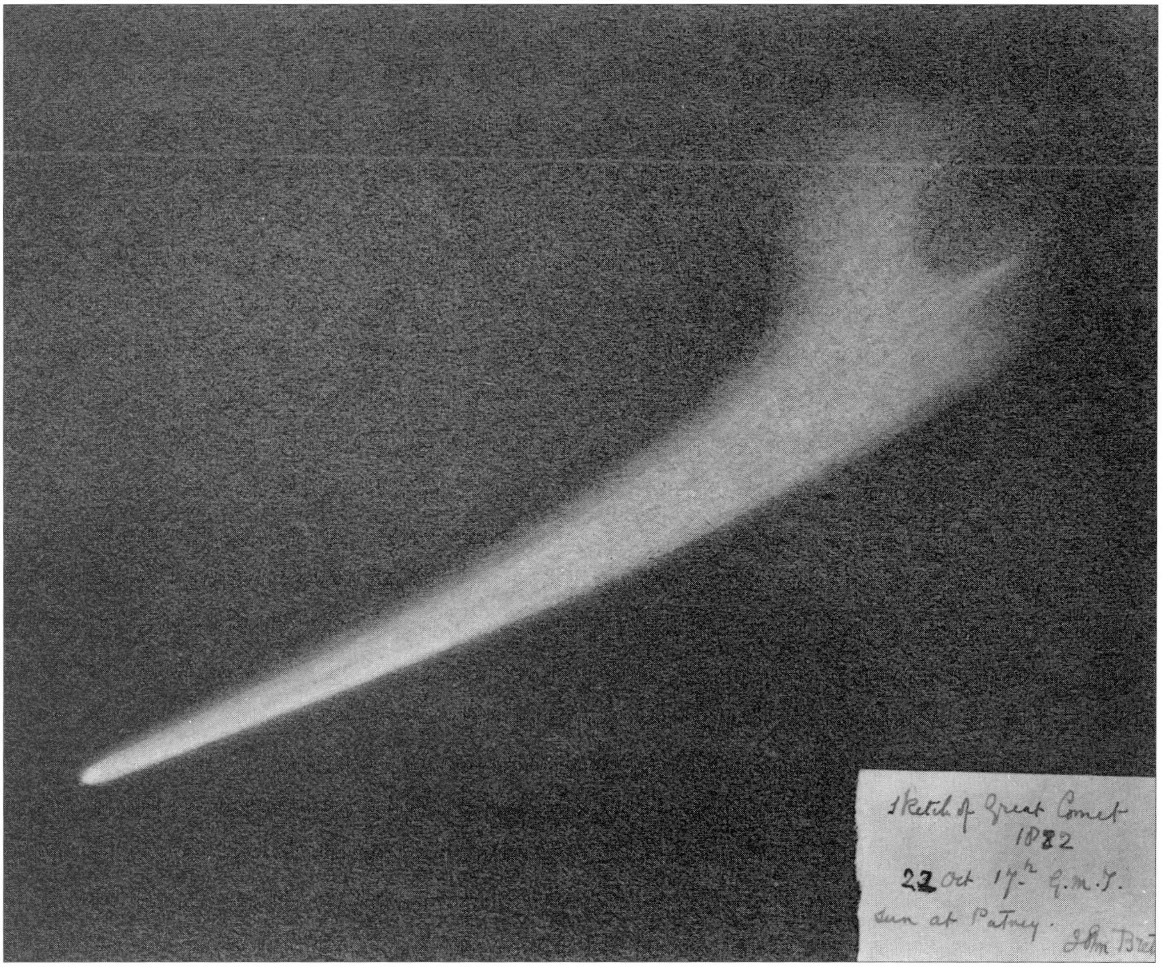

Figure 144. John Brett, *The Great Comet (1882 II) on 22 October* (published in the *Monthly Notices*, vol. 43, opposite p. 288), a photograph of an original in black and white chalk (?), Royal Astronomical Society Library, London.

in Keswick Road, Putney.[72] In his lunar observations, Brett followed in the footsteps of his artist predecessor John Russell,[73] and in several of his paintings, responding to Ruskin's encouragement of exacting Pre-Raphaelite ideals, he demonstrated better than perhaps any other artist how truth to nature meant geological as well as botanical accuracy.[74] Nevertheless, in his painted oeuvre, one sees a greater appreciation of meteorological rather than astronomical phenomena.

Brett's sketch of the Great Comet of 1882 (C/1882 R1, formerly 1882 II), published in the *Monthly Notices*, most successfully combines his two fields of interest: art and astronomy. It reveals both an acute sense of observational accuracy and an artistic ability in capturing the effects of cometary illumination [Fig. 144]. The now-lost depiction was published as a photographic plate with a handwritten caption on a torn piece of paper at the lower right. It notes that Brett observed the comet at 17:00 on 22 (originally the date was given as 23 but the digit 3 was changed to 2) October at Putney.[75]

Two of Brett's 70 sketchbooks contain additional studies of the Great Comet of 1882 as well as sketches of an earlier comet in 1881, which he identifies as Pollie's Comet (there is no mention of a Pollie in standard astronomical sources), whose time frame matches that of Tebbutt's Comet of 1881. Brett's first analytical graphite sketch [Fig. 145] records his telescopic studies of this comet's envelopes and nucleus as well as the variations in its tail over time.[76] The second may very well be a preparatory study for his larger chalk drawing of 1882 [Fig. 146].[77] It shows the entire comet with its center of condensation and sweeping tail at the same angle as in the more finished published work. Moreover, the date of its inscription, 23 October at 17h (5:00 P.M.), agrees with the one tacked onto the photographed drawing, where it is apparent that the 23 was modified to a 22.

Although it seems that some of the theories and ideas which Brett proposed at meetings of the Royal Astronomical Society were

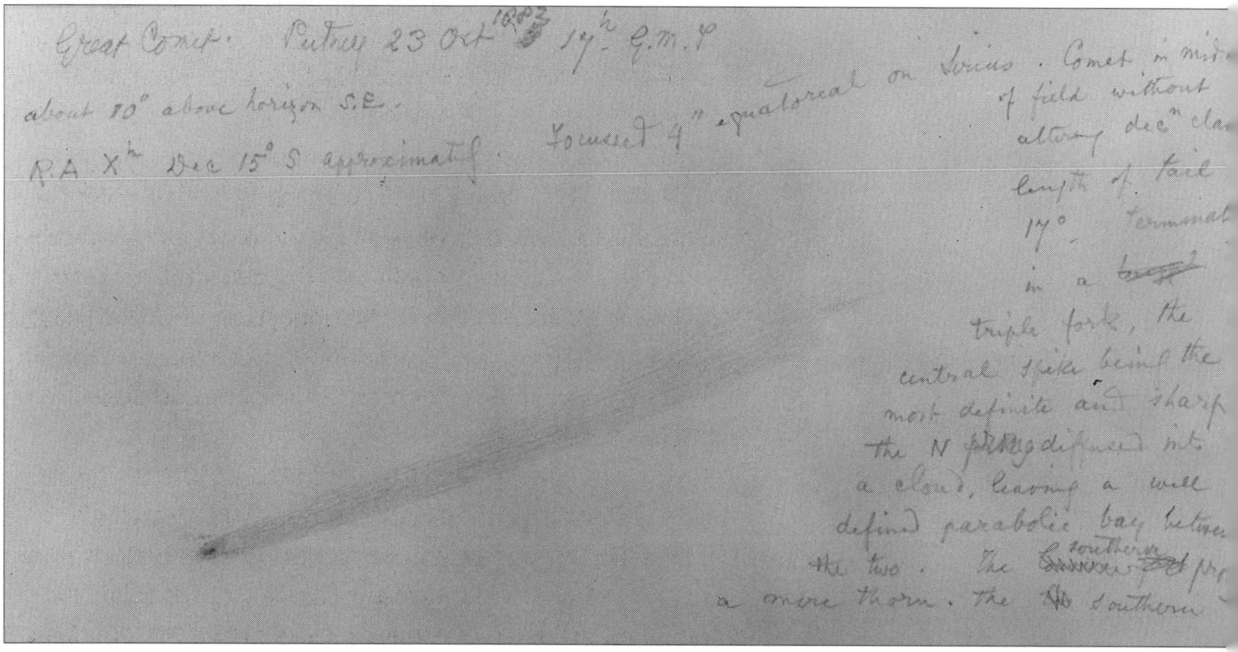

Figure 145. *(Opposite top)* John Brett, *"Pollie's" Comet (1881 III)*, Sketchbook 53 page 0001r, 1881, graphite, 82 × 175 mm, National Maritime Museum, Greenwich.

Figure 146. *(Opposite bottom)* John Brett, *Great Comet of 1882*, Sketchbook 48 page 9881v, 1882, graphite, 89 × 177 mm, National Maritime Museum, Greenwich.

Figure 147. Edward Poynter, *Ides of March*, signed and dated 1883, oil on canvas, 153 × 112.6 cm, Manchester City Art Galleries.

maligned by fellow members, some of them proved to be valid and ahead of his time. In the long run, as reported by the writer of Brett's obituary, many were verified by others and he was partially vindicated.[78] Brett's immersion in astronomical pursuits is buttressed by the fact that he made his own telescopes and remained active in the Royal Astronomical Society until his death.[79]

Poynter's and Hardy's Visual and Literary Comets

Brett's work and specializations contrast with those of Sir Edward John Poynter (1836-1919), who has been characterized as a natural Victorian polymath.[80] As opposed to the Pre-Raphaelites'

Figure 148. Edward Poynter, *The Great Comet of 1882 Over Loch Luichart*, graphite, 180 × 340 mm, Private Collection, New York.

preference for medieval subjects, he specialized in themes from the classical world (as well as biblical illustrations), lavishing much decorative detail on them. (Taken together, the examples of these two artists demonstrate how rich and multiple were the cultural interests of the Victorians.) It comes as no surprise, therefore, that Poynter painted a scene from one of the most famous stories from Roman history, that of Julius Caesar. He chose not the dramatic mob scene of the assassination, but rather the pivotal earlier event when Caesar could have avoided his destiny. In Poynter's scene, Caesar and his wife Calpurnia stand in a loggia looking out at a landscape illuminated by the tail of a very large comet. Entitled "*The Ides of March*," the composition is known in at least three versions, the largest of which was exhibited at the Royal Academy in 1883 [Fig. 147].[81] A replica in pastel, dated 1916, has a larger, more mystical and disembodied comet as well as less detail.[82] Further, two earlier preparatory studies have been recorded, reportedly dated 1881 and 1882. The 1882 study is known to exist, while the one supposedly dated 1881 may actually be the same study (it has not been reproduced). It is telling that the 1882 example, which differs from the Manchester picture in a number of minor ways, has one major variation: the frightening comet portent hovering in the sky is tiny and barely discernible.[83] If the earlier study, purported to date from 1881, exists, it most likely does not have a dominant comet.[84]

It seems logical that the appearance of the Great Comet of 1882 may have inspired Poynter to darken the sky that is lit by a nearly unidentifiable source in the 1882 study and to illuminate his 1883 canvas with the almost piercing light reflected by the tail of a huge comet. In fact, the angle and profile of Poynter's comet in the 1883 canvas looks very much like that of the Comet of 1882 as photographed by David Gill [Fig. 135]. A graphite drawing made by Poynter in Scotland on 23 October 1882 proves that the artist acutely observed this naked-eye comet [Fig. 148], which served as both a model and a catalyst for the brighter, more frightening comet tilted in a more vertical direction in the final historical painting.[85] The drawing preserves not only the comet's stunning profile, but also the wonder Poynter experienced while viewing it.

At the very least, the Great Comet of 1882 served as an emphatic, physical reminder of one of the most salient features of the painting's textual source: William Shakespeare's *Julius Caesar*, Act II, 2. In this scene, Calpurnia implores her husband not to go to the Senate House, where he will be murdered. Shakespeare wrote the dialogue that Poynter later painted when Caesar defies the warning to beware the Ides of March, stating:

> Yet Caesar shall go forth: for these predictions
> Are to the world in general as to Caesar,

and his wife Calpurnia answers:

> When beggars die there are no comets seen:
> The heavens themselves blaze forth the death of princes.

In all known versions of the theme by Poynter, Calpurnia points to the omen in the sky. In both the preparatory study and the painting (but not the replica), the bust of Caesar's friend and rival, Brutus, at the left casts an ominous shadow. This vertical shadow resembles a dark comet with a tail, and literally foreshadows Caesar's assassination by Brutus.[86]

Poynter and Shakespeare depended on ancient traditions in which a comet could signal both the birth and death of a king, designate a person as significant, or function as a prescient sign of doom. Both poet and painter have exploited the tension inherent in this ambiguous symbolism, on the one hand designating Caesar as significant and on the other indicating his impending demise. Moreover, Caesar's identity was so connected with a contemporary comet (the great daylight Comet of 44 B.C.) that the Roman coin commemorating him features a large comet on its reverse.[87] Thus, the appearance of the Great Comet of 1882 could have encouraged this historically minded artist to remember Shakespeare's text and to make more prominent the comet in his painting. The fact that the comet is less clear in the replica, as the memory of the spectacular apparition faded, argues that Poynter did use the actual apparition of 1882 II, recorded in his graphite drawing, as his model for the 1883 painting.

The strong link between visual and literary comet images, together with their inherently powerful potential as symbols, appears not only in the writings of Thomas Hardy (1840-1928) but also in his art work. This novelist and poet, who was obsessed with life forces, sexual relationships, and nature, peppered his literary output with dramatic comet symbolism. For Hardy, comets were ineffable, positive, and inspirational. For example, in the *The Woodlanders* (1887), a comet becomes a simile of the human soul:

> Then the unhappy Grace regarded him by the light of the candle. There was something in his look which agonized her, in the rush of his thoughts, accelerating their speed from minute to minute. His soul seemed to be passing through the universe of ideas like a comet; erratic, inapprehensible, untraceable.

In his earlier novel *Two on a Tower* (1882), Hardy's fictional protagonist is an astronomer who was saved by a comet.

> The strenuous wish to live and to behold the new phenomenon supplanting the utter weariness of existence...gave him a new vitality. The crisis passed.... The comet had in all probability saved his life. The limitless and complex wonders of the sky resumed their old power over his imagination....[88]

Hardy's comet in this context signifies resurrection and its mesmerizing power, inspiration. Certainly, the apparition of the Great Comet of 1882 coincided with, if not actually triggered, Hardy's conceit.

Hardy observed actual comet apparitions during his lifetime and even designed and drew the image of a powerfully primordial comet poised above the silhouette of a cropped city skyline for

Figure 149. Thomas Hardy, *Headpiece Illustration for "A Sign-Seeker,"* c. 1898, black ink wash and white body color, 57 × 173 mm, City Museums and Art Gallery, Birmingham.

one of his *Wessex Poems*, "A Sign-Seeker," published in an edition of 1898 that was illustrated with 30 drawings by the writer [Fig. 149].[89] Hardy's comet headpiece accompanies the following passage and visualizes one of his "eccentric orbs":

> I learn to prophesy the hid eclipse,
> The coming of eccentric orbs;
> To mete the dust the sky absorbs,
> To weigh the sun and fix the hour each planet dips.

Hardy's ability in drawing and design derived from his training as an architect and his work on church restoration in Dorset. He even won the prize for design given by the Architectural Association in 1863, the same year in which he was awarded the medal of the Royal Institute of British Architects for his essay on colored brick and terracotta architecture. It is in Hardy's creative work where both the potential verbal and visual power of comets is demonstrated by one individual.

British Astronomy Near the Close of the Century

John Russell Hind (1823-1895), who played a major part in late nineteenth-century British astronomy, contributed to the on-going study of comets. Although Hind is now chiefly remembered for discovering in 1852 what is called "Hind's Variable Nebula," and was famous in the nineteenth century as the most prolific asteroid discoverer, with ten found,[90] he was also an early participant in the difficult task of distinguishing between comets and asteroids. In addition to dozens of articles on comets, Hind published two historical books about them.[91] He also discovered three comets, leading Lancaster Brown to cite him as one of the "great comet hunters of the nineteenth century."[92]

The most prolific author about meteors in Victorian Britain was William F. Denning. From his first article in 1877 through his columns in the 1920s, Denning justified his reputation as one of the greatest amateur astronomers of modern times. His interest was whetted by the great Leonid meteor shower of 1866, and positions of the radiants of many meteor showers, that is, the points in the sky from which the meteors appear to come. He also discovered four comets (in 1881, 1890, 1892, and 1894), joining Hind as one of the only three British subjects to discover comets since the time of Caroline Herschel.[93]

The third nineteenth-century British discoverer of comets was Edwin Holmes. He found a short-period comet, 17P/Holmes, near the Andromeda Galaxy in 1892. Then, after Denning's 1894 final discovery, no comet was discovered in the United Kingdom until George Alcock found a comet in 1959.

In 1890, yet another astronomical society was founded, the British Astronomical Association, testifying to further specialization in the profession and the participation of amateurs. The group was formed primarily for those who found the subscription to the Royal Astronomical Society too high or the papers too technical, or who, being women, were excluded. It still survives today. The impetus for this organization started with a letter written by W.H.S. Monck to a popular magazine, the *English Mechanic*. One of the most enthusiastic supporters was Elizabeth Brown, Director of the Solar Section of the Liverpool Astronomical Society, which had been founded in 1881 but was in serious decline by 1890. The first officers and council of the British Astronomical Association included such well-known figures as the astronomer William and Mrs. Huggins, Isaac Roberts, the Earl of Rosse, Agnes Clerke, and Sir Howard Grubb. Membership spread beyond Great Britain out into the Empire. For example, the New South Wales Branch, founded in 1894, had as its first president John Tebbutt and is the only one to survive to the present day.[94]

Figure 150. Walter Crane, *Shakspere Calendar, 1897*, whereabouts unknown.

Comets and Meteors of the Illustrators, an Escape into the Imagination

Nevertheless, comets and meteors were still employed in the visual tradition as the work of the illustrator, painter, and designer Walter Crane (1845-1915) demonstrates. Crane, who was connected to the latter day Pre-Raphaelites and William Morris, illustrated the famous meteor quotation mentioned previously from Shakespeare's *A Midsummer's Night Dream*, Act II, scene I, as the

Figure 151. Arthur Rackham, *"To Hear the Sea-Maid's Music," from Shakespeare's "A Midsummer Night's Dream," Act II, Scene 2*, for Heinemann edition, signed and dated 1908, pen and black ink and watercolor, 382 × 268 mm, The Victoria and Albert Museum, London.

cover of his *Shakspere Calendar* [Fig. 150].[95] In its central roundel, Crane placed a mermaid with a harp seated on a fantastic sea-monster-like dolphin.[96] In the background, as the text indicates, stars are falling from the heavens in a meteor shower. Certainly the annual August Perseids, near the middle of the hottest part of summer, would have signaled to Crane's audience, as well as to

Shakespeare's, the humid steamy nights of summer. It is an image fraught with synesthesia.

It was the illustrators, those visual magicians of the imagination, who conjured up comet and meteor imagery at the end of the century, rather than the landscape or genre painters. Perhaps the advent of photography was influential on this trend, although the highly symbolic, aesthetic, and escapist mood of the late decades (the *fin de siècle*) and the early years of the next century (*Belle Epoque*) demanded this type of response, not just in England but throughout the western world. In Great Britain, illustrators enjoyed halcyon days during the renaissance of book illustration that began with William Morris and the Kelmscott Press and continued into the twentieth century.

For example, Arthur Rackham (1867-1939) again visualized the same Shakespearean image in his preparatory watercolor for a 1908 edition of *A Midsummer's Night Dream* [Fig. 151].[97] Like his predecessors who also illustrated this passage [Figs. 94, 150], Rackham punctuated his evocative scene with meteor showers (see chapter III):

> ...And heard a mermaid on a dolphin's back
> Uttering such dulcet and harmonious breath,
> That the rude sea grew civil at her song.
> And certain stars shot madly from their spheres.
> To hear the sea-maid's music.

However, Rackham's illustration is more complicated than the earlier two and includes additional arcane symbolic elements from the following lines.

> That very time I saw – but thou couldst not –
> Flying between the cold moon and the earth,
> Cupid all armed; a certain aim he took
> At a fair vestal throned by the west.
> And loosed his love-shaft smartly from his bow
> As it should pierce a hundred thousand hearts:
> But I might see young Cupid's fiery shaft
> Quenched in the chaste beams of the watery moon,
> And the imperial votaress passed on,
> In maiden meditation, fancy-free:
> Yet marked I where the bolt of Cupid fell:
> It fell upon a little western flower....

They include the cupid in the roundel at the top and the Elizabethan lady holding a rainbow in the lower roundel inscribed *NON SINE SOLE IRIS*, which may allude to the queen herself. Shakespeare and Rackham herein preserve the association from Antiquity between Aphrodite, the goddess of love, and meteorites, and their worship in her temples.[98]

Rackham employed meteor showers again in an amusing illustration of *Hercules Supporting the Sky in Place of Atlas* [Fig. 152], reproduced as a color plate in the 1922 edition of Nathaniel Hawthorne's *A Wonder Book* with the caption: "Hercules gave a great shrug of his shoulders. It being now twilight, you might have seen two or three stars tumble out of their places."[99] Herein, the meteor showers signify the great cosmic strain during Hercules's struggle to support Atlas's burden. As part of his eleventh labor, the Greek hero persuades the giant Atlas to fetch the apples of the Hesperides (symbolic of immortality), temporarily holding the globe on his shoulders for Atlas. In Rackham's illustration, Hercules also holds up the heavens, a mythological alternative, which in the sixteenth century functioned as an allegory of astronomy. (According to ancient mythographers, Atlas taught Hercules astronomy.)[100]

Just after the turn of the century, Henry John Stock (1853-1930) illustrated the book of Revelation in a series of dramatic watercolors, two of which feature meteoric images.[101] Stock's watercolors harken back to traditional apocalyptic representations, but were most strongly influenced by Blake. The first one to be discussed [Fig. 153 color] illustrates Revelation 9:1: "And the fifth angel blew his trumpet and I saw a star fall from heaven unto earth; and to him was given the key to the bottomless pit; and there arose a smoke out of the pit, as the smoke of a great furnace; and the sun and air were darkened...."[102] Stock's falling star resembles a huge flaming fireball but also formally recalls one of Blake's illustrations for *Jerusalem* [Fig. 58]. A second, later watercolor for the series has a meteor shower [Fig. 154 color].[103] It illustrates Revelation 6:12-13: "And I beheld when he had opened the sixth seal, and, lo, there was a great earthquake; and the sun became black as sackcloth of hair, and the moon became as blood; And the stars of heaven fell unto the earth, even as a fig tree casteth her untimely figs, when she is shaken of a mighty wind."

Figure 152. Arthur Rackham, *Hercules Supporting the Sky In Place of Atlas* (or *Falling Stars*), for Nathaniel Hawthorne's *A Wonder Book*, 1922, watercolor and pen and black ink, 380 × 277 mm, whereabouts unknown, formerly Christie's, New York.

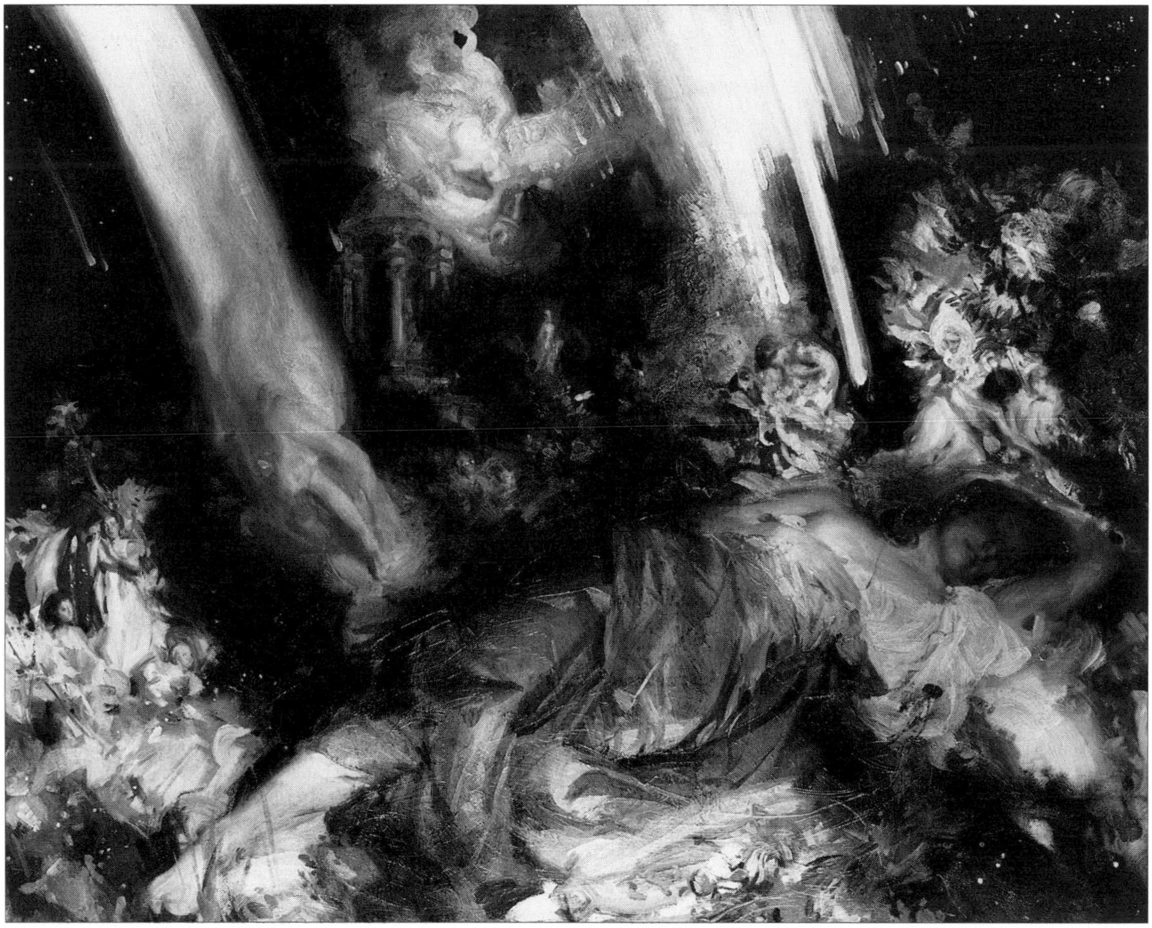

Figure 155. Charles Sims, *Portrait of Julia MacWhirter*, after 1909?, oil on canvas, 71 × 91.5 cm, whereabouts unknown, formerly Sotheby's, London.

Charles Sims (1873-1928) included a meteor shower in an unusual portrait, known in at least two versions, of his future wife Julia [Fig. 155].[104] (Julia MacWhirter was the daughter of the Royal Academician and Scottish painter John MacWhirter.) The work supposedly commemorated their honeymoon, while illustrating a poem by the seventeenth-century poet Robert Herrick "A Night Piece to Julia." This four-stanza paean to Herrick's beloved Julia, part of the "Hesperides," includes the lines utilized by Sims for the meteor imagery in his canvas:

Her eyes the glow-worm lend thee,
The shooting stars attend thee,
The elves also, whose little eyes glow
Like sparks of fire, befriend thee.

Figure 156. Maxwell Ashby Armfield, *Morning*, 1937, tempera and gold leaf on canvas laid on shaped panel, 71.2 × 33.7 cm, whereabouts unknown, formerly Sotheby's, New York.

The canvases are luscious, tender Belle Epoque celebrations of the artist's passion for Julia. Both are set in a sensual garden and feature meteor showers, symbolic of love and associated with the goddess of love and gardens, Aphrodite/Venus. Meteor showers are potent visual symbols because like sexual passion, they are transient and intense, while their phallic shape and fiery nature are emphatically evocative.

That comets and meteors survived in the British illustrative tradition well into the twentieth century is born out by Maxwell Ashby Armfield's (1881-1972) anachronistic panel *Morning* [Fig. 156].[105] Armfield – a writer, illustrator, and fierce advocate of tempera painting, used a comet (or shooting star) to symbolize hope and the dawning of a new day. The comet appears adjacent to the morning star, hanging above an archaically dressed and bearded poet or musician wearing a laurel wreath and playing a lyre. Presumably, the artist intended the celestial form to be a comet, since early morning is one of the two times when comets are highly visible and because comets can function as positive portents, here of poetic inspiration. Armfield's composition is part of a diptych and its upper left corner (masked in the reproduction) is inscribed *MORN* and the right *ING*. Its pendant, *Evening*, is punctuated by stars, a crescent moon, and lightning above the female figure.[106] These allegorical panels date from 1937 and embody Armfield's attachment to both tempera, which was reinforced by periodic trips to Italy, and archaic symbols.

The last decades of the nineteenth century witnessed sweeping changes in taste and artistic styles. They were linked to the current pan-European trend for escapist, Symbolist art. Since the 1880s, increasing numbers of British artists were drawn to Paris as the dominant artistic center, where they studied, facilitating the spread of international ideas and styles. Also, it is interesting to note that from the 1860s, the British government began to draw back from subsidizing high culture, perhaps partly because of an uncertainty about taste.[107] Artists were encouraged, therefore, to indulge their highly personal visions and to turn to new avenues of patronage. It was in this artistic climate, which due to the perfection of photography lacked an impetus for mimesis, that comet and meteor images became once again highly symbolic.

The Fate of Astronomical Illustration

Until the perfection of astronomical photography to the point where photographs could be widely used as illustrative material, astronomers drew equally on technical, scientific, and artistic skills in both the field and the observatory. They were not willing to sacrifice beauty and accuracy for mechanical autonomy. Then, during the 1870s drawings and astronomical photographs began to share a more equal status, except in the case of comet illustrations, which were not part of this trend until later. At times, astronomers chose to publish composite drawings that combined the best features of several photographs and drawings. These were eventually issued as engravings requiring a close collaboration between astronomers and engravers. This approach assumed that the informed judgments of artists and scientists could compensate for the drawbacks in the originals. At its best, the engraver became a translator of the artist and not a mere copyist. It comes as no surprise, therefore, that there were running debates over which printing processes produced the best results.[108] A case in point is represented by lithography, whose stones tend not to give as many clear images in mass printings as engravings, but which has other assets. By the 1880s and 1890s, however, astronomers abandoned this heterogeneous mix of skills and technologies in favor of the increasingly mechanical system of making and copying photographs (for example, half-tone and photogravure).[109] At the same time, the emergence of collotyped reproductions began to undermine the engravers, lithographers, and the old system of reproductive illustration, resulting in a state of turmoil in the whole reproductive process, including the illustration of works of art. Some illustrators still printed idealized engravings of works of art, others set out to reproduce illusionistically oils, while still others turned to engraving facsimiles of photographs. Books reflected this confusion and were illustrated by engravings of photographs as well as by printed photographs of engravings or drawings.[110] John Brett's drawing of the Comet of 1882 [Fig. 144] is an example of the latter technique.

It was at this juncture that comet photography came of age and supplanted the old system of representation, making illustrations by artists and illustrators less in demand. Finally, as Pang observes, professional expeditions equipped with large cameras "drove amateurs and drawing out of the field and photomechanical printing technologies delivered the *coup de grace*, offering astronomers, increasingly suspicious of human judgment and observation, a vision of a fully mechanical means of imaging, a closed system free of human frailty that would transmit pictures from the field to the pages of the journal."[111] Parallel problems and solutions to those of scientific illustrations occur in the processes to reproduce works of art.[112] Both became increasingly high-tech in their orientation.

Conclusion

During the last decades of the century, specialization helped to separate the interests of artists and astronomers so that with the exception of a few individuals, such as Holiday and Brett, there was a reduction in the cross currents between the two fields. Further, astronomy no longer held the same interest for the masses. Astronomy was being revolutionized by major technological developments (in instrumentation, for example) and by the emergence of astrophysics as a distinct discipline, especially in countries not participating in the *Carte du Ciel*, notably the United States. This situation created a gulf between professional astronomers and amateurs, two groups that had previously thrived in a symbiotic relationship. In an even more general sense, the changes in both intellectual climate and technology resulted in an increasing separation between the arts and the sciences. Moreover, it seems that after the 1860s, English artists began to turn away from the English tradition of landscape painting and the dialogue between art and nature fostered by its artists, first toward an historicism and then toward an aestheticism that was sweeping other countries of Europe.[113] By the same token, beginning in the 1870s genre painting, which had enjoyed an undiminished vigor for two centuries, suffered a marked decline.[114] Finally, the advent of photography removed the impassioned interest in mimesis or realism and pushed artists further into escapist realms of symbols, imagination, and fantasy, which were fed by a disillusionment with visual "reality." These factors are among the many which precipitated the end of the intense British artistic and astronomical involvement in comets and meteors, which was encouraged by a strong literary tradition and exciting astronomical discoveries. Nevertheless, the appearances of bright comets and meteors continued to galvanize British observers, who were poised for the appearance of Halley's Comet in 1910. But never again, was there the sense that artists and astronomers, as well as the culture at large, were riveted by these celestial visitors as we have seen was the case between c. 1780 and c. 1882.

Notes

1 See Pang, 1994, 249-274.

2 Smyth, 1843, 75. Discussion about drawing comets continues today (J.E. Bortle and C.S. Morris, "The Visual Observation of Comets," *International Comet Quarterly*, 16, 1994, 129-130; S.J. Edberg and D.H. Levy, *Observing Comets, Asteroids, and the Zodiacal Light*, Cambridge, 1994, 39-63; J. Shanklin, "Estimation of DC – an Experiment," *International Comet Quarterly*, 17, 1995, 85-87).

3 Kidger, 231-238, believes that this enhancement was due to the number of Kreutz group objects that returned over a comparatively short period. He also notes that there has not been a lack of bright comets in the twentieth century, although the number of bright comets was sharply reduced in the 1980s and 1990s. This situation has changed with comets Hyakutake and Hale-Bopp.

4 R. Bhathal and G. White, *Under the Southern Cross: A Brief History of Astronomy in Australia*, Kenhurst, Australia, 1991, 39.

5 Tebbutt discovered it in Windsor in New South Wales thirty miles south of Sydney. Vsekhsyvatskii, 206-216; Kronk, 1984, 49-53. Three issues of *The Illustrated London News* cover the event with illustrations (vol. 39, 6, 12, 20 July 1861). The 20 July issue (p. 49) cites J.R. Hind's announcement that on 30 June the earth had passed through the comet's tail. It is followed in the supplement (p. 65) with an engraving of the comet's path, together with "Relics of the Past," on historical attitudes towards comets that quotes Shakespeare's *Henry VI*:

> Comets, importing changes of time and states.
> Brandish your crystal tresses in the sky.

6 See De la Rue, 1862, 78-84, pls. VII-VIII, figs. 1-9, engravings after the drawings; Olson, 1985b, 100; chap. VI.

7 Tebbutt's success led him to construct his own observatory. He turned down the offer to be Government Astronomer and head the Sydney Observatory for a number of reasons, including to be unimpeded in his own work, which centered on comets.

8 Australia's southern location augmented by personal interests in comets led to many discoveries, a situation extending to the present day. Bhathal and White, 40-49. See also R. Bhathal, *Australian Atronomer John Tebutt*, 1993, pp. 21-22, 46, 54-64.

9 Holiday was not a fellow of the Royal Astronomical Society, a fact checked by Peter Hingley. See Reynolds, 1966, 65, 69, 71, 72; Staley, 1973, 175.

10 H. Holiday, *Reminiscences of My Life*, London, 1914, 19-20, 73.

11 *Ibid.*, 209, fig. 13.

12 A.C. Ranyard, ed., "Observations made during total Solar Eclipses," *Memoirs of the Royal Astronomical Society*, 41, 1879, 63.

13 Holiday, 119.

14 *Ibid.*, 169-170. Opp. p. 170 are three of his drawings of the moon's surface.

15 See, for example, V. Surtees, *The Paintings and Drawings of Dante Gabriel Rossetti (1828-1882)*, 1, Oxford, 1971, 73-74, nos. 117, 117A; 2, figs. 179, 180.

16 Among the numerous nineteenth-century treatments of this popular theme in European art and literature was the tragedy by Silvio Pellico, *Francesco da Rimini* (1815), which Rossetti saw in 1850. See R.J.M. Olson, *Ottocento. Romanticism and Revolution in 19th-Century Italian Painting*, exh. cat., New York and Florence, 1992, 247-248, no. 98, for a discussion of the subject in nineteenth-century art.

17 Inv. no. 3056 in the Tate Gallery, London, in watercolor measuring 248 × 445 mm. It is inscribed: "*Quanti dolci pensier, Quanto disio;*" "*Menò costoro al doloroso passo;*" "*O lasso!*" Surtees, 1, 36-37, no. 75. The second version or first replica is inv. no. P.548/P.M. 25 in the Cecil Higgins Art Gallery, Bedford, in watercolor measuring 330 × 610 mm, monogrammed and dated 1862 (*ibid.*, 1, 38-39, no. R.1; vol. 2, fig. 88). The third or the second replica is inv. no. 3266/4 in the National Gallery of Victoria, Melbourne, in watercolor measuring 400 × 330 mm, monogrammed and dated 1867 (*ibid.*, 1, 39, no. R. 2). See *ibid.*, 37-38, nos. 75A-75D, for the preparatory studies; C. Gizzi, *Dante Gabriel Rossetti*, exh. cat., Milan, 1984, 145-146; 170-171, nos. 22-27.

18 Surtees, 1, 37.

19 In the Gerald H. Morris Collection, London, a chromolithograph measuring 90 × 135 mm, from A. Giberne, *Sun, Moon, and Stars: A Book for Beginners*, London, 1884, opp. p. 218, published by Seeley and Co. For other illustrations of this shower, see D. Hughes, "A Mysterious Woodcut," *Sky and Telescope*, 74, 1987, 252; *idem.*, "The World's Most Famous Meteor Shower Picture," *Earth, Moon, and Planets*, 68, 1995, 311-322.

20 See also F.L. Fennell, *Dante Gabriel Rossetti. An Annotated Bibliography*, New York and London, 1982; A.C. Foxon, *Dante Gabriel Rossetti*, New York, 1989. It is pertinent that Jean-François Millet c. 1846-49 painted a canvas *The Shooting Stars*, in the National Museum of Wales, Cardiff, which depicts two couples, each representing a personification of a shooting star, wound inextricably together in erotic embraces as they float against the backdrop of a starry sky, rather like Paolo and Francesca (Olson, 1985b, fig. 78). Meteors and their showers were used as metaphors for physical love, perhaps a meteoric and spectacular passion, in literature as well.

21 Inv. no. 1943.672 in the Fogg Art Museum, Harvard University, Cambridge, MA, in pen and black ink on vellum measuring 156 × 192 mm. See Cambridge, MA, The Fogg Art Museum, *Paintings and Drawings from the Pre-Raphaelites and Their Circle*, exh. cat., Cambridge, MA, 1946, 24-25, no. 5; M. Harrison and B. Waters, *Burne-Jones*, New York, 1973, fig. 44, whose caption dates it 1858, but on pp. 38-41 it is dated 1858-59; D.N. Mancoff, *The Arthurian Revival in Victorian Britain*, New York and London, 1990, 164-165, who believes that it was begun as a tribute to Alfred Lord Tennyson.

22 See G. Burne-Jones, *The Memorials of Edward Burne-Jones*, 1, London, 1904, 178; P. Fitzgerald, *Edward Burne-Jones. A Biography*, London, 1975; E. Burne-Jones, *The Little Holland House Album*, North Berwick, 1981.

23 Tennyson's later *Idylls of the King* and Sir Thomas Malory's *Morte d'Arthur* were influential on Burne-Jones, as well as on his teacher, Rossetti, and his roommate in Oxford, William Morris. Apparently Burne-Jones shared a passion for Malory with Morris, who in 1855 bought a copy of the 1817 Robert Southey edition of Malory's *Morte d'Arthur* as a gift for Jones. Cambridge, MA, The Fogg Art Museum, *Paintings and Drawings from the Pre-Raphaelites and Their Circle*, 24. For Tennyson, see R.B. Martin, *Tennyson: The Unquiet Heart*, Oxford and London, 1980.

24 For example, the horse resembles the steed in Carpaccio's *St. George*, while its saturnine expression recalls that of the horse in Dürer's

engraving *Knight, Death, and the Devil*, later given to Jones by the critic John Ruskin, another frequenter of the Prinseps's circle.

25 Harrison and Waters, 37-41. They mention the possible influence of Morris's long poem *The Defense of Gueneverre* (1857) and its section, "Sir Galahad: A Christmas Mystery," wherein Galahad is described as riding, unlike in Tennyson's *Idylls of the King*.

26 Mancoff, 1990, 164.

27 See D.N. Mancoff, "'Because my Heart is Pure': Sir Edward Burne-Jones and Tennyson's *Sir Galahad*," *Avalon to Camelot*, 1:1, 1983, 24-26, for the significance of Sir Galahad to Victorian society and to the artist. She believes that this drawing depicts not the otherworldly mystic of Malory's *Morte d'Arthur* but the new incarnation of the chaste Grail knight created by Tennyson. Galahad fuses chivalry with religion according to N.J. Lacey, *The Arthurian Encyclopedia*, New York and London, 1986, 203.

28 See *ibid.*, 257-260. For the Arthurian craze, see R. Simpson, *Camelot Regained. The Arthurian Revival and Tennyson 1800-1849*, Woodbridge, Suffolk, 1990.

29 G. Burne-Jones, 1, 77. Burne-Jones continued to be involved with the story of the Grail in later works.

30 See Olson, 1979. It is also telling that Morris's poem "Sir Galahad: A Christmas Mystery" makes the link with Christmas: "It is the longest night in all the year, Near on the day when the Lord Christ was born."

31 Less convincingly, Orloff has allied Burne-Jones's comet to light forms described by Tennyson in *The Holy Grail*, wherein they are associated with Galahad:

> ...and then stream'd thro' my cell a cold and silver beam,
> and down the long beam stole the Holy Grail,
> Rose-red with beatings in it as if alive....
>
> A beam of light seven times more clear than day,
> And down the long beam stole the Holy Grail
> All cover'd with a luminous cloud....
>
> ...and never yet
> ...this Holy Thing fail'd from my side, nor come
> Cover'd but moving with me night and day,
> Fainter by day, but always in the night
> Blood red....
>
> ...and from the star there shot
> A rose-red sparkle to the city, and there
> Dwelt, and I knew it was the Holy Grail.

S. Orloff, "Images de comètes: analogies et analyses iconographiques," *Bulletin de la Classe des Beaux-Arts*, 70, 1988, 229- 230. She states that the poem was first published in 1842, when, in fact, it was written in 1868-69 and not published until 1869, and later included in *Idylls of the King*. We thank Christopher Ricks for confirmation of this point. For Tennyson's poetry, see A. Tennyson, *The Poems of Tennyson*, ed. C. Ricks, London and New York, 1969. Nevertheless, it is tempting to think that Burne-Jones knew about Tennyson's ideas prior to the publication of *The Holy Grail*, because certain lines echo ideas associated with comets. For example, earlier Tennyson claims the Holy Grail was "blood-red," a traditional description of a comet, and that it moved with the knight during the evening and day, an image that conjures up the Star of Bethlehem associated with the Magi's visit to the Christ Child (Olson, 1979).

32 See.B. Brightwell, *A Concordance to the Entire Works of Alfred Tennyson*, New York, 1970, 73, for "comet;" 173, for "meteor."

33 Orloff, 229, believes that Comet Brorsen in June of 1857 (5D/1857 F1, formerly 1857 II; Kronk, 1984, 239-40) piqued the artist's interest.

34 G. Burne-Jones, 1, 58; Orloff, 229-230, notes the fact that the 1851 catalogue of the artist's library lists A. von Humboldt's *Cosmos. Essai d'une description physique du monde*, 1, Brussels, 1852, 76-88, with a chapter on comets.

35 G. Burne-Jones, 2, 303-304.

36 Inv. no. 1912.33 in the Manchester City Art Galleries, in graphite and grisaille watercolor measuring by sight 708 × 480 mm. It is signed with the monogram S that resembles a snake on an escutcheon. There is a smaller illustration for the scene in the same museum (inv. no. 1912.28), in pen and black ink with wash and white heightening measuring 121 × 78 mm. This unsigned version (E. Mills, *The Life and Letters of Frederic Shields*, London, 1912, pl. 2; Manchester City Art Galleries, *Masterpieces From the Manchester City Art Galleries*, exh. cat., Manchester, 1993, no. 82) has variations, such as the orientation of the comet's tail, which runs from the lower left to the upper right with its head (in a nocturnal scene this orientation is not proper). The comet in the larger study is more accurate. Manchester, City Art Galleries, *Concise Catalogue of British Watercolours and Drawings*, 1, Manchester, 1984, 159, lists another "sketch" for "Solomon Eagle" (inv. no. 1912.34, in watercolor and body color measuring 575 × 295 mm).

37 In his diaries, Shields notes on 29 June 1863 that he "Traced Solomon Eagle." Mills, 77.

38 Shields was reputedly one of the best Pre-Raphaelite illustrators, and his designs for the *History of the Plague* are generally regarded as his masterpiece in this field. They drew enthusiastic praise from both Ruskin and Rossetti, but in the opinion of the artist the woodcuts were ruined in the cutting. To preserve the images Shields had the drawings on wood photographed before their cutting; they were shown at the Memorial Exhibition in London in the autumn of 1911. However, in his diary for 29 January, Shields notes: "Sold my Plague sketches to Mr. Barrett £5;" this entry must refer to the artist's preparatory sketches. According to Mills, the edition of Defoe's classic tale with Shields's illustrations was published in a cheap, paper-covered series, entitled *Laurie's Shilling Entertainment Library*, engraved by Swain and Morton and published in 1863. Mills (*ibid.*, 78) writes that the book is not in the British Library, "and is apparently not to be procured." See nn. 39, 42 below, which suggests a copy is in The New York Public Library.

39 D. Defoe, *History of the Plague of London 1665*, published by Thomas Laurie, 28 Paternosters Row, n.d. It may indeed be the same item cited in n. 38 because the cover and title page are inscribed with the name of the series: *Laurie's Entertaining Library*.

40 *Ibid.*, 22-23.

41 *Ibid.*, opp. p. 24.

42 D. Defoe, *History of the Plague of London (1665)*, published in London by Longman, Green, 1863. The Boston Public Library owns a copy of the

same volume, edited by J.S. Laurie, in "The shilling entertaining library," according to the *National Union Catalogue*, 136, 558.

43 Defoe, 1863, 101-102. The illustration appears opp. p. 101.

44 The dominant comet in Shield's illustration of this scene serves as a foil for an earlier painting of the same subject by Paul Falconer Poole (1807-1879) in the Sheffield Art Gallery (*Solomon Eagle exhorting the people to repentance, during the plague of the year 1665*) that was exhibited in 1843 at the Royal Academy. While Poole does not actually depict the comet, he implies its existence in the sky. See Leeds Art Gallery, *Great Victorian Pictures: Their Paths to Fame*, exh. cat., 1978, 68-69, no. 45.

45 Satis House, Tuxedo Park, in oil on board measuring 110 × 305 cm.

46 J. Smetham, *Letters of James Smetham*, eds. S. Smetham and W. Davies, London, 1892, 154-155.

47 S. Casteras, *James Smetham. Artist, Author, Pre-Raphaelite Associate*, Aldershot, Hants, 1994, 108.

48 Mancoff, 1990, 195, discusses depictions of the death of Arthur.

49 Casteras, 108.

50 Olson, 1985b, 104-105.

51 Smetham's interest in astronomical forms seems to be for their poetic, symbolic power rather than for their scientific import, as exemplified by a drawing (1854, in the Smetham Family Collection) of an "Orrery of personal responsibility," whose concentric circles literally chart his friendships (Casteras, fig. 15).

52 D.H. Levy, "Poet and Observer: Gerard Manley Hopkins and Some Mid-19th Century Comets," *Journal of the Royal Astronomical Society of Canada*, 75:3, 1981, 139-150. See Vsekhsvyatskii, 220-221; Kronk, 1984, 54.

53 In the Watts Gallery, Compton, Guildford, in oil on canvas measuring 147 × 66 cm. It is a replica of a version (424 × 201 cm), designed c. 1868 and completed in 1872, which Watts presented to the Royal Academy as his diploma work. The Watts Gallery canvas was part of Watts's illustrations to the story of Genesis and was intended to be included in his later, modified cosmic scheme of *The House of Life*. There are two related sketches: one in monochrome (150 × 69 cm; Sotheby's, London, 9/VI/95, no. 348) and one in color (in the Castle Museum, Norwich). See London, Arts Council, *George Frederick Watts 1817-1904*, exh. cat., London, 1954, 36-37, no. 5; London, The Whitechapel Gallery, *G.F. Watts. A Nineteenth Century Phenomenon*, exh. cat., London, 1974, no. 25; Minneapolis, The Minneapolis Institute of Arts, *Victorian High Renaissance*, exh. cat., Minneapolis, 1978, 81-83, no. 24.

54 *Ibid*. See also Mrs. R. Barrington, *G.F. Watts*, New York, 1905.

55 Vsekhsvyatskii, 240-242; Kronk, 1984, 60-61.

56 Ms. Add. 161/1-3 in the Royal Astronomical Society Library, London, in white wash on paper prepared with black ink and backed by a single board. All three are signed *G.H. With*. With published an article on his observations, "Observations of Coggia's Comet," *Monthly Notices of the Royal Astronomical Society*, 36:6, 1876, 277-278, in which he relates that he was also including copies of three "sketches" made within a few hours of the time of each observation on 8, 11, and 14 July. A reproduction opp. p. 278 records the comet on 12 July, a date mentioned in neither the article nor noted on the three RAS images, although its style is similar to the three reproduced in this study which suggests that With rendered an even larger series.

57 See n. 56; it measures 162 × 114 mm and is inscribed and signed:
Coggia's Comet *8½ inch silvered glass Newtonian*
July 5th 1874 [bracket] *Power 125 ±* *G.H. With*
10:30 p.m.
Its verso is inscribed: *To be viewed in a north light/at a distance of 6 feet from/the eye G. With/No 1.*

58 See n. 56; it measures 161 × 114 mm and is inscribed and signed:
Coggia's Comet *12¼ speculum.*
July 11: 1874 [bracket] *6 ft 6 n focus* [bracket] *G.H. With*
10:30 p.m. *Power 125 ±*
Its verso is inscribed: *To be viewed in a north/light; at a distance of 8 feet from the eye./G. With/No 2.*

59 See n. 56; it measures 162 × 116 mm and is inscribed and signed:
Coggia's Comet *8½ Speculum*
July 14:1874 [bracket] *5 ft. focus* [bracket] *G.H. With*
10:30 to 11:30 p.m. *Power 300 ±*

60 From these inscriptions, we know that With used a telescope whose mirror was 8½ inches across for the first and third images, and one with a 12¼-inch mirror for the second. Since the light-gathering power is proportional to the mirror's area, the advantage of the larger mirror is the ratio of the squares of the diameters, $12^2/8^2 = 144/64$, or more than a factor of two. Thus, the larger mirror collects over twice as much light and gives a brighter image. In the third image, a shorter focal-length eyepiece was used than in the first, to give a magnification of 300 instead of 125, because the magnification of a telescope is the ratio of the focal length of the main mirror to the focal length of the eyepiece.

61 See Bond, 1862.

62 Vsekhsvyatskii, 252-253; Kronk, 1984, 64-65. It was also observed from Australia by Tebbutt.

63 Vsekhsvyatskii, 258-260; Kronk, 1984, 66-67. It was followed by another bright, naked-eye comet (C/1881 N1, formerly 1881 IV), again observed by Tebbutt.

64 Its long visibility allowed a thorough study that revealed several unusual characteristics: its single nucleus appears to have split into two elongated ones; six to eight smaller objects were noticed nearby; and beginning 16 October an anomalous sunward tail was visible to the naked eye. Vsekhsvyatskii, 265-267; Kronk, 1984, 69-70.

65 Chap. VI, n. 44.

66 D. Gill, "Observations of the Great Comet of 1882, made at the Royal Observatory, Cape of Good Hope," *Annals of the Royal Observatory, Cape of Good Hope*, 2:1, 1886, 2-19. See also W.L. Elkin, "Observations with the Heliometer," *Annals of the Royal Observatory, Cape of Good Hope*, 2:1, 1886, 19-25.

67 L.J. Robinson, "Book review of *Celestial Objects for Common Telescopes*," *Sky and Telescope*, 26, 1963, 95.

68 A.C.R., "Obituary: Thomas William Webb," *Monthly Notices of the Royal Astronomical Society*, 46, 1886, 198-201.

69 Clerke, 326, citing p. 306 of T.W. Webb, "On the Great Comet of 1861," *Monthly Notices of the Royal Astronomical Society*, 22, 1861-62, 305-314. For additional information, see, Robinson, 95-96; J. Ashbrook, "Astronomical Scrapbook," *Sky and Telescope*, 47, 1974, 26-27; K.G. Jones, ed., *Webb Society deep-sky observer's handbook*, Short Hills, NJ, 1979.

70 In the Royal Astronomical Society Library, London, not catalogued but given a provisional designation, RAS Archive Webb 5A, fol. 103r, in pen and brown ink measuring 196 × 155 mm, with an inserted drawing in watercolor and body color with oxidized white heightening on blue paper measuring 62 × 116 mm.

71 Thanks to Peter Hingley for the suggestion that Mrs. Webb may have been the scribe. For information on the Webbs, see *Webb Weekend 1990 – Nov. 24th-25th*, Hereford, 1990.

72 See *Compact Edition of Dictionary of National Biography*, 1, London, 1975, 1191. A considerable part of Brett's strange house, which he built for himself, was designed totally on utilitarian principles and devoted to astronomical observation. On the roof were mounted an equatorial telescope, resting on a solid brick pier going down to the foundation level, and an alt-azimuth reflector. For a list of notices of reports and articles and the publication of one drawing of the comet C/1882 R1, formerly 1882 II, by Brett in the *Monthly Notices*, see Anonymous, *General Index to volumes XXX to LII of the Monthly Notices of the Royal Astronomical Society 1869-1892*, London, 1896, 12. Much of Brett's artistic material is unknown and has remained in the possession of his descendants. See Cummings and Staley, 327-329; Staley, 1973, 124, n. 2.

73 As in a Brett drawing in black chalk with white heightening of Gassendi's crater in a private collection, New York. For the latter, see London, J. and S. Maas Ltd., *John Ruskin and His Circle*, sale cat., London, 1991, 37, no. 63. It is in black chalk heightened with body color measuring 216 × 159 mm and is inscribed *M 5 [or S] 84*, indicating a date in 1884.

74 See A. Bowness, "The Victorians," in *The Genius of British Painting*, ed. D. Piper, New York, 1975, 267, as well as Staley, 1973, 124-137.

75 It appears to have been drawn in black and white chalk and no measurements are known. The inscription reads: *Sketch of Great Comet/1882/22 [23 originally] Oct. 17h G.M.T./seen at Putney./John Brett.* J. Brett, "Sketch of the Great Comet (1882 b)," *Monthly Notices of the Royal Astronomical Society*, 43, opp. p. 288.

76 Inv. no. PR86-55, sketchbook no. 53, for 1881-82, page 0001r, in the National Maritime Museum Greenwich, in graphite measuring 82 × 175 mm. It is inscribed: *Comet/June 23, 81/11P.M./3 inch Cooke/P [power] ≥ 50.* According to the computer program Voyager II, at the time and place Brett observed the comet, it would have been in Auriga near the star Capella. Brett also studied it on nine successive pages until the end of June in views of the comet over a landscape as well as telescopic studies of its head.

77 Inv. no. PR86-55, sketchbook no. 48, page 9881v, in the National Maritime Museum Greenwich, in graphite measuring 89 × 177 mm. It is inscribed: *Great comet. Putney 23 Oct 1882 17* h*-G.M.T. /about 10° above horizon S.E./R.A.* X*h Dec 15° S approximately. Focussed 4" equatorial on Sirius. Comet in middle of field without altering dec*n*. clamp [.] Length of tail 17°-terminating in a triple fork, the central spike being the most definite and sharp [.] the north prong diffused into a cloud, leaving a well defined parabolic bay between the two. The southern prong a mere thorn. The southern edge of tail denser and better defined than N. and the general direction of parallel to the Ecliptic, but there is a sweeping curve in its form suggesting that it is part of an annulus seen in very oblique perspective. The nucleus is an oval equally difficult all round, and very undefined.*

78 H.H.T., "Obituary: John Brett," *Monthly Notices of the Royal Astronomical Society*, 62, 1902, 238-241.

79 Brett also participated in expeditions to study solar eclipses, where according to the writer in *ibid.*: "His trained eye was a most effective astronomical weapon." For example, he observed a total solar eclipse in December 1870, reported in J. Brett, "Extract of a report of the Eclipse Expedition of 1870 at Augustea, Sicily," *Monthly Notices of the Royal Astronomical Society*, 31, 1870, 163-165, which includes drawings. Moreover, his observations on planetary detail called attention to several points of great importance.

80 Reynolds, 1966, 119. Poynter was a practicing artist, who was also a connoisseur, collector, and administrator. He served as President of the Royal Academy as well as Director of the National Gallery (Charles Eastlake was the only other individual to hold these two posts simultaneously.)

81 Inv. no. 1883.18 in the Manchester City Art Galleries, in oil on canvas measuring 153 × 112.6 cm. Signed with a monogram and dated *18 EJP 83*.

82 Whereabouts unknown, in pastel on paper laid down on canvas measuring 76.2 × 55.9 cm. It is signed with a monogram and dated *19EJP16*. See Christie's, London, sale cats., 14/VI/91, no. 11, 19/VII/88, no. 110.

83 Whereabouts unknown, in oil on canvas measuring 40.5 × 28 cm. It is signed with a monogram and dated 1882. See Christie's, London, sale cat., 13/III/92, no. 94; Sotheby's, Belgravia, sale cat., 27/VI/78, no. 50.

84 Whereabouts unknown, measuring c. 40.6 × 28 cm. It is supposedly signed with a monogram and dated 1881 with a provenance from the collection of Lady Pease, Pinchinthorpe, as the other work in n. 83. See Christie's, London, sale cat., 29/VII/77, no. 189. One wonders whether these are really the same work and the date was mistranscribed in the 1977 sale catalogue.

85 In a private collection, New York, in graphite measuring 180 × 340 mm. It is inscribed: *Loch Luichart October 23.82. 4½ a.m.* See M. Bell, *Drawings of Sir E.J. Poynter Bart. P.R.A.*, London and New York, n.d., pl. 33.

86 When the Manchester painting was reviewed in the May 1883 issue of *The Atheneum*, its critic wrote: "...the pathos of the picture is due to the strangeness of the light, reversed from the natural order, and shining from below on the polished shafts, sculptured panels, and gilded mouldings, among which all sorts of weird shadows seem to lurk as if they were about to assume even more monstrous shapes."

87 Olson, 1985b, 28-29, fig. 23. The obverse has a profile portrait, while the reverse features an eight-pointed comet-star with a flaming tail inscribed *DIVVS IVLIVS*, signifying Divine Julius. Shakespeare was responding to ancient historians, like Suetonius, who in *The Lives*, wrote: "He...was numbered among the gods, not only by a formal decree, but also in the conviction of the vulgar. For at the first of the games which Augustus gave in honor of his apotheosis, a comet shone for seven successive nights...and was believed to be the soul of Caesar, who had been taken to heaven...." See also J.T. Ramsey and A.L. Licht, *The Comet of 44 B.C. and Caesar's Funeral Games*, Atlanta, 1997.

88 Certainly in this love story, Hardy was referring to George Eliot's novel of 1860, *Mill on the Floss*, where one character remarks: "I suppose it's all astronomers [who hate women] because, you see, they live up in high towers, and if the women came there they might talk and hinder them from looking at the stars."

89 Inv. no. 81'11 in the City Museums and Art Gallery, Birmingham, in black ink wash and white body color measuring 57 × 173 mm. It is

signed in white with Hardy's monogram *TH*. It is a headpiece to the poem on p. 129 in Hardy's *Wessex Poems and Other Verses*, London and New York, 1898, published by Harper and Brothers Publishers. In the preface, Hardy relates that while the poems date from various periods, the illustrations were all done "recently."

90 John Herschel named this asteroid Irene after the Greek Hour who opened the gates of Heaven and Mt. Olympus and was the personification of peace. Schmadel quotes Hind as stating that "You will readily discover that this name, properly Eirene (peace), has some relation to this event (the Great Industrial Exhibition) which is now filling our Metropolis with the talent of art and science, in which all...must feel an interest." L.D. Schmadel, *Dictionary of Minor Planet Names*, Berlin, 1993 (second edition), 20-23. Hind was a Fellow of the Royal Astronomical Society and its President for a two-year term. See also "Obituary: John Russell Hind," *Monthly Notices of the Royal Astronomical Society*, 56, 1895-96, 200-205; Dreyer and Turner, 250; O. Struve and V. Zebergs, *Astronomy of the 20th Century*, New York, 1962, 403-404.

91 See Anonymous, *A General Index to the First Twenty-Nine Volumes of the Monthly Notices of the Royal Astronomical Society, 1827-1869*, London, 1870, 104-107; Anonymous, *General Index to Volumes XXX to LII of the Monthly Notices of the Royal Astronomical Society, 1869-1892*, 63-65. His books on comets include: *Recent Comets and the Elements of their Orbits* ("Obituary: John Russell Hind," 204); *On the Expected Return of the Great Comet of 1264 and 1556 with a History of Former Appearances, compiled from various authors and ephemerides, for facilitating its Re-discovery*, London, 1848; *The Comets: A Descriptive Treatise Upon those Bodies. With... a Table of all the Calculated Comets, from the Earliest Ages to the Present Times*, London, 1852. He also analyzed the orbit for the comet of 1556 (Halley's Comet 1P) to determine whether it would return in 1858 or 1860 (*Sweeping ephemerides for the comet of 1556*, London, 1857; *The Comet of 1556; being popular replies to every-day questions...*, London, 1857). See Yeomans, 186-187.

92 P. Lancaster-Brown, *Comets, Meteorites and Men*, London, 1973, 31-32.

93 C.P. Oliver, "Obituary: William Federick Denning," *The Observatory*, 54, 1931, 276-283.

94 See London, the British Astronomical Association, "The British Astronomical Association: The First Fifty Years," *Memoirs of British Astronomical Association*, 36:2, 1948, 1-132 (reprinted as *idem.*, "The British Astronomical Association: The First Fifty Years," *Memoirs of the British Astronomical Association*, 42:1, 1989, 1-132). Thanks to Jacqueline Mitton for this information.

95 It was published by Marcus Ward & Co. Ltd., London, Belfast, and New York; see O. von Schleinitz, *Walter Crane*, Bielefeld and Leipzig, 1902, 95. Unfortunately, we have not been able to locate a copy of the work, either in England or the United States, so that the reproduction is from Schleinitz. See also P.G. Nonody, *The Art of Walter Crane*, London, 1902; I. Spencer, *Walter Crane*, London, 1975; G. Smyth and S. Hyde, *Walter Crane. Artist, Designer and Socialist*, exh. cat., London, 1989.

96 It is inscribed: *AN APPROPRIATE QUOTATION FOR EVERY DAY IN THE YEAR*, followed by a passage from Shakespeare on the border of the roundel: *I HEARD A MERMAID ON A DOLPHIN'S BACK UTTERING SUCH DULCET & HARMONIOUS BREATH THAT THE RUDE SEA GREW CIVIL AT HER SONG AND CERTA....* It continues on banderoles: *TO HEAR THE SEA MAID'S MUSIC; STARS SHOT MADLY FROM THEIR SPHERES.*

97 Inv. no. P.34-1917 in the Victoria and Albert Museum, London, in pen and black ink and watercolor measuring 382 × 268 mm. It is signed and dated: *Arthur Rackham 1908*. See L. Lambourne and J. Hamilton, *British Watercolours in the Victoria and Albert Museum*, London, 1980, 305. It was published in William Shakespeare, *A Midsummer-Night's Dream*, London, 1908, opp. p. 48, by William Heinemann. D. Hudson, *Arthur Rackham. His Life and Work*, London, 1960, 168, notes that a deluxe edition limited to 1,000 copies was signed by Rackham.

98 See Burke, 220-221.

99 Whereabouts unknown, in watercolor and pen and black ink measuring 380 × 277 mm; see Christie's, New York, sale cat., 30/X/85, no. 480. It is signed and inscribed: *Wonder Book* on the verso *54/no 82* and *Falling Stars* (scratched out); on the verso of the mount it is inscribed: *Hercules Supporting the Sky in Place of Atlas (Hawthorne's Wonder Book)*. *A Wonder Book* was published by Hodder & Stoughton in London and the illustration was reproduced opp. p. 130 (measuring 182 × 126 mm). Hudson, 171, notes a deluxe edition of 600 signed by Rackham.

100 J. Hall, *Dictionary of Subjects and Symbols in Art*, New York, 1979, 150.

101 Stock is one of the less well-known but highly individual, albeit sometimes eclectic, figures of the latter day English Romantic tradition. He painted portraits for his bread and butter, but his main interest was illustrating imaginative subjects from the Bible or literature or symbolist subjects. He identified with Blake, who had also been born in Soho and who later had lived at Felpham, where Stock settled in 1909.

102 Whereabouts unknown, in graphite and watercolor with touches of white heightening measuring 333 × 232 mm. It is signed and dated *H.J. Stock./1902*, numbered on the verso *No 3*, and inscribed on an old label on its backing: *And, I saw a Star fall/from heaven unto the/Earth./Rev. IX.1./Henry J. Stock*. See Christie's, New York, sale cat., 4/XI/94, no. 58.

103 Whereabouts unknown, in graphite and watercolor heightened with body color measuring 330 × 232 mm; see Christie's, New York, sale cat., 4/XI/94, no. 62. It is signed and dated *H.J. Stock./1910* and inscribed on its backing: *And the whole moon became as blood-/and the stars of heaven fell unto the earth.*

104 Whereabouts unknown, in oil on canvas measuring 71 × 91.5 cm; see Sotheby's London, sale cats., 25/XI/87, no. 87, 3/XI/93, no. 247. In 1909 Sims, who became an Associate of the Royal Academy in 1908, submitted for exhibition at the Royal Academy a version of the composition with minor variations (in tempera and oil on canvas measuring c. 71 × 91.5 cm; illustrated in *Pall Mall Magazine Pictures of 1909*, on p. 4; *The Studio*, 2, 1909, on p. 34; C. Sims and A. Sims, *Picture Making. Technique and Inspiration*, London, 1934, pl. 18, discussed on pp. 80-81). The Frick Art Reference Library, New York, notes it inscribed on a scroll:

To Julia
Her eyes the glow worm lend thee
The shooting stars attend thee
The elves also Sims 09

It also notes it in the George A. Hearn Collection sale, American Art Galleries, New York, 25/II/11, no. 157, sold to J.C. Elms. See H. Speed, "Charles Sims, R.A.," in *The Old Water-colour Society's Club 1928-1929*, London, 1929, 48; Hutchison, 127.

105 Whereabouts unknown, in tempera and gold leaf on canvas laid on shaped panel measuring 71.2 × 33.7 cm. It is signed with the artist's monogram. See Sotheby's, New York, sale cat., 20/VII/94, no. 238; Christie's, London, sale cat., 8/IX/96, no. 44.

106 Both are reproduced in M.A. Armfield, *Tempera Painting Today*, London, 1946. Pl. XV is *Evening*, identified as "Prayer or the Sacrifice of Obedience (Reception)," illustrating Psalm 40. Pl. XVI is *Morning*, identified as "Prayer of the Sacrifice of Praise (Expression)," illustrating a verse from Hebrews 13 ("Let us offer the sacrifice of praise to God: the fruit of our lips."). *Morning* was in the exhibition London, Fine Arts Society, *Homage to Maxwell Armfield*, exh. cat., London, 1970, no. 112.

107 Cunningham, 321. Earlier he noted (p. 291) that plutocracy began to supplant aristocracy. Certainly that had an impact on cultural expression with a levelling of culture and entertainment. There was, however, much optimism about the changes. Anthony Trollope in 1868, for example, expressed confidence that England possessed the largest and wealthiest leisure class in Europe with time and money to spend. The Universal Exposition of 1851 already had testified to this attitude.

108 Pang, 1994, 249ff.

109 *Ibid.*, 255. See also: W. Ivins, *Prints and visual communication*, Cambridge, MA, 1973; G. Bridson and G. Wakeman, *Printmaking and picture printing: A bibliographical guide to artistic and industrial techniques in Britain 1750-1900*, Williamsburg, VA, 1984.

110 Pang, 1994, 250, who also cites the instructions of Brett and Holiday for draughtsmen written in 1870 and 1871 for the Royal Astronomical Society eclipse observations.

111 *Ibid.*, 269. Pang (p. 255) remarks:

> In the field, large telescopes and photographic plates replaced astronomers equipped with modest instruments and sketch-pads, and at home photomechanical printing technologies replaced lithography and steel engraving. Astronomers came to distrust composite drawing, while seeing photographs as flawed but potentially flawless systems for representing astronomical reality. In terms of technology, practice and expectation, the 1880's were to witness profound changes in observing practices in astronomy.

Pang (1995, 67) points out that not everyone was willing to give up astronomical drawing and the British Astronomical Association, which sponsored four expeditions for amateurs, was at the forefront of an effort to reform drawing styles and to set drafting standards.

112 See L. Goldschmidt, "Tangible fact, poetic interpretation," in *The truthful lens: A survey of the photographically illustrated book 1844-1914*, ed. L. Goldschmidt and W.J. Naef, New York, 1970, 3-7; C. Fox, "The engravers' battle for professional recognition in early nineteenth-century London," *London Journal*, 2, 1976, 3-32; A. Dyson, *Pictures to print: The nineteenth-century engraving trade*, London, 1984.

113 Staley, 1968, 30.

114 Johnson, 1986, 260ff, discusses these changes and the emphasis on Art for Art's sake, as well as the change in the patronage system with the entry of large and powerful galleries, which acted as agents and provided exhibition space.

VIII

Comets and the New Century

The turn of the century wiped the nineteenth-century slate clean and called forth the dawn of a new age in many people's minds. It was, therefore, a time of assessing the past and predicting the future. This feeling was true for many artists, a goodly number of whom embarked on experimental paths toward non-representational art, as well as for astronomers.[1]

Century of Progress: Nineteenth-Century Comet Astronomy Appraised

From time to time during the eighteenth and nineteenth centuries, people attempted to survey how astronomy had changed in the previous interval, often after one hundred years. For example, in 1850 Elias Loomis, a professor of mathematics and natural philosophy then in New York City and later at Yale, wrote about "recent progress," with the implicit statement that progress had been made.[2] In Scotland, Hector MacPherson, Jr., wrote about "progress," in the previous hundred years,[3] as did Reginald Waterfield in England.[4] All three of these authors demonstrate how internationally widespread was the interest in comets. Nevertheless, Great Britain was no longer perceived as being in the forefront of comet and meteor studies.

McPherson, writing in 1906, framed his review of the century's astronomy in terms of Sir William Herschel, whose interests advanced from the solar system to beyond the external star system. Nevertheless, there was still tremendous interest in solar system objects, and the beginning of the century had seen the discovery of asteroids (a name invented by Herschel).[5] In his fifteen-page

W HEATH ROBINSON

Figure 157. William Heath Robinson, *Searching for Halley's Comet At Greenwich Observatory*, from *The Sketch*, 17 November 1909, 1909, engraving, 290 × 170 mm, National Maritime Museum, Greenwich.

chapter on comets, McPherson discusses several bright ones that had appeared during the previous century.[6] He suggests that comets had already ceased to be objects of superstition, and that this modern understanding dates from the Great Comet of 1811. In addition, suggesting the international nature of astronomy, the only British astronomer he mentions is Huggins, the first to try to photograph a stellar spectrum.

Waterfield also stresses the importance of William Herschel's work.[7] In his comet chapter, he discusses mainly theories rather than the individuals who made advances.[8] Again, the scientists whose work he cites are widely distributed around the world.

Halley's Comet 1910

When these assessments of astronomy were written, the world was changing rapidly, and skies were no longer as dark as they had been in prior centuries. It would have been rare for anyone to make an observation like that Edward Pigott, who wrote in 1793 about a comet:

> "Walking home last night I accidentally cast my eye on a Comet; as it is equal in brightness to stars of the 3^{d} magnitude, possibly it may have been already discovered. . . ."[9]

As urbanization and electrification took control, light pollution continued to grow, robbing the public of the spectacle of the heavens and greatly diminishing an individual's chance that he or she would discover a comet or even see a spectacular one.

But at the turn of the twentieth century, on 1 January 1901, astronomers, artists, and the public knew that they could soon expect the return of Halley's Comet, due back in 1910. At the Royal Greenwich Observatory, Philip H. Cowell (1870-1949) and Andrew C.C. Crommelin (1865-1939) recalculated the positions for the recovery of Halley's Comet. In works published between 1907 and 1910, they used a variety of methods to include the effects of gravity of the relevant planets.[10] The expected cometomania was satirized in a caricature by W. Heath Robinson, which shows "astronomers" at the Observatory searching for the comet, using a magnifying glass, binoculars, a spyglass, and eyeglasses to augment their ridiculous jury-rigged telescope aimed at a comic, snake-like comet [Fig. 157].[11] Another illustration by Lewis

Brammer in *Punch*, which depicts not the comet itself but a thief breaking into a house during the spectacle, testifies not only to the public's interest in viewing the comet, but also to its relative dispassion toward the event.[12] Apparent in both is also the lack of interest in rendering the comet itself, the antithesis of the attitudes prevalent in the eighteenth and nineteenth centuries.

However, there were astronomical discoveries to be made during this apparition. What made the earth's encounter with Halley's Comet in 1910 so interesting and unusual was that our planet was positioned in such a manner that the comet passed precisely between the earth and the sun. At that time, earth was closer to the comet but as the two swept by one another travelling in opposite directions, the comet passed directly in front of the sun. Because it could not be seen as a dot as it passed over the sun's surface, astronomers decreased estimations of the size of its nucleus. In addition, cyanogen, a poisonous gas, was discovered in its coma by spectroscopy. Needless to say, the prospect of cyanogen poisoning the earth, when the tail brushed the planet, caused consternation among nonscientists. Despite assurances that the comet posed no threat, numerous stories worldwide testify to the public's panic. The blatant exploitation of the fright and the fascination by the media, sometimes playful, led to the enjoyment of the event by those staging comet parties or composing music, such as "The Comet Rag," or producing an endless flow of ephemera with depictions of this famous comet during the last gasps of the Belle Epoque.[13]

International cooperation for studying Halley's Comet in 1910 was vital, since the comet usually was observable only during the time just after sunrise or just before sunset, making it important to have observations from locations scattered along all the earth's longitudes. In the United States, the Astronomical and Astrophysical Society of America[14] set up a committee to coordinate forthcoming observations.[15] Edward Barnard of the Yerkes Observatory in Chicago, an expert on astronomical photography, was on the committee. He was especially interested in obtaining wide-field photographs of the tail. When the Comet Committee's appeal was published, it was badly received in England. Only American manufacturers of telescopes and photographic plates had been listed. The editors of *The Observatory*, an independent

British journal of astronomy still publishing today, wrote, "May we who live on this side of the Atlantic be permitted to remark, that though we occasionally have to regret seeing art treasures from this side pass over to America, we are not disposed to give up our rights in Halley's Comet just yet."[16]

As the historian of astronomy John Lankford describes, British astronomers had "sometimes resented the rise of U.S. astronomy to world-class rank. (Transcripts of Royal Astronomical Society meetings indicate members tended to be inordinately skeptical of American research.)"[17] With hindsight, the center of gravity of astronomy was indeed changing, especially with new telescopes in the United States on the mountain sites of the Lick Observatory (on Mount Hamilton) and the Mount Wilson Observatory. It was decades before the British would move their main telescopes to mountain sites, in spite of Piazzi Smyth's pioneering study of Tenerife.

The British also began preparing for Halley's return with great fervor. Photographs were taken by British astronomers in 1909 in an early search for the comet. They included observations made by Harold Knox-Shaw in Helwan, Egypt; interestingly, the telescope used to guide the tracking of the photographic telescope on that occasion was one of those that John Herschel had used to observe Halley's Comet from the Cape of Good Hope in 1836. Although some of these photographs contained images of Halley's Comet, the comet was not noticed before its official recovery by Maximilian Wolf (1863-1932) in Germany on 11 September 1909. Cowell and Crommelin had made preliminary calculations about the date of its reappearance, although their predictions had not included the effects of non-gravitational forces, that is, the jet effects of gases being sublimed off the comet nucleus under solar heating. Amazingly, Cowell and Crommelin had been off by less than 3 days in predicting its perihelion. Their mistake is understandable since the effects of the nongravitational forces can easily account for that small deviation.[18] Even though hundreds of photographs were taken of Halley's Comet in 1910, when it came so close that the earth went through the outer parts of its tail, not a single British photograph appears in the book of photographs of Halley 1910 edited by Bertram Donn, Jürgen Rahe, and John C. Brandt,[19] nor are any in the collection of the

Royal Observatory Edinburgh.[20] Indeed, the photographs of Halley's Comet available at the Old Royal Observatory at Greenwich , now a museum, are from Mt. Wilson.

Subsequently, British astronomers tried to retain their observational capabilities. In the 1950s, the British built a 2.6-m (104-inch) telescope yet put it at Herstmonceux Castle in Sussex, where the Royal Observatory had been moved, a low, foggy site that greatly impeded its usefulness. (It has since been moved to the Canary Islands, and the Royal Observatory relocated to Cambridge.) More profitably, the British teamed with the Australians to build in the 1950s the Anglo-Australian Siding Spring Observatory, Australia. A large reflecting Anglo-Australian Telescope (1975) and the U.K. Schmidt Telescope (1973), a wide-field telescope, are on the site.[21] Some of the best photographs of Halley's Comet in its 1985-86 apparition [Figs. 158, 160] were taken with those telescopes and processed by Brian H. Hadley of the Royal Observatory Edinburgh and David F. Malin of the Anglo-Australian Observatory. British observations of this apparition were also taken with the U.K. Infrared Telescope, which was erected on Mauna Kea in Hawaii (1979) and the Isaac Newton Telescope now on La Palma in the Canary Islands. In 1996, SOHO (the Solar and

Figure 158. UK Schmidt Telescope in Australia, *Halley's Comet 10 March 1986 with a tail disconnection event*, processed at the Royal Observatory Edinburgh.

Heliospheric Observatory), a joint ESA-NASA program, captured a comet hitting the sun in a coronagraph [Fig. 159 color], which occults not only the sun (shown by a white circle) but also some region around it. SOHO discovers about a comet a week. They are almost all in the family of sungrazers and do not emerge from the sun; presumably they have vaporized or, in some cases, hit the sun when hidden from view behind the occulting disk.

Sadly, it was announced in 1997 that the Royal Observatories will be no more. Many of the functions of the Royal Greenwich Observatory, most recently in Cambridge, and the Royal Observatory Edinburgh, have been combined in Edinburgh under the name of the Astronomy Technology Centre. The historic buildings in London are now known as the Royal Observatory Greenwich, and are part of the National Maritime Museum.

Notes

1 For twentieth-century comet art, see Olson, 1985b, 108ff.

2 E. Loomis, *The Recent Progress of Astronomy; especially in the United States*, New York, 1850.

3 H. MacPherson, Jr., *A Century's Progress in Astronomy*, Edinburgh and London, 1906.

4 R.L. Waterfield, *A Hundred Years of Astronomy*, London and New York, 1938.

5 MacPherson, 97-102.

6 *Ibid.*, 123-127.

7 Waterfield, 153-154.

8 *Ibid.*, 457-473.

9 No. L&PX.29, p. 1 in the Archives of the Royal Society, London.

10 Yeomans, 258, 456. H-H. Altfeld, *Bibliographical Guide for Cometary Science*, Munich, 1983, 58-59, lists nine joint papers by them on the perturbations of Halley's Comet through 1907 and three summary papers. For other references to their work on comets, see Freitag, 519.

11 An engraving measuring 290 × 170 mm, from *The Sketch*, 68, 17 November 1909, p. 175.

12 Illustrated *Punch, or The London Charivari*, 138, 1 June 1910, p. 403.

13 For representations of comets around the 1910 apparition and the commercial approach of advertising using comets, see R. Etter and S. Schneider, *Comet Halley. Memories of 1910*, New York, 1985; M. Littmann and D.K. Yeomans, *Comet Halley: Once in a Lifetime*, Washington, D.C., 1985, 37ff; J. Metz, *Halley's Comet, 1910: Fire in the Sky*, St. Louis, MO, 1985.

14 Now the American Astronomical Society.

15 J. Lankford, "How Not to Run an International Halley Watch," *Sky and Telescope*, 70, 1985, 530-533.

16 *Ibid.*, 532.

17 *Ibid.*

18 Yeomans, 258.

19 B. Donn, J. Rahe, and J. C. Brandt, eds., *Atlas of Comet Halley 1910 II*, Washington, D.C., 1986.

20 B.H. Hadley, personal communication, 1994.

21 See Pasachoff, 1995, 60-62.

EPILOGUE

Comets and the New Millennium

BY COLIN T. PILLINGER

This book has concentrated on comets as they were observed in Britain or studied by British scientists. In this epilogue, let me discuss British contributions to cometary science in the past and future. Some of these contributions are through the European Space Agency (ESA), of which Britain is a member.

Halley's Comet 1985-1986

In 1985, when the ESA launched its Giotto mission to fly-by comet Halley, the prime contractor for engineering the spacecraft was British Aerospace. This spacecraft, however, was not the earliest British Aircraft industry product related to comets and meteors. Earlier in the century, it had produced two notable engineering firsts named after rapidly moving heavenly bodies. One of these innovations was the speedy Gloster Meteor, the second jet fighter aircraft in service, used by the Royal Airforce just at the end of the Second World War in the battle against V1 missiles. The other was the De Havilland Comet, the first jet passenger aircraft to grace the world's airways. The two names carry just the right amount of symbolism – the meteor flashy, instantaneous, and with a connotation of destruction needed for its military context, and the comet conveying more serenity and glamour for commercial use. As we have seen in previous chapters, the two words have been used since the eighteenth century for mechanical and transportation items that usually involved speed.

Names that make a connection to mythology or classical periods of history are very popular with space mission teams. In

Figure 160. Halley Multicolour Camera aboard the European Space Agency's *Giotto* spacecraft, *Nucleus of Halley's Comet with Jets of Water Vapor and Dust*, 1986, Courtesy of H.U. Keller, © 1986 Max-Planck-Institut für Aeronomie, Lindau/Harz, FRG.

calling its Halley spacecraft "*Giotto*," the ESA opened up a connection to the art world, since it was Giotto di Bondone who painted the 1301 apparition of the comet in his Adoration of the Magi fresco in the Scrovegni Chapel, Padua.[1]

Adopting an artistic image as a logo experienced a new twist with a British experiment called "DIDSY" (Dust impact detection system) included as part of the Giotto payload. DIDSY was an international collaboration led from the University of Kent at Canterbury by Tony McDonnell and John Zarnecki, with engineering input from Ray Turner at the Rutherford Appleton Laboratory. Its objective was to characterize the flux of particles from the comet during the mission. The Canterbury connection with comets, particularly Halley, can be traced back to its 1145 apparition and to another work of art: a marginal drawing of Halley's Comet by the cathedral monk Eadwine in a Psalter (a manuscript collection of the Psalms) that he was illustrating at the time.[2] This manuscript, which has the earliest British depiction of a comet known, was removed from Canterbury when Thomas Neville decided to take it with him to Trinity College, Cambridge, on being appointed master in 1615. It would be interesting to know if Trinity's greatest scholar, Isaac Newton, ever saw Eadwine's "doodle." Even if he did, neither he nor Halley would have recognized that was part of the pattern for Halley's momentous prediction concerning the return of this comet in 1758.

Beside Eadwine's picture is a legend which refers to the fact that comets appear infrequently and that they are portents; it certainly did not act as an evil omen for DIDSY, which was one of the huge successes of *Giotto* when it passed approximately 600 kilometers from the nucleus on 14 March 1986, the closest of five spacecraft to make the trip [Fig. 160].[3] Despite being hit by 1000 impacts, some of particles over 10 milligrams in mass travelling at 68 kilometers per second, the experiment survived and went on to fly past another comet, Grigg-Skjellerup, as part of the Giotto extended mission (GEM). Other experiments were not so lucky. The Multicolour Camera was disabled during the encounter, but not before its team, led by Uwe Keller of the Max Planck Institut für Aeronomie and including British astronomer David Hughes, had obtained images never before possible.[4] The Giotto mission included another important British experiment, led by Alan Johnstone of the Mullard Space Science Laboratory, to study the plasma that constitutes the enormous display associated with comets. This experiment, too, participated in the extended mission. While there was lots of activity in space, there was still room for contributions from the ground, and several British astronomers were involved in the International Halley Watch, a worldwide monitoring program. A notable achievement was the wide-field photography of the comet's tail carried out with the Anglo-Australian Telescope at Siding Spring, Australia [Fig. 158].

Rosetta

The future involvement of British scientists in cometary research is now guaranteed, insofar as any space launch can be, well into the twenty-first century through the ESA's Cornerstone mission to the comet 46P/Wirtanen. This mission goes by the name "Rosetta," after the Rosetta stone, because of the potential it holds for understanding the events occurring in the early solar system. The Rosetta stone, from ancient Egypt and now in the British Museum, led to the deciphering of Egyptian hieroglyphics because it carries the same inscription not only in hieroglyphics but also in two other languages that were understood: Demotic script and Greek.

Originally Rosetta was conceived as a mission with NASA to return cometary samples to earth, but because of funding difficulties in the USA, it became a rendezvous activity. The mother spacecraft and its instruments will orbit a comet while another package of experiments will be released to land on the nucleus close to aphelion. The spacecraft will then travel through the solar system with the comet as it evolves and grows its coma and tail. The beauty of this scenario is that Rosetta will both investigate the pristine chemical and physical nature of the materials constituting the inactive solid body and document the processes of the cometary phenomenon that have fascinated mankind since time immemorial. The drawback is the time it is going to take. Launch is scheduled for the year 2003, the encounter in 2011, and the culmination with perihelion passage in 2013.

Originally several comets were potential targets, but mass and power limitations have narrowed the choice down to one object, the 5.5-year-period comet 46P/Wirtanen. A photograph used in the selection process was taken by the British astronomers Iwan Williams and Alan Fitzsimmons, using the 1-m Jacobus Kapteyn telescope on the Canary Islands. Their picture of Wirtanen was acquired almost accidentally, when a break in their planned observations meant they had a few spare minutes. During 1996, Wirtanen passed through the inner solar system and was the subject of a very extensive observing program, since it was a known target for Rosetta. The size of its nucleus is probably no greater than 1.2 kilometer in diameter and the maximum rate at which it gives off molecules is 1.5×10^{27} molecules per second (that is, 15 followed by 26 zeroes). Since by far the most abundant species in a cometary coma is water (80 percent), Andrew Coates of the Mullard Space Science Laboratory has come up with a very descriptive analogy. The display developed by Wirtanen is the equivalent of throwing a bath full of water into space every second. Extending this unit of measurement to C/1995 O1 (Hale-Bopp) suggests it was shedding a small swimming pool every second at its maximum activity during its 1997 apparition.

The actual logistics of landing on comet Wirtanen, since no one knows the consistency of its surface – rock hard ice, delicately soft snowdrifts, or a crust of mineral and organic refractories (high melting-point substances) left after sublimation – is not a simple one. Problems, such as anchoring to a body with a c. 1 kilometer diameter that has virtually no gravity, have to be addressed. The thermal regime – from the −200 °C of deep space to temperatures that might be several hundred degrees on dark, absorbing areas when the comet is closest to the sun – has to be considered for survival. The spacecraft involved will be coated with solar cells, which produce about 10 watts to power the experiments when they are 3 astronomical units from the sun. Building its lander is a collaborative effort among many countries, including Great Britain, with the lead coming from Germany and the chief designer, Helmut Rosenbauer of the Max Planck Institut für Aeronomie, Lindau.

The name "Rosetta," selected by the ESA for this cometary mission, continues the links between science and art. The Rosetta stone itself was found in 1798 by Pierre François Xavier Bouchard, an officer in the army of cometophile Napoleon Bonaparte at Rashid (Rosetta) in Egypt and deposited in the National Institute in Cairo. When this city was threatened by British military operations, the stone was moved to Alexandria which, in turn, was attacked and capitulated. The Rosetta tablet by this time was packed and hidden in the luggage of the defeated commander General Menou. Although he claimed it was his personal property, he was forced to hand it over. Two men with strong links to the subject of meteoritics were eye witnesses to the events that unfolded during this confrontation in the streets of Alexandria. One was Sir William Hamilton FRS – sometime President of the Society of Antiquaries and British Ambassador to Naples and the uncle of Charles Greville.

Hamilton's money had allowed his nephew to build up his mineral and meteorite collection that was valued at £14,000 when purchased for the nation by the British Museum. Sir William also provided the sample of the Siena meteorite seen by King and analyzed by Howard. The other witness was Edward Daniel Clarke, a

British traveler, famed as an undergraduate for flying a hydrogen-filled balloon, piloted by the Jesus College Cambridge cat! Clark, who became the first professor of mineralogy at Cambridge in 1808, was a meteorite fanatic, giving lectures on the subject of stones from the sky which attracted large audiences. To illustrate his lectures he constantly badgered James Sowerby to loan him the Wold Cottage Meteorite,[5] though he had little success.

One might have hoped that hostilities between Britain and France concerning the Rosetta stone would have ceased long ago. But it was not to be, because argument continues over the translation and who was responsible for it. The British feel that Thomas Young, the polymath Foreign Secretary of the Royal Society, a man easily recognizable as a scientist (although the term was not coined until 1833), played a key part and should have been given substantial credit for the translation. The French, on the other hand, argue that Jean François Champollion, a scholar of linguistics, was responsible. The truth is that both played a part, as did a number of others. However, Young should be credited with first having demonstrated that the hieroglyphics inside elongated ovals or cartouches were proper names. He correctly transcribed *Ptolemy* from the Rosetta stone and identified *Berenice* (an Egyptian Queen) and *Tuthmosis* (another King) in other locations. By 1819 he published, in the *Encyclopaedia Britannica*, an assigned meaning for some 200 of the known 700 hieroglyphics, which he had correctly realized were a mix of ideograms and phonograms. He also communicated directly his information to Champollion who, as late as 1821, was maintaining that hieroglyphs were signs, not sounds. Suddenly, in 1822, Champollion published a full and accurate translation of the Rosetta stone; he had found the cipher, the name Cleopatra, ironically provided on a tablet discovered by a British archaeologist, W.J. Bankes. Champollion never admitted in print to any contribution by Young but to some extent Young was partially himself to blame, because he frequently published his work anonymously so that he could not be accused of neglecting his duties as a medical practitioner.

Because of Young's prodigious interests it is perhaps inevitable that he should impinge, if only briefly, on comets and meteors, and show some artistic talent. At a very early age it was recognized that he was specially gifted when he mastered seven languages. As a student, he was a prodigy, studying at three universities: Cambridge, Edinburgh, and Göttingen. After qualifying as a physician, he turned physicist; his name is immortalized in that field by the measure of elasticity known as Young's Modulus. His main research interests, however, concerned light and vision, which certainly fall within the spectrum of both art and science. Among the enormous legacy he left were major contributions concerning how the eye works, the first diagnosis of astigmatism, and the three-receptor theory of color vision. Young also demonstrated the undulating nature of light via the classic double-slit experiment and in doing so solved the problem of Newton's rings, which had defeated Newton himself. While he was never an artist in the true sense, his draftsmanship is amazing, as can be seen from the notes to lectures which he gave as Professor of Natural Philosophy at the newly formed Royal Institution and which were published in 1807. Moreover, he presented at least one regular lecture on comets and meteorites, which he illustrated with Newton's description of Halley's Comet during its perihelion passage and the Lord Paisley diagrams of the comet of 1723 [Fig. 22] redrawn by himself. As another demonstration of his drafting skills, there is a note in the Combination Room Book of Emmanuel College, Cambridge, which acknowledges that Young won a bet for being able to rule 100 lines to the inch. On top of all these accomplishments, Young published explanations for solar/lunar haloes, explained the tides, and gave the proper definition of energy, among many other things.

Young's work on hieroglyphics enables us to identify possible inscriptions related to meteors, meteorites, and comets from Egyptian sources. The word for star is used sometimes in phrases such as "stars falling from heaven," which must mean shooting stars. Likewise, *ssd* (pronounced seshed) means brilliant or flashing light; it could be related again to shooting stars or a comet. Finally, the words *biz-n-pt* (pronounced bia-en-pit) could simply be iron from the sky. The literal translation is a bronze or copper mineral of heaven. An interpretation of this strange association is that copper is equated with blue (the sky), and the meteorite was a piece which broke off.[6] For many years it was used as a logo on the front of *Meteoritics*, the Meteoritical Society journal.

Given their roles in unravelling the inscription on the Rosetta stone, it seems highly appropriate that this ESA space mission, which hopes to decipher the composition of comets, should honor Champollion and Young. Indeed, French scientists chose to give the name Champollion to the package of instruments they intended to land on the comet. The French lander, however, was abandoned because NASA collaborators were unable to commit to funding on the time scale needed. The French instruments thus were incorporated into the German-led lander. Young is commemorated extensively in the British experiment, which is called "MODULUS" after his most well known concept; MODULUS stands for Methods Of Determining and Understanding Light elements from Unequivocal Stable isotope compositions. It is a very ambitious project in that it employs chemical reactions in space for the first time. In order to obtain clear cut measurements of isotopes of various elements (D/H, $^{13}C/^{12}C$, $^{15}N/^{14}N$ and $^{18}O/^{17}O/^{16}O$) at high precision, it is necessary to remove some problems that occur in low resolution mass spectrometers. The problem, simply stated, is that unless a mass spectrometer is very large, which automatically would make it too massive for a space instrument, it cannot tell the difference between the molecule ^{16}OH and the atom ^{17}O, since both have the same integer mass (that is, 17 atomic mass units) though both are actually very slightly different in total mass. The difficulty can be overcome in an elegantly simple fashion by converting all molecules to their elemental constituents. These changes can be effected by carrying the reagents required in a solid state and switching them on by applying heat. MODULUS also incorporates the necessary means (gas chromatography) of resolving complicated mixtures; for example, carbon dioxide, carbon monoxide, formaldehyde, and nitrogen all have ions at mass 28 and 29 and so must be separated before quantitative isotope work can be attempted. MODULUS also will be able to identify unknown molecules from their fragmentation pattern (analytical mass spectrometry) and will provide absolute and relative abundances. Another refinement is the possibility of continuously recalibrating the instrument *in situ*, essential after a ten-year dormancy during the flight to the comet.

MODULUS consists of two sensors, one on the lander (*Ptolemy*) and one on the orbiter (*Berenice*); the link to the key words recognized by Young on the Rosetta stone is obvious. *Ptolemy*'s goal is to identify the ground-truth state of molecules in the comet nucleus and to this end it will analyze samples collected by drilling. It will provide information about the building blocks of the solar system; the isotopic compositions may identify primordial matter not mixed in or homogenized with the rest of the material, which went to form the solid planets or the asteroids that are the parent bodies of most meteorites. *Ptolemy* will also address other fundamental questions about whether the earth's oceans would have been derived from comets like Wirtanen. *Berenice* will monitor the changes in composition, both in terms of compounds and isotopes, as the comet evolves. Its objective is to unravel the mechanisms that drive the cometary phenomenon, the rates and timings of processes which have made comets such awe-inspiring manifestations throughout history. According to Young in one of his lectures, a comet's coma was of great interest in this regard: "It is possible that, on account of the intense cold to which comets are subjected in the greatest part of their revolutions, some substance may be retained in liquid or even in a solid form until they are disengaged by the sun's heat." Today, measurements of isotopic compositions of species like water and carbon dioxide, the primary aim of the MODULUS experiment, are exceedingly informative when investigating processes driven by thermal evaporation on earth, such as the hydrogeological cycle, the interaction between the atmosphere and the oceans, and the sedimentary record.

Even though very few isotopic measurements exist for comets, since it is very difficult to make high-precision measurements from the ground by spectroscopy; those available trace their origins back to studies by British scientists in the mid-nineteenth century.[7] Carbon isotopes were first discovered in 1929, when it was recognized that carbon-rich stars gave spectra which could be interpreted in respect of the existence of ^{12}C-^{12}C, ^{12}C-^{13}C and even ^{13}C-^{13}C bonds. The data were seized upon by N.T. Bobrovnikov of the Lick Observatory to explain bands that he had seen in the spectrum of Halley's Comet during its 1910 passage. Many comets, in fact, show the same bands, called "Swan bands" after the British scientist William Swan. They are interpreted as coming from a carbon molecule of as yet unknown parentage. They were named for the Professor of Natural Philosophy at St. Andrews University in Scotland by the amateur astronomer Sir William Huggins in 1868.[8] The first spectroscopic observations of a comet had been those of C/1864 N1 (Tempel), formerly 1864 II on 5 August 1864. These were followed in 1866 by cometary studies of Huggins, who matched the emission features seen by Donati and himself with results obtained by Swan from electrical discharges in ethylene vapor.

Not only does it seem that a Briton may have made the first photograph of a comet, but also there has never been any question about the first photographic spectrum. It was obtained by Huggins on 24 June 1881, from C/1881 K1 (Comet 1881 III). His measurements revealed strong features subsequently proved to be due to CN, in addition to Swan bands. Over a hundred years later, only these two species in comets are well-enough understood for them to be used to give ground-based estimates for the $^{13}C/^{12}C$ ratio, which resembles that of earth though within very large uncertainties.

The mid-nineteenth century was a vital time for spectroscopy; another scientist who made a huge contribution was Sir Norman Lockyer. Again inspired by the spectacular meteor display of 1866 and the apparition of a comet (he was invited to view Coggia's Comet of 1874 through a 25-inch refractor at Gateshead), Lockyer turned his attentions to the links between these bodies. Since radioactivity or nuclear fusion had not yet been discovered, Lockyer believed that the sun was fueled by in-falling meteorites. In fact his ideas went further than that, as summed up in his own words: "All self-luminous bodies in the celestial spaces are composed of either swarms of meteorites or masses of meteoritic vapor produced by heat." Thus he was fascinated by comparing the spectra of the sun, stars, and comets with meteorites heated in vacuo in the laboratory, in an attempt to prove what became known as his Meteorite Hypothesis, a statement of the results of a spectroscopic inquiry into the origin of cosmical systems.[9] In the event, he actually obtained the right answer for the wrong reason, because our best estimate of the composition of the sun comes not from stellar observations but from the careful chemical analysis of the most primitive meteorites, a type known as CI carbonaceous chondrites. Lockyer also discovered and named helium, which composes around 25 percent of the sun but was not known on earth until thirty years later when it was found in 1895 as a radioactive decay product. His most lasting creations were the Science Museum, which began with the loan of equipment from him, and the journal *Nature*, which he founded in 1869.

Lockyer is worth mentioning in connection with Rosetta, because it is his philosophy for investigating meteorites that MODULUS will apply on the surface of 46P/Wirtanen. It has been used successfully to study the isotopic composition of trace constituents in carbonaceous meteorites over the last twenty years. Lockyer studied every kind of meteorite he could obtain and gently heated them with an electric current gradually increasing the temperature. He then went on to use a Bunsen burner and an oxy-coal gas flame, noticing the changes in the spectra. He rationalized that the constituents would appear not in their order of abundance but according to volatility, thereby inventing stepped heating, which is now used by virtually everyone currently seeking traces of interstellar grains in meteorites. The current method is to recognize anomalous isotopic composition in carbon, nitrogen, and noble gases as a function of extraction temperature, and as a result minerals like diamond and silicon carbide have been recognized as interstellar. Such techniques will undoubtedly be employed on the Rosetta Lander to seek out interstellar grains because they are likely to be abundant in comets. Interstellar grains also will be studied intensively by the grain-analysis experiment on the Rosetta Orbiter, whose principal investigator, Jochen Kissel, possibly detected them in Halley with a version of the equipment flown on *Giotto*. The Rosetta Lander, in addition to exploring carbon and its compounds, will measure the major, minor, and trace-element composition using a mass spectrometer; it will, of course, test Lockyer's hypothesis and may reveal comets to be more primitive than the most primitive meteorites and, therefore, a better model for the relative abundances of the elements in the sun.

With the lander playing such a fundamentally important role in the Rosetta mission, it is a pity it is not more inspiringly named: its appellation RoLand is supposedly short for Rosetta Lander but the more mischievous among the team of experimenters believe it has something to do with the main designer (that is, Rosenbauer Lander). Yet it is possible to equate RoLand with much more heroic and romantic events and to make a connection to cometary imagery. Roland, the man, was a knight in the army of Charlemagne (Karl der Gros), killed in battle at Roncesvalles during a rearguard action protecting the withdrawal of the Holy Roman Emperor's forces from Spain. His exploits are featured in many European illustrations and sculptures, in particular in northern Germany, where he is viewed as the symbol of imperial justice. But most important, his deeds are described in the epic poem, the *Chanson de Roland* (*Rolandslied*), which was reputedly penned by Turold, the future Abbot of Peterborough, to inspire the forces of William the Conqueror before their invasion of Britain at Hastings in 1066. Thus a link can be forged to the most famous cometary image of all in British historical terms: the *Bayeux Tapestry*. Why not have such a tenuous link? There is nothing more inspirational in art or science than a comet, which is very tenuous, almost ethereal, in the extreme.

The Future

Just as past and recent comets have inspired works of art as well as collaborations between artists and scientists,[10] we are confident that bright comets of the future, so rare in this century until recently,[11] will be inspirational muses in these astro-humanistic endeavors. That Comet Hale-Bopp [Figs. 161, 162 both color], discovered in 1995, became as bright in the spring of 1997 as the Comet of 1811, dazzling people around the world with its beauty even to the unaided eye, augurs well for the future of widespread interest in comets among astronomers, artists, and the public at large. Its appearance, with a broad, sweeping dust tail and a narrower, blue, structured gas tail, resembled reports of Comet Donati and the drawings by Bond.[12] In the new millennium, before Rosetta arrives at comet 46P/Wirtanen in 2011, the New Millennium Deep Space 1 mission will fly past Comet 76P/West-Kohoutek-Ikemura on 1 June 2000; the Stardust spacecraft will fly past Comet 81P/Wild 2 on 2 January 2004, and will bring a dust sample back to Earth; and the New Millennium Deep Space 4 mission will land upon Comet 9P/Tempel 1 in January 2006 and return a surface sample to Earth in June 2010. Yet, it is difficult to believe that there ever will be a similarly integrated fascination and involvement with comets and meteors among scientists, artists, and the general culture like the one in late eighteenth- and nineteenth-century Britain.

Notes

1 The theory first was advanced by Olson, 1979, 160-170. See also Olson, 1985b, 15-17; Olson and Pasachoff, 1986; Olson, 1994.

2 See Olson, 1985b, 15, fig. 14.

3 See Pasachoff, 1998, 272-291, for other illustrations.

4 H.U. Keller et al., "Comet P/Halley's nucleus and its activity," *Astronomy and Astrophysics*, 187, 1987, 807-823 (reprinted in M. Grewing, F. Praderie, and R. Reinhard, eds., *Exploration of Halley's Comet*, Berlin and Heidelberg, 1988, 807-823).

5 Pillinger and Pillinger, 589-605.

6 Thanks to Penny Wilson for this information.

7 For the history of stable isotope research as initiated from the study of meteorites, see C.T. Pillinger, "Light element stable isotopes in meteorites – from grams to picograms," *Geochimica et Cosmochima Acta*, 48, 1984, 2739-2766.

8 Huggins began as a microscopist and turned to astronomy as an amateur with a telescope he built at his house. He served as President of the Royal Astronomical Society (1876-1878) and the Royal Society (1901-1906); both gave him their gold medals. See chap. VI, n. 42.

9 J.N. Lockyer, *The Meteoric Hypothesis. A Statement of the Result of a Spectroscopic Inquiry into the Origin of Cosmical System*, London, 1890.

10 For example, a conference on "The Inspiration of Astronomical Phenomena," held in 1994 at the Villa Mondo Migliore, Rocca di Papa, Italy, included discussions on literature, art, myth and religion, history and politics (see articles in *Vistas in Astronomy*, 39, 1995, 381-733, and *Leonardo*, 29:2, 1996, 123-154). R.J.M. Olson and J.M. Pasachoff were involved in a session at the 1996 annual meeting of the College Art Association in Boston entitled "Galaxies, Black Holes, Dark Matter: Contemporary Interpretations of Outer Space," organized by Susan Schwab.

11 M. Gingerich, 86-88; Kidger, 231-238.

12 See chaps. V, VI. See also Dewhirst and Hoskin, 1996, 277-278, who reproduce both one of Bond's drawings of the head of Donati's Comet, showing the structure in the head that was also visible through telescopes about 140 years later in C/1995 O1 (Hale-Bopp), and also a wide-field drawing showing multiple tails. In the latter comet, a third tail of neutral sodium was also discovered intermediate between the dust and gas tails.

APPENDIX I

Comets and Meteors discussed in the *Philosophical Transactions of the Royal Society*

Bibliography compiled by year and grouped by phenomena.

Because the style of headings changed several times during the course of the long run of these articles, we have chosen to be consistent by giving the titles from the articles whenever they give sufficient information and in using index entries otherwise, or in combining information from the two. Furthermore, we are consistently using the full name or initials as they appear with the articles. When titles are given for individuals, it is because their first names or initials do not appear.

Anonymous. "The Motion of the late Comet predicted." [Indexed as "The motions of the Comet of Decemb. 1664. predicted], *Phil. Trans.* 1, #1, 1665-66, 3-8.

Anonymous. "Extract of a Letter written from Rome, concerning the late Comet, and a New one." *Phil. Trans.* 1, #2, 1665-66, 17-18.

Anonymous. "Extract of another Letter from Paris, containing some Reflexions on the precedent Roman Letter." *Phil. Trans.* 1, 1665-66, #2, 1665-66, 18-20.

Anonymous, "The Motion of the Second Comet predicted, by the same Gentleman, who predicted that of the former." [Indexed as "Comet of March, April 1665"] *Phil. Trans.* 1, #3, 1665-66, 36-40.

Hevelius, J. "An Account of Hevelius his Prodromus Cometicus together with some Animadversions made upon it by a French philosopher." *Phil. Trans.* 1, #6, 1665-66, 104-108.

Anonymous. "On the Judgement of some of the English Astronomers, touching the difference between the two named men, about an Observation made of the First of the Late Comets." *Phil. Trans.* 1, #9, 1665-66, 150-151.

Hevelius, J. "Descriptio Cometæ." *Phil. Trans.* 1, #17, 1665-66, 301-304.

Cassini, G.D. "Some Observations Concerning the Comet, that hath lately appear'd in forain parts, communicated from Italy and Portugal. The Italian Account, given by Gio. Domen. Cassini, we thus English." *Phil. Trans.* 3-4, #25, 1668-69, 683-684.

Mutoli, Pietro. "Del movimento della Cometa apparsa il mese di Dicembre 1664, da Pietro Maria Mutoli, in Pisa, in 4°." [Book review]. *Phil. Trans.* 3-4, #53, 1668-69, 1069-1070.

Bartholinus, Erasmus. "Erasmi Bartholini de Cometis An. 1664 & 1665. Opusculum; Ex Observationibus Hafniæ habitis adornatum. Hafniæ in 4°." *Phil. Trans.* 3-4, #53, 1668-69, 1071.

Hevelius, J. "Extract of a letter of Monsieur Hevelius from Dantzick written to the Publisher in Latin, March 9. (st. nov.) 1672; giving some Accompt of a New Comet, lately seen in that Country: Englished as followeth." [Indexed as "A Comet seen March and April 1672. by Monsieur Hevelius at Dantzick."]. *Phil. Trans.* 7, #81, 1672, 4017-4018.

Cassini, G.D. "Observations of a New Comet, made at Paris in the Royal Observatory." [Indexed as "The motions observed by Sign. Cassini at Paris"]. *Phil. Trans.* 7, #82, 1672, 4042-4050.

Estancel, Valent. "Observations concerning the Comet that was seen in Brazil. An. 1668. in March." *Phil. Trans.* 9, #105, 1674, 91-93.

Flamsteed, John. "Answer to the former three Letters [of Signor Cassini], containing also some celestial Observations." *Phil. Trans.* 11, #123, 1676, 565-567.

Wallis, John. "An Extract of Two Letters... concerning a considerable Meteor seen in many distant places of England at the same time." *Phil. Trans.* 12, #135, 1677, 863-864.

Wallis, John. "A Third Letter from the same hand, concerning the same Meteor." *Phil. Trans.* 12, #135, 1677, 864-866.

Cassini, G.D., Hevelius, J., Flamstead [sic], J. "Letters about the Comet seen in April last 1677. from Sign. Cassini, M. Hevelius and Mr. Flamstead." 12, #135, 1677, 868, 869-873, 873-875.

Hevelius, Johannes. "Observations by him made at Dantzick, of the Comet which began there to appear, Aug. 16, 1682." *Phil. Trans.* 13, #143, 1683, 16-19.

Hevelius, Johannes. "Cometæ Anni 1683." *Phil. Trans.* 13, #154, 1683, 416-424.

Ciampini, Senior [sic]. "Concerning a late Comet seen at Rome." Hevelius, Johannes. "Cometæ Anni 1683." *Phil. Trans.* 15, #169, 1685, 920-921.

Anonymous. "An Account of a Comet seen at Lipsick, Sept. 1686." *Phil. Trans.* 16, #186, 1687, 256-258.

Vans, R. listed as "Extract of a Letter." [Indexed as "An Account of an Extraordinary Meteor, or kind of Dew resembling Butter, that fell last Winter and Spring."]. *Phil. Trans.* 19, 1696, 224. [Not a meteor.]

Anonymous. "Situs novi Cometæ mense Februario, Anni 1699. in Regio Observatorio Parisiensi Observati." *Phil. Trans.* 21, #250, 1699, 79-81.

Halley, Edmond. "Astronomiæ Cometicæ Synopsis, Autore Edmundo Halleio apud Oxonienses Geometriæ Professore Saviliano, & Reg. Soc. S." *Phil. Trans.* 24, #297, 1704-5, 1882-1899.

Ray [Wray], John. "Observations made at Rome, by the late Reverend Mr. John Ray, of the Comet which appeared Anno 1664." *Phil. Trans.* 25, #309, 1706-07, 2350-2352.

Derham, W. "An Account of some Inundations, Monstrous Births, Appearances in the Heavens, and other Observables he received from Ireland. With his Observations on the Eclipse of the Sun, Sept. 3 and of the Moon, Sept. 18, 1708." *Phil. Trans.* 26, #320, 1708-09, 308-311.

Thoresby, Ralph. "An Account of a Meteor, which was seen in Yorkshire, and other Neighbouring Countries, upon May 18, 1710." *Phil. Trans.* 27, #331, 1710-12, 322-323.

Halley, Edmund. "An Account of several extraordinary Meteors or Lights in the Sky." *Phil. Trans.* 29, #341, 1714-16, 159-164.

Kirch, G. "Observationes quædam accuratæ infignis Cometæ sub finem anni 1680 visi, Coburgi Saxoniæ." [Indexed as "Accurate observations of the remarkable comet seen at the end of the year 1680, at Coburg in Saxony."]. *Phil. Trans.* 29, 1714-16, 170-172.

Halley, Edm. "An Account of a small Telescopical Comet seen at London on the 10th of June 1717." *Phil. Trans.* 30, #354, 1717-19, 721-723.

Anonymous. "Cometae Berolini" ["Comet observed at Berlin"]. *Phil. Trans.* 30, #357, 1717-19, 820-821.

Barham, Henry. "A Relation of a fiery Meteor seen by him, in Jamaica, to strike into the Earth; with Remarks on the Weather, Earthquakes, &c. of that Island." *Phil. Trans.* 30, #357, 1717-19, 837-838.

Halley, Edmond. "An Account of the Extraordinary Meteor seen all over England, on the 19th of March 171. With a Demonstration of the uncommon Height thereof." *Phil. Trans.* 30, #360, 1717-19, 978-990.

Cotes, Roger. "A Description of the great Meteor which was on the 6th of March, $171\frac{5}{6}$." *Phil. Trans.* 31, #365, 1720-21, 66-70.

Kirchius, Christfridus. "Cometae Berolini, *anno* 1718" [Comet observed at Berlin]. *Phil. Trans.* 32, #375, 1723, 238-240.

Bradley, Rev. "Observations upon the Comet, that appear'd in the Months of October, November, and December, 1723." *Phil. Trans.* 33, 1724-5, 41-49.

Paisley, Lord [ms. under Hamilton, James]. "With some curious Figures of the same Comet." *Phil. Trans.* 33, 1724-5, 50-51.

Bianchini, F., and Carbone, P.J.B. "Observatio ejusdem Cometæ." *Phil. Trans.* 33, 1724-5, 51-52.

Saunderson, W. "Observations upon the Comet that appeared in October 1723, made at Bombay." *Phil. Trans.* 34, 1726, 213-214.

Lynn, G. "A Method for determining the Geographical Longitude of Places, from the Appearance of the common Meteors called Falling Stars." *Phil. Trans.* 35, 1727-8, 351-353.

Greenwood, I. "A new method for composing a natural history of meteors." *Phil. Trans.* 35, about 1728, 390-402.

Dereham, Thomas. "Of the Meteor called the *Ignis Fatuus,* from Observations made in England." *Phil. Trans.* 36, #411, 1729-30, 204-214.

Dove, J. "A letter...containing an Account of a Comet seen on February 29, 173½; From on Board the Monmouth, in Table Bay." *Phil. Trans.* 37, 1731-2, 393-394.

Bradley, J., et al. "A collection of Observations communicated to the Royal Society, relating to the Comet that appeared in the Months of January, February, and March 1736-7":

at Oxford, J. Bradley, "Observations upon the Comet that appear'd in the Months of January, February, and March 1737, made at Oxford," *Phil. Trans.* 40, 1737-8, 111-118.
at Rome, D. de Revillas, "Cometes Romæ ex Monte Aventino observatus mense Februario 1737," *Phil. Trans.* 40,1737-8, XL 118-119.
at Philadelphia, Dr. Kearsley, "Observations upon the Comet seen in January and February 1736-7. and of an Eclipse of the Sun, Feb. 18, 1736-7. made at Philadelphia in Pennsylvania," *Phil. Trans.* 40, 1737-8, 119-121.
at Jamaica, R. Fuller, "Part of a letter...mentioning a Comet seen in Jamaica, in January 1736-7," *Phil. Trans.* 40, 1737-8, 122.
at Madras, Mr. Sartorius, "Extract of a Letter...concerning the Comet," *Phil. Trans.* 40, 1737-8, 122.
at Lisbon, G.R. Vanbrugh, "An Observation of the Comet seen at Lisbon...on board the Burford Man of War," *Phil. Trans.* 40, 1737-8, 123.

Crocker, Mr. "An Account of a Meteor seen in the Air in the Day-time, on Dec. 8. 1733." *Phil. Trans.* 41, 1739-41, 346-347.

Breitnall, J. "Notices of some Meteors observed at Philadelphia in North-America." *Phil. Trans.* 41, 1739-41, 359-360.

Short, Thomas. "An Account of several Meteors." *Phil. Trans.* 41, #459, 1739-41, 625-630.

Zanotti, E. "The parabolic Orbit for the Comet of 1739, observed at Bologna." *Phil. Trans.* 41, 1739-41, 809.

Beauchamp, Lord. "An Account of the Fire-ball seen in the Air, and of the Explosion heard Dec. 11, 1741, near London." *Phil. Trans.* 41, 1739-41, 870-871.
In Sussex–Fuller, J. *Phil. Trans.* 41, 1739-41, 871-872.
In Kent–Gostling, W. *Phil. Trans.* 41, 1739-41, 872-873.

Mason, Christopher. "Extract of a letter...concerning a Fire-ball seen in the Air, and a great Explosion heard, Dec. 11, 1741." *Phil. Trans.* 42, #462, 1742, 1.
Cooke, Benjamin. "Giving an Account of the Fire-ball seen Dec. 11, 1741." *Phil. Trans.* 42, #462, 1742-43, 25-27.
Gordon, William. "A Letter from Capt. William Gordon to Capt. Samuel Mead, F.R.S. inclosing an Account of the Fire-ball seen Dec. 11, 1741." *Phil. Trans.* 42, #463, 1742-43, 58-60.

Gostling, William. "Concerning the Fire-ball seen Dec. 11 last, and the Mock-Suns seen the 19th of the same Month." *Phil. Trans.* 42, #463, 1742-3, 60-61. [mock-sun]

Milner, T. "An Account of a meteor seen at Peckham, Dec. 11, 1741." *Phil. Trans.* 42, #464, 1742-43, 138-139.

Lovell, Thomas Lord. "An Account of a Meteor seen near Holkon in Norfolk in August 1741." *Phil. Trans.* 42, #465, 1743, 183-184.

Frantz, R.P. "Observationes Cometæ." ["Observations of a comet made by R.P. Frantz in the month of February of the year 1743 in Vienna, Austria."] *Phil. Trans.* 42, #470, 1742-43, 457-458.

Cradock, Zach. "The Appearance of a fiery Meteor." *Phil. Trans.* 43, #473, 1744-45, 78. [numbering begins again with #473]

Betts, Joseph. "Observations on the late Comet made in Sherborn and Oxford; with the elements for computing its Motions." *Phil. Trans.* 43, #474, 1745, 91-100.

Costard, G. "Concerning a fiery Meteor seen in the Air on July 14, 1745." *Phil. Trans.* 43, 1744-5, 522-524.

M[ortimer], C. "[Fireball]: December 16, 1742." [Indexed as "Account of a fiery meteor seen Dec.16, 1742."]. *Phil. Trans.* 43, 1744-5, 524-525.

Hodgson, J. "Via Cometæ: Observations on the comet of March 1742, by the Jesuits at Pekin [in Latin]." *Phil. Trans.* 44, 1746, 264-266.

Stryck, N. [Latin: Struijck, N.]. "Viæ Cometarium: secundùm hypothesin quæ statuit illos cursu suo Parabolam circa Solem describere." *Phil. Trans.*, 46, 1749-50, 89-92.

Barker, T. "An Account of an extraordinary Meteor seen in the County of Rutland, which resembled a Water-Spout." *Phil. Trans.* 46, 1749-50, 248-249. [not a meteor]

Hallerstein, A. "Epistola una cum Observationibus Cometæ visi Pekini 1748." *Phil. Trans.* 46, 1749-50, 305-315.

Gaubil, A. "Observatio Cometæ, Anno 1748." *Phil. Trans.* 46, 1749-50, 316-319.

Smith, William. "Concerning a Fire-ball, seen in the Air July 22, 1750." *Phil. Trans.* 47, 1751-52, 1-2.

Baker, Henry. "An Account of the same Meteor." *Phil. Trans.* 47, 1751-52, 3-4.

Dunthorne, Richard. "Concerning Comets." *Phil. Trans.* 47, 1751-52, 281-288.

Hirst, William. "An Account of a Fire-Ball, seen at Hornsey." *Phil. Trans.* 48, Part II, 1754, 773-776.

Barker, Thomas. "Concerning the Return of the Comet, expected in 1757, or 1758." *Phil. Trans.* 49, Part I, 1755, 347-350.

Bradley, J. "Observations upon the Comet that appeared in the Months of September and October 1757, made at the Royal Observatory [Greenwich]." *Phil. Trans.* 50, 1757-8, 408-415.

Klinkenberg, D. "Observations on the late Comet in September and October 1757; made at the Hague." *Phil. Trans.* 50, 1757-8, 483-488.

Bevis, J. "An Account of the Comet seen in May 1759." *Phil. Trans.* 51, Part I, 1759, 93-94.

Munckley, Nicolas. "An Account of the same Comet." *Phil. Trans.* 51, Part I, 1759, 94-96.

Pringle, John. "Several Accounts of the fiery Meteor, which appeared on Sunday the 26th of November, 1758, between Eight and Nine at Night." *Phil. Trans.* 51, Part I, 1759, 218-259.

Pringle, John. "Some Remarks upon the several Accounts of the fiery Meteor (which appeared on Sunday the 26th of November, 1758), and upon other such bodies." *Phil. Trans.* 51, Part I, 1759, 259-274.

Forster, Richard. "An Account of a Meteor seen at Shefford, in Berkshire, on Saturday, October 20th, 1759; with some Observations on the Weather of the preceding Winter." *Phil. Trans.* 51, Part I, 1759, 299-300.

Colebrooke, Josiah. "An Account of the same Meteor, seen at Bath." *Phil. Trans.* 51, Part I, 1759, 301-302.

Dutton, William. "An Account of the Meteor seen at Chigwell Row, in Essex, on the 20th of October 1759." *Phil. Trans.* 51, Part I, 1759, 302-303.

Short, James. "Observations on the Comet Seen in January 1760." *Phil. Trans.* 51, Part II, 1760, 465.

Michell, Rev. John. "Observations on the Same Comet." *Phil. Trans.* 51, Part II, 1760, 466-467.

Munckley, Nicolas. "An Account of the Same Comet." *Phil. Trans.* 51, Part II, 1760, 467-468.

Day, Mark. "An Account of the Said Comet." *Phil. Trans.* 51, Part II, 1760, 469.

De la Caille, Abbé. "[index: Observations of the comet seen at Paris, Feb. 1760]." *Phil. Trans.* 51, Part II, 1760, 635.

Winthrop, J. "An Account of a Meteor seen in New England, and of a Whirlwind felt in that Country." *Phil. Trans.* 52, 1762, 6-16.

Swinton, J. "An Account of a remarkable Meteor seen at Oxford [Sept 21, 1760]." *Phil. Trans.* 52, 1762, 99-102.

De la Lande, Mons. "An Account of the Comet seen at Paris, in June 1762." *Phil. Trans.* 52, 1762, 581.

Gabry, Petrus. "Observatio cometæ, qui mense Maio, A. 1759 apparuit, facta Haggœ-Comit." ["An observation of the comet which appeared in the month of May 1759, made in the Hague."]. *Phil. Trans.* 53, 1763, 3-4.

Gabry, Petrus. "Observatio cuiusdam Meteori igniti instar Chasmatis, facta Hagœ-Comit." ["An observation of a certain fiery meteor like a chasm (a rending of the firmament), made in the Hague on the 21 Dec. 1758 New Style."]. *Phil. Trans.* 53, 1763, 5-6.

Dunn, Samuel. "An Account of a remarkable Meteor." *Phil. Trans.* 53, 1763, 351-352.

Messier, Charles. "A Table of the Places of the Comet of 1764 discovered at the Observatory of the Marine at Paris, the 3d of January, about 8 o'clock in the Evening, in the Constellation of the Dragon, concluded from its Situation with regard to the Stars." *Phil. Trans.* 54, 1764, 151.

Winthrop, John. "An Account of several fiery Meteors seen in North America." [Letter to John Pringle.] *Phil. Trans.* 54, 1764, 185-188.

Cawley, James. "Account of Three Meteors seen in North America." *Phil. Trans.* 54, 1764, 188-191.

Sullivan, John, and Mark of Thomas + Frog. (sailors' testimonial). *Phil. Trans.* 54, 1764, 191-192.

Swinton, J. "An account of a remarkable meteor seen at Oxford, March 5, 1764." *Phil. Trans.* 54, 1764, 326-329. "It may not be improper here to remark, that a meteor, called an Aurora Borealis, was seen. . . ."; so this paper is not about a meteor. It lasted 4 hours. Reference to: Martyn, J., "An Account of an Aurora Australis seen Jan. 23, 1749-50." *Phil. Trans.* 46, 1756, 319-321.

Swinton, J. "An Account of a remarkable Meteor seen at Oxford, April 23, 1764." *Phil. Trans.* 54, 1764, 332-336. [Luminous arch; not a meteor]

Messier, C. "A Memoir, containing the History of the Return of the famous Comet of 1682, with Observations of the same, made at Paris." *Phil. Trans.* 55, 1765, 294-325.

Messier, Charles. "A Discovery, with Observations, of two new Comets in the Marine Obvatory [sic] at Paris." Trans., M. Maty. *Phil. Trans.* 56, 1766, 57-65; followed by "Mr. Pingré's remarks on the two comets of this year, " 65.

Bryce [Brice], Alexander. "An Account of a Comet seen by him." *Phil. Trans.* 56, 1766, 66-67.

Swinton, J. "Description of a Meteor seen at Oxford, Oct. 12, 1766." *Phil. Trans.* 57, 1767, 108-110. [Another sky phenomenon; not a meteor]

Winthrop, John, communicated by Franklin, Benjamin. "*Cogitata de Cometis.*" *Phil. Trans.* 57, Part I, 1767, 132-137.

Swinton, J. "An Account of very remarkable Meteor seen at Oxford, Oct. 24, 1769." *Phil. Trans.* 60, 1770, 532-535.

Messier, M. "Extract of two letters on a new comet observed Jan. 10, 1771, at Paris [index title]." *Phil. Trans.* 61, 1771, 104-106.

Brydone, Patrick. "An Account of a fiery Meteor, seen on the 10th of February last; and also of some new electrical Experiments." *Phil. Trans.* 63, Part I, 1773, 163-170.

Lexell, J. A. "Disquisitio de tempore Periodico Cometæ anno 1770 observati." ["A disquisition of the periodical time of the comet of 1770."] *Phil. Trans.* 69, 1779, 68-85.

Herschel, William. "Account of a Comet [in 1781]." *Phil. Trans.* 71, Part II, 1781, 492-500. [Actually the discovery of Uranus]

Pigott, Edward. "Extract of a letter...containing the Discovery of a Comet [19 Nov. 1783]." *Phil. Trans.* 74, Part I, 1784, 20.

Cavallo, Tiberius. "Description of a Meteor, observed Aug. 18, 1783." *Phil. Trans.* 74, Part I, 1784, 108-111.

Aubert, Alex. "An Account of the Meteors of the 18th of August and 4th of October, 1783." *Phil. Trans.* 74, Part I, 1784, 112-115.

Cooper, William. "Observations on a remarkable Meteor seen on the 18th of August, 1783." *Phil. Trans.* 74, Part I, 1784, 116-117.

Edgeworth, Richard Lovell. "An Account of the Meteor of the 18th of August, 1783." *Phil. Trans.* 74, Part I, 1784, 118.

Blagden, Charles. "An Account of some late fiery Meteors; with Observations." *Phil. Trans.* 74, Part I, 1784, 201-232.

Pigott, Nathaniel. "An Account of an Observation of the Meteor of August 18, 1783, made on Hewit Common near York." *Phil. Trans.* 74, Part II, 1784, 457-459.

Pigott, Edward. "Observations of the Comet of 1783." *Phil. Trans.* 74, Part II, 1784, 460-462.

Morgan, William. "Electrical Experiments...." *Phil. Trans.* 75, Part I, 1785, 272-278.

Maskelyne, Nevil. "Advertisement of the expected Return of the Comet of 1532 and 1661 in the Year 1788." *Phil. Trans.* 76, Part I, 1786, 426-431.

Herschel, Caroline. "An account of a new Comet [in 1786]." *Phil. Trans.* 77, Part I, 1787, 1-3.

Herschel, W. "Remarks on the new Comet." *Phil. Trans.* 77, Part I, 1787, 4-5.

Wollaston, Francis. "Observations on Miss Herschel's Comet." *Phil. Trans.* 77, Part I, 1787, 55-60.

Herschel, William. "Observations on a Comet [in 1789]." *Phil. Trans.* 79, Part II, 1789, 151-153.

Herschel, William. "Account of a comet [in 1791]." *Phil. Trans.* 82, Part I, 1792, 23-24.

Gregory, Edward. "Extracts of two Letters containing an Account of the Discovery of a comet, with Observations thereon." *Phil. Trans.* 83, Part I, 1793, 50-54.

Maskelyne, Nevil. "Observations on the Comet of 1793." *Phil. Trans.* 83, Part I, 1793, 55.

Herschel, Caroline. "An account of the Discovery of a Comet." *Phil. Trans.* 84, Part I, 1794, 1.

Herschel, Caroline. "Account of the Discovery of a new Comet." *Phil. Trans.* 86, Part I, 1796, 131-132.

Herschel, William. "Additional Observations on the Comet." *Phil. Trans.* 86, Part I, 1796, 133-134.

Howard, E. "Experiments and Observations on certain stony and metalline Substances, which at different Times are said to have fallen on the Earth; also on various Kinds of native Iron." *Phil. Trans.* 92, 1802, 168-212; includes indexed articles:

Williams, J. L. "Account of the Explosion of a Meteor, near Benares, in the East Indies; and the falling of some Stones at the same Time, about 14 Miles from that City." *Phil. Trans.* 92, 1802, 175-179;

Howard, E. indexed as "Of one seen in [South] America." *Phil. Trans.* 92, 1802, 202.

Herschel, William. "Observations on the two lately discovered celestial bodies." *Phil. Trans.* 92, 1802, 213-232. [Comparing new bodies with comets].

Herschel, William. "Observations on the Nature of the new celestial Body discovered by Dr. Olbers, and of the Comet which was expected to appear last January in its return from the Sun." [Indexed as "Additional observations on the comet."]. *Phil. Trans.* 97, Part II, 1807, 260-266.

Herschel, William. "Observations of a Comet, made with a View to investigate its Magnitude and the Nature of its Illumination. To which is added, an Account of a new Irregularity lately perceived in the apparent Figure of the Planet Saturn." *Phil. Trans.* 98, Part II, 1808, 145-163.

Herschel, W. "Observations on the Nature of the new celestial Body discovered by Dr. Olbers, and of the Comet which was expected to appear last January inits return from the Sun." *Phil. Trans.* 98, Part II, 1808, 260-266.

Herschel, William. "Observations of a Comet, with Remarks on the Construction of its different Parts." *Phil. Trans.* 102, Part I, 1812, 115-143.

Herschel, William. "Observations of a second Comet, with Remarks on its Construction." *Phil. Trans.* 102, Part II, 1812, 229-237.

Ivory, James. "A new Method of deducing a first Approximation to the Orbit of a Comet from three Geocentric Observations." *Phil. Trans.* 104, Part I, 1814, 121-186.

Hall, Basil. "Extract of a letter containing observations of a comet seen at Valparaiso." *Phil. Trans.* 112, Part I, 1822, 46-49.

Brinkley, J. "Elements of Captain Hall's Comet." *Phil. Trans.* 112, Part I, 1822, 50-63.

Rumker, Mr. "Observations of comets, made at Paramatta." [A. The Comet of Encke in 1822; B. Comet of September and October 1822, in Ophiucho; C. Comet in the Lion, July 1824; D. Comet in the Lion, July 1825; E. Great Comet of 1825; F. Comet in Orion, Sept. 1826; G. Re-appearance of the Comet of Encke in 1828]. *Phil. Trans.* 119, Part III, 1829, 54-69.

Robertson, Lieutenant (now Captain) William, R.N. "Observations of the Second Comet of 1822, made at Rio de Janeiro." *Phil. Trans.* 121, Part I, 1831, 1-8.

Henderson, Thomas. "Observations of the Comet of Encke, made in June 1832." *Phil. Trans.* 123, Part II, 1833, 549-558.

Maclear, Thomas. "An Account of the Fall of a Meteoric Stone in the Cold Bekkeveld, Cape of Good Hope." *Phil. Trans.* 129, Part I, 1839, 83-85.

Faraday, Michael. "Chemical Account of the Cold Bokkeveld Meteoric Stone." *Phil. Trans.* 129, Part I, 1839, 86-88.

Palgrave, Sir Francis. "An Account of the Shooting Stars of 1095 and 1243." *Phil. Trans.* 130, 1840, 175-176.

Maclear, Thomas. "Further particulars of the Fall of the Cold Bokkeveld Meteorolite." *Phil. Trans.* 130, 1840, 177-182, with a plate facing 177.

APPENDIX II

Selected Significant Comets, 1577-1910

Comets are in upright type; *meteors are in italics.*

1577, C/1577 V1: Comet of 1577. Observed by Tycho Brahe, who used it to show that comets were outside the earth's atmosphere.

1618, C/1618 W1: First bright comet of the century. Observed by John Milton as a boy. In Haddington's portrait. C/1618 Q1 preceeded and C/1618 V1 followed C/1618 W1 in this year. [Chapter I]

1664 and 1665, C/1664 W1 and C/1665 F1: Associated with the plague and fire, respectively. [Chapter I] Hevelius's *Cometographia* (1668), the first systematic account of past comets. [Introduction] Described in Defoe's *History of the Plague of London* (1722). [Chapter VII]

1680, C/1680 V1: Great Comet of 1680-81. Observed by Newton and Halley. First telescopic discovery of a daylight comet. Visible for 4 months. Drawn and last observed in 1681 by Newton. [Chapter I]

1682, 1P/1682 Q1: Halley's Comet, visible for only 31 days. Observed by Newton and Halley. Later had its orbit identified by Halley with the orbits of the comets of 1531 and 1607, and, when it reappeared on schedule in 1758 (1758 Y1), was called "Halley's Comet." Recorded in Wedgwood and Hoare portraits of Newton. [Chapter I]

1744, C/1743 X1: Brighter than all stars. Spectacular just after perihelion throughout Europe. May have inspired Hogarth. Drawn from London. [Chapter II]

1758, 1P/1758 Y1: Halley's Comet, on its first predicted return. [Chapter I]

1760, C/1760 B1: Messier's Comet, the first comet to be discovered telescopically as part of an organised search.

1781: "Comet" described by William Herschel that soon turned out to be the discovery of the planet Uranus. [Chapter I]

1783: fireball of 18 August 1783, widely viewed and drawn from Windsor and elsewhere. Extensively described in Philosophical Transactions of the Royal Society. Recorded by the brothers Sandby and many other artists and scientists. [Chapter II]

1786, 1788, 1790, 1790, 1791-92, 1794, 1795, 1797; C/1786 P1, formerly 1786 II ; 35P/1788 Y1, formerly 1788 II; C/1790 A1, 1790 I; C/1790 H1, formerly 1790 III ; C/1791 X1, formerly 1792 I ; C/1793 S2, 1793 I (an independent discovery of a comet first discovered by Messier 10 days earlier); 2P/1795 V1, formerly 1795; C/1797 P1, 1797: Comets discovered by Caroline Herschel. [Chapter II]

1795: meteorite fell to earth on 13 December 1795 weighing 56 stone at Wold Cottage in the East Riding of Yorkshire. President of the Royal Society, Sir Joseph Banks observed it and asked Edward Charles Howard to examine several chondrites.

1795, 2P/1795 V1: Encke's Comet, sketched by William Herschel. [Chapter I]

1801, C/1801 N1: Simultaneously discovered by several observers in France; drawn on Durs Egg's flintlock fowling piece. [Chapter III]

1807, C/1807 R1: Bright comet, sketched by Caroline Herschel. [Chapter III]

1811, C/1811 F1: Great Comet of 1811-12: A daylight comet, visible in the nighttime sky for seventeen months and with the naked eye for over nine months, the longest naked-eye apparition recorded. Carefully drawn by William and Caroline Herschel as well as by Linnell. Subject of caricatures, like those by Rowlandson and Gillray. May have inspired Blake. Linked to Napoleon. [Chapter III]

1811 II, C/1811 W1: Second comet of 1811, visible from November 1811 through January 1812. [Chapter III]

1812, 12P/1812 O1: Pons-Brooks Comet, first recorded appearance, naked eye to Napoleon's troops invading Moscow and to the Russians as well. [Chapter III]

1835, 1P/1835 P1: Halley's Comet, predicted and extensively observed in both northern and southern hemispheres. Perhaps influential on Turner's watercolor. [Chapter IV]

1843, C/1843 D1: Spectacular comet. Visible in daylight. Sungrazer. Brighter in southern hemisphere but very visible also in northern hemisphere. Painted by Piazzi Smyth from South Africa. [Chapter IV]

1853, C/1853 R1: Used by George Cruikshank as a vehicle for a caricature. Daumier inspired by it. [Chapter IV]

1858, C/1858 L1: Donati's Comet. One of the most beautiful and spectacular comets of all time. Subject of first comet photographs. Period about 2000 years. Large curving dust tail. Subject of George P. Bond's major study. Many artists painted, such as Palmer, Dyce, and Turner of Oxford. [Chapter V]

1861, C/1861 J1: Tebbutt's Comet, Great Comet of 1861. A bright naked-eye object discovered in Australia and later visible in the northern hemisphere. Subject of De la Rue's unsuccessful attempts to photograph it. [Chapter VII]

1864, C/1864 N1: The comet for which spectra were first obtained, by Donati. [Chapters V, VII]

1874, C/1874 H1: Coggia's Comet. Spectrum observed. Painted in watercolor by Piazzi Smyth from Edinburgh. [Chapters IV, VII]

1880, C/1880 C1: Large Southern Comet, discovered by Tebbutt. [Chapter VII]

1881, C/1881 K1: Comet Tebbutt, subject of first generally cited comet photographs, by Janssen and Draper. [Chapters VI, VII]

1882, C/1882 E1: Swift's Comet. Changes in the tail were photographed, though the comet was not prominent to the naked eye. [Chapter VI]

1882, C/1882 R1: Great September Comet. Magnificent for several months. Daylight comet, photographed at Cape of Good Hope by Gill, which led to the *Carte du Ciel* project. At perihelion, observed very near the sun. Spectrum photographed. Sketched by Webb, Poynter, and Brett. [Chapters VI, VII]

1882: Comet discovered near the sun during a total solar eclipse. No orbit known so no number is assigned. [Chapter VI]

1893, C/1893 U1: Brooks's Comet. Changes seen in the tail were photographed. [Chapter VI]

1910, 1P/1909 R1: Halley's Comet, its third return since its first viewing by Halley in 1682 (with 1758 and 1835 as intermediate apparitions). Widely viewed and photographed. [Chapter VIII]

APPENDIX III

Astronomers Royal and Gresham Professors

Astronomers Royal

John Flamsteed	1675-1719
Edmond Halley	1720-1742
James Bradley	1742-1762
Nathaniel Bliss	1762-1764
Nevil Maskelyne	1765-1811
John Pond	1811-1835
Sir George Biddell Airy	1835-1881
Sir William Christie	1881-1910
Sir Frank Dyson	1910-1933
Sir Harold Spencer Jones	1933-1955
Sir Richard Woolley	1956-1971
Sir Martin Ryle	1972-1982
Sir Francis Graham Smith	1982-1990
Arnold Wolfendale	1991-1994
Sir Martin Rees	1994-

Astronomers Royal for Scotland

Thomas Henderson	1834-1846
Charles Piazzi Smyth	1846-1888
Ralph Copeland	1888-1905
Frank Dyson	1905-1910
[became Astronomer Royal]	
Ralph Sampson	1910-1938
William M.H. Greaves	1938-1957
Herman A. Brück	1957-1975
Vincent C. Reddish	1975-1987
Malcolm S. Longair	1987-1990
John C. Brown	1995-

Gresham Professors of Astronomy

Edward Brerewood	1596-1597
Thomas Williams	1613
Edmund Gunter	1619-1620
Henry Gellibrand	1626-1627
Samuel Foster	1636-1637
Mungo Murray	1637
Samuel Foster	1641
Lawrence Rooke	1652
Christopher Wren	1657
Walter Pope	1660-1661
Daniel Man	1687
Alexander Torriano	1691
John Machin	1713
William Romaine	1751
William Cockayne	1752
R.P. Sandiford	1795
Joseph Pullen	1833
Edmund Ledger	1875
S.A. Saunder	1908
Arthur R. Hinks	1913
Lectures in abeyance	1939-1945
William Herbert Steavenson	1946
Sir John Carroll	1964
Sir Martin Ryle	1968
R.J. Taylor	1969
Martin Rees	1975
David W. Dewhirst	1976-1980
M. Rowan-Robinson	1981
A.C. Fabian	1982
Raymond Hide	1984
Lord Porter	1990
Heather Couper	1993
C.T. Pillinger	1996

Selected Bibliography

All eighteenth- and nineteenth-century articles on comets and meteors in the *Philosophical Transactions of the Royal Society* are found in Appendix I, organized chronologically, rather than in the bibliography. The title is abbreviated as *Philosophical Transactions* in the footnotes.

The *Monthly Notices of the Royal Astronomical Society* is abbreviated in the bibliography only to *Monthly Notices*.

Abetti, G. *The History of Astronomy*. Trans. B.B. Abetti. New York, 1952.

Abley, M., ed. *The Parting Light. Selected Writings of Samuel Palmer.* Manchester, 1985.

A.C.R. [A. Cowper Ranyard]. "Obituary: Thomas William Webb." *Monthly Notices*, 46, 1886, 198-201.

Adams, E. *Francis Danby – Varieties of a Poetic Landscape*. New Haven/London, 1973.

Altfeld, H.-H. *Bibliographical Guide for Cometary Science*. Munich, 1983.

Alexander, A.F.O'D. *The Planet Uranus: A History of Observation, Theory and Discovery*. London, 1965.

Alvarez, L.W. "Mass extinctions caused by large bolide impacts." *Physics Today*, 40, 1987, 24-33.

Alvarez, L.W., Alvarez, W., Asaro, F., and Michel, H.V. "Extraterrestrial Cause for the Cretaceous-Tertiary Extinction." *Science*, 208, 1980, 1095.

Alvarez, W. *T. Rex and the Crater of Doom*. Princeton, 1997.

Alvarez, W., Kauffman, E.G., Surlyk, F., Alvarez, L.W., Asaro, F., Michel, H.V. "Impact Theory of Mass Extinctions and the Invertebrate Fossil Record." *Science*, 223, 1984, 1135-1141.

Andrews, K., and Brotchie, J.R. *Catalogue of Scottish Drawings*. Edinburgh, 1960.

Anonymous. "Death of Mr. W. Usherwood." *Dorking and Leatherhead Advertiser*, November, 1916.

Anonymous. *A General Index to the First Twenty-Nine Volumes of the Monthly Notices of the Royal Astronomical Society, 1827-1869*. London, 1870.

Anonymous. *General Index to Volumes XXX to LII of the Monthly Notices of the Royal Astronomical Society, 1869-1892*. London, 1896.

Anonymous. "Giant meteor shatters Dutch calm." *New Scientist*, 126, 21 April, 1980, 21.

Anonymous. "How Mr. De La Rue Photographed the Moon." *The British Journal of Photography*, 25, 1868, 256-257, 270-271, 279-280.

Anonymous. "Observations of Donati's Comet made at the Observatory of Edward J. Cooper, Esq., with the large Equatorial." *Monthly Notices*, no. 35, published in *Memoirs of the Royal Astronomical Society*, 29, 1859-60, n.p.

Anonymous. "Report of the Council." *Monthly Notices*, 19, 1858, 138-139.

Arago, F. "Le Daguerréotype." *Compte Rendu des Séances de l'Academie des Sciences*, 9, 1839, 250-267.

Arago, F. *Popular Astronomy*. 2 vols. Trans. W.H. Smyth and R. Grant. London, 1855.

Armfield, M.A. *Tempera Painting Today*. London, 1946.

Armitage, A. *William Herschel*. London, 1962.

Ashbrook, J. "Astronomical Scrapbook." *Sky and Telescope*, 47, 1974, 26-27.

Ashworth, W.J. "The calculating eye: Baily, Herschel, Babbage, and the business of astronomy." *British Journal for the History of Science*, 27, 1994, 409-441.

Bachrach, A.G.H. "The Field of Waterloo and beyond." *Turner Studies*, 1:2, 1981, 5-13.

Bailey, M.E., Clube, S.V.M., and Napier, W.M. *The Origin of Comets*. Oxford, 1990.

Bailey, S.I. *The History and Work of Harvard Observatory, 1839 to 1927*. New York and London, 1931.

Ball, J. *Paul and Thomas Sandby, Royal Academicians*. Bath, 1985.

Balston, T. *John Martin 1789-1854. His Life and Works*. London, 1947.

Banister, J. "A Postscript to the Barnard Ledgers." *Proceedings of the Silver Society*, 3, 1983, 39.

Barger, S.M. "The Moon, 6 August 1851." *1989 Annual Report, The Christian A. Johnson Memorial Gallery*. Middlebury, VT, n.d., n.p.

Barker, P. "The Optical Theory of Comets." *Physis (Firenze)*, 30, 1993, 1-25.

Barkla, H.M. "Cometarium made by R. Fidler, early 19th century." Unpublished manuscript, 1994.

Barkla, H.M. "Ingenuity Misdirected." Unpublished manuscript, n.d.

Barnard, E.E. "The Development of Photography in Astronomy." *Popular Astronomy*, 6:8, 1898, 425-455.

Barnard, E.E. "The Development of Photography in Astronomy." *Science*, 8, no. 195, 1898, 341-395.

Baroni, I. "Note Scientifiche: Le Comete. Storia – Fantasia – Realtà." *Emporia*, 31, 1910, 212-304.

Barrington, Mrs. R. *G.F. Watts*. New York, 1905.

Baskett, J., and Snelgrove, D. *The Drawings of Thomas Rowlandson in the Paul Mellon Collection*. New York, 1977.

Bauer, H.H. *Beyond Velikovsky: The history of a public controversy*. Urbana, IL, 1984.

Bayard, J. *Works of Splendor and Imagination: The Exhibition Watercolor, 1770-1870*. Exh. cat. New Haven, 1981.

Beckett, R.B. *John Constable's Correspondence*, 3 (*Suffolk Records Society*, 7). Ipswich, 1965.

Beech, M. and Youngblood, R. "That Which We Call a Meteorite." *The Observatory*, 114, 1994, 312.

Bell, L. *The Telescope*. New York, 1922 (1981 ed. with intro. by J.M. Pasachoff).

Bell, M. *Drawings of Sir E.J. Poynter Bart. P.R.A.* London and New York, n.d.

Bennett, J.A. "Catalogue of the Archives and Manuscripts of the Royal Astronomical Society." *Memoirs of the Royal Astronomical Society*, 85, 1978, 2-90.

Bennett, J.A. "'On the power of penetrating into space': The telescope of William Herschel." *Journal for the History of Astronomy*, 7, 1976, 75-108.

Berendzen, R., Hart, R., and Seeley, D. *Man Discovers the Galaxies*. New York, 1976.

Bhathal, R. *Australian Astronomer John Tebutt*. Kenthurst, Australia, 1993.

Bhathal, R., and White, G. *Under the Southern Cross: A Brief History of Astronomy in Australia*. Kenthurst, Australia, 1991.

Bindman, D., ed. *Blake's Illustrated Books*. 5 vols. Princeton, 1991-93.

Bindman, D., ed. *The Complete Graphic Works of William Blake*. New York, 1978.

Bindman, D., ed. *William Blake. His Art and Times*. Exh. Cat. London, 1982.

Binyon, L. *Catalogue of the Drawings by British Artists and Artists of Foreign Origin working in Great Britain preserved in the Department of Prints and Drawings in the British Museum*, 2. London, 1900.

Birmingham, Museum and Art Gallery. *Canaletto and England*. Exh. cat. Birmingham, 1993.

Bishop, R.L. "Newton's Telescope Revealed." *Sky and Telescope*, 59, 1980, 207.

Black, D.J. *An Historical Account of the Lands and Mansion formerly called HERMITS AND TERMITS now known as the Coalyard House....* Edinburgh, 1972.

Bode, U., et al. *George Cruikshank 1792-1878*. Exh. cat. Stuttgart, 1983.

Bond, G.P. *Account of the Great Comet of 1858* (*Annals of the Astronomical Observatory of Harvard College*, 3). Cambridge, MA, 1862.

Bond, G.P "An Account of Donati's Comet of 1858." Cambridge, MA, 1858. (Reprinted from "An Account of the Comet of Donati. 1858." *Mathematical Monthly.*, 1:2, 1858, 61-67, and "An Account of the Comet of Donati. 1858. Part II." *Mathematical Monthly*, 1:3, 1859, 88-116.)

Bortle, J.E. "How to Observe Comets." *Sky and Telescope*, 61, 1981, 210-214.

Bortle, J.E., and. Morris, C.S. "The Visual Observation of Comets." *International Comet Quarterly*, 16, 1994, 129-130.

Bowness, A. "The Victorians." In *The Genius of British Painting*. Ed. D. Piper. New York, 1975, 249-290.

Brett, J. "Extract of a report of the Eclipse Expedition of 1870 at Augustea, Sicily." *Monthly Notices*, 31, 1870, 163-165.

Brett, J. "Sketch of the Great Comet (1882 b)." *Monthly Notices*, 43, facing 288.

Bridson, G., and Wakeman, G. *Printmaking and picture printing: A bibliographical guide to artistic and industrial techniques in Britain 1750-1900.* Williamsburg, VA, 1984.

Brightwell, D.B. *A Concordance to the Entire Works of Alfred Tennyson*. New York, 1970.

Brion, M. *Art of the Romantic Era*. New York, 1966.

Broughton, P. "Arthur Storer of Maryland: His Astronomical Work and His Family Ties with Newton." *Journal for the History of Astronomy*, 19, 1988, 77-96.

Brown, D.B. *Samuel Palmer 1805-1881. Loan Exhibition from the Ashmolean Museum Oxford.* Exh. cat. London, 1982.

Brown, D.B. *Turner and Byron*. Exh. cat. London, 1992.

Brück, H.A., and Brück, M.T. *The Peripatetic Astronomer: The Life of Charles Piazzi Smyth*. Bristol, 1988.

Bryant, M., and Henage, S. *Dictionary of British Cartoonists and Caricaturists 1730-1980*. Aldershot, Hants, 1994.

Bryan's Dictionary of Painters and Engravers. 5 vols. London, 1904.

Buchanan-Brown, J. *The Book Illustration of George Cruikshank*. North Pomfret, Vermont, 1980.

Burke, J.G. *Cosmic Debris. Meteorites in History*. Berkeley/Los Angeles/ London, 1986.

Burne-Jones, E. *The Little Holland House Album*. North Berwick, 1981.

Burne-Jones, G. *The Memorials of Edward Burne-Jones*. 2 vols. London, 1904.

Bury, S. *Jewellery 1784-1910: The International Era*. 2 vols. London, 1991.

Butlin, M. "Blake, the Varleys, and the Graphic Telescope." In *William Blake: Essays in honour of Sir Geoffrey Keynes*. Eds. M.D. Paley and M. Phillips. Oxford, 1973, 294-304.

Butlin, M. *The Blake-Varley Sketchbook of 1819 in the Collection of M.D.E. Clayton-Stamm*. 2 vols. London, 1969.

Butlin, M. *The Paintings and Drawings of William Blake*. 2 vols. New Haven/ London, 1981.

Butlin, M. *Tate Gallery Collection 5: William Blake 1757-1827*. London/Seattle, 1990.

Butlin, M. *William Blake*. Exh. cat. London, 1978.

Buttmann, G. *The Shadow of the Telescope: A Biography of John Herschel*. New York, 1970. (Reprint, London, 1974.)

Calvert, H.R. *Scientific Trade Cards in the Science Museum Collection*. London, 1971.

Cambridge, MA, The Fogg Art Museum. *Paintings and Drawings from the Pre-Raphaelites and Their Circle*. Exh. cat. Cambridge, MA, 1946.

Campbell, M. *David Scott 1806-1849*. Edinburgh, 1990.

Campbell, M.J. *John Martin Visionary Printmaker*. Exh. cat. York, 1992.

Carey, A.M. *English, Irish, and Scottish Firearms Makers*. London, 1954.

Carina Software. Voyager II. San Leandro, CA, 1993. Computer program.

Carter, H.B. *Sir Joseph Banks, 1743-1820*. London, 1988.

Casteras, S. *James Smetham. Artist, Author, Pre-Raphaelite Associate*. Aldershot, Hants, 1994.

Cavallo, T. *A complete treatise of electricity in theory and practice; with the original experiments*. London, 1777.

Cecil, D. *Visionary and Dreamer. Two Poetic Painters: Samuel Palmer and Edward Burne-Jones*. Princeton, NJ, 1969.

Chancellor, E. B. "London in the Works of Canaletto and Scott." *Connoisseur*, 37, 1913, 127-132.

Chapman, A. "The Astronomical Work of Thomas Harriot (1560-1621)." *Quarterly Journal of the Royal Astronomical Society*, 36, 1995, 97-107.

Chew, K., and Wilson, A. *Victorian Science and Engineering Portrayed in The Illustrated London News*. Phoenix Mill, Gloucestershire, 1993.

Chladni, E.F.F. *Uber den Ursprung der von Pallas gefundenen und anderer ihr ahnlicher Eisenmasse und über einig damit in Verbindung stehende Naturerscheinungen*. Riga, 1794.

Chladni, E.F.F. *Uber Feure-Meteore und über die mit denselben herangefallen Massen*. Vienna, 1819.

Christianson, G. *In the Presence of the Creator: Isaac Newton and His Times*. New York and London, 1984.

Clerke, A.M. "Charles Piazzi Smyth." *Compact Dictionary of National Biography*, 2, London, 1975, 2460. (Reprinted from the *Dictionary of National Biography*, New York and London, 1891.)

Clerke, A.M. *A Popular History of Astronomy During the Nineteenth Century*. London, 1902.

Cliver, E.W. "Was the Eclipse Comet of 1883 a Disconnected Coronal Mass Ejection?" *Solar Physics*, 122, 1989, 319-333.

Combs, H.C., and Sullens, Z.R., eds. *A Concordance to the English Poems of John Donne*. Chicago, 1940.

Common, A.A. "Photography as applied to Comet b 1881." *The Observatory*, 4, 1881, 232-233.

Compact Dictionary of National Biography. 2 vols. London, 1975. (Reprinted from *Dictionary of National Biography*, New York and London, 1885-1901.)

Cook, H.J. "The Practice of Science: A Behaviorist's View." *American Scientist*, 83, 1995, 271-273.

Copeland, R. "Obituary: Charles Piazzi Smyth." *Monthly Notices*, 61, 1901, 189-196.

Cowley, R.L.S. *Hogarth's Marriage A-La-Mode*. Ithaca, 1983.

Crookshank, A., and the Knight of Glin. *The Watercolours of Ireland: Works on Paper in Pencil, Pastel and Paint c. 1600-1914*. London, 1994.

Crowther, J.G. *Men of Science*. New York, 1936.

Culme, J. *The Directory of Gold and Silversmiths, Jewellers and Allied Traders, 1838-1914*. 2 vols. Woodbridge, Suffolk, 1987.

Cummings, F. "Romanticism in Britain, 1760-1860." In F. Cummings and A. Staley. *Romantic Art in Britain. Paintings and Drawings 1760-1860*. Exh. cat. Philadelphia, 1968, 17-24.

Cummings, F., and Staley, A. *Romantic Art in Britain. Paintings and Drawings 1760-1860*. Exh. cat. Philadelphia, 1968.

Cunningham, H. "Leisure and culture." In *The Cambridge Social History of Britain*, 2. Ed. F.M.L. Thompson. Cambridge, 1990, 279-339.

Cunnington, C.W., and Cunnington, P. *Handbook of English Costume in the Eighteenth Century*. London, 1957.

Crouan, K. *John Linnell: A centennial exhibition*. Exh. cat. Cambridge, 1982.

Daumas, M. *Scientific Instruments of the Seventeenth and Eighteenth Centuries and their Makers*. Trans. and ed. M. Holbrook. London, 1992.

Davies, C.D.P. "Herschel's 18¾ in. Speculum (the '20 ft')." *Monthly Notices*, 84, 1923-24, 23-26.

Davies, G. *The Early Stuarts 1603-1660*. Oxford, 1949.

Davies, J.M.Q. *Blake's Milton Designs: The Dynamics of Meaning*. West Cornwall, Conn., 1993.

Dawes, W.R. "Observations of Donati's Comet made at Haddenham, Bucks." *Monthly Notices*, 19, 1859, 88-91.

De Grazia, A., ed. *The Velikovsky Affair*. London, 1966.

De la Rue, W. "The Bakerian Lecture. – On the Total Solar Eclipse of July 18, 1860, observed at Rivabellosa, near Miranda de Ebro, in Spain." *Philosophical Transactions*, 52, 1862, 333-416.

De la Rue, W. "Observations on Donati's Comet. 6: Observations by Mr. Warren De la Rue." *Memoirs of the Royal Astronomical Society*, 30, 1862, 78-84.

Delteil, L. *Le Peintre-Graveur Illustré*, 20-29: *Honoré Daumier*. New York, 1969. (Reprint of Paris, 1925-30 ed.)

De Vaucouleurs, G. *Astronomical Photography. From the Daguerreotype to the Electron Camera*. Trans. R. Wright. London, 1961.

Dewhirst, D. "Christmas Stars and Celestial Portents." Exh. labels. Cambridge University Library, 1985.

Dewhirst, D., and Hoskin, M. "The Memory of Starlight: The Rise of Astrophysics." In *Cambridge Illustrated History of Astronomy*. Ed. M. Hoskin. New York and Cambridge, 1996, 256-343.

Dickinson, H.T. *The Caricature and the Constitution 1760-1832*. Ed. M. Duffy. Cambridge, 1986.

Digges the Elder, L. *A Prognostication Everlasting of Ryght Good Effecte*. London, 1556 (second impression).

Donn, B., Rahe, J., and Brandt, J.C., eds. *Atlas of Comet Halley 1910 II*, Washington, D.C., 1986.

Donnelly, I. *Ragnarok: The Age of Fire and Gravel*. New York, 1887.

Draper, H. "Note on Photographs of the Spectrum of Comet b 1881." *The Observatory*, 4, 1881, 252-253.

Draper, H. "Photographs of Comet b 1881." *The Observatory*, 4, 1881, 239.

Dreyer, J.L.E., and Turner, H.H., eds. *History of the Royal Astronomical Society, 1820- 1920*. London, 1923. (Reprint, Oxford, 1987.)

Duffy, M. *The Englishman and the Foreigner*. Ed. M. Duffy. Cambridge, 1986, 328-329.

Dunbar, P. *William Blake's Illustrations to the Poetry of Milton*. Oxford, 1980.

Dyson, A. *Pictures to print: The nineteenth century engraving trade*. London, 1984.

Eaves, M. *The Counter-Arts Conspiracy. Art and Industry in the Age of Blake*. Ithaca/London, 1992.

E.B.K. [E.B. Knobel.] "Obituary: Warren De la Rue." *Monthly Notices*, 50, 1889-90, 155-164.

Edberg, S.J., and Levy, J.D. *Observing Comets, Asteroids, and the Zodiacal Light*. Cambridge, 1994.

Edgerton, J. *Wright of Derby*. Exh. cat. London, 1990.

Edgerton, S.Y., Jr. *The Heritage of Giotto's Geometry*. Ithaca/London, 1991.

Edinburgh, The Crawford Collection. *Catalogue of the Crawford Library in the Royal Observatory Edinburgh*. Edinburgh, 1890.

Edwards, G. *International Guide to Nineteenth Century Photographers*. Boston, 1988.

Einberg, E. *Manners and Morals: Hogarth and British Painting 1700-1760*. Exh. cat. London, 1987.

Elkin, W.L. "Observations of the 1898 Leonid meteors and discovery of a comet (Chase, Nov. 14) at the Yale Observatory." *Astronomical Journal*, 19, 1898, 151-152.

Elkin, W.L. "Observations with the Heliometer." *Annals of the Royal Observatory Cape of God Hope*, 2:1, 1886, 19-25.

Elkin, W.L. "Report of Dr. Elkin." *Report for the Year 1898-99 of the Observatory of Yale University*. New Haven, 1899, 4-5.

Ellery, R.L.J. "Observations of the Great Comet (b), 1882 made at the Melbourne Observatory." *Monthly Notices*, 43, 1883, 29-30.

Erdman, D.V., et al., eds. *William Blake's Designs for Edward Young's Night Thoughts*. 2 vols. Oxford, 1980.

Erffa, H. von, and Staley, A. *The Paintings of Benjamin West*. New Haven and London, 1986.

E.S.H. [Edward S. Holden.] "Notices from the Lick Observatory: Photographs of Donati's Comet in September 1858." *Publication of the Astronomical Society of the Pacific*, 9, 1897, 89.

Essick, R.N. *The Separate Plates of William Blake. A Catalogue*. Princeton, 1983.

Essick, R.N. *William Blake's Commercial Book Illustrations. A Comprehensive Study of the Plates Engraved by Blake after Designs by Other Artists*. Oxford, 1991.

Essick, R.N., and Pearce, D., eds. *Blake in His Time*. Bloomington/London, 1978.

Essick, R.N., and Viscomi, J., eds. *Blake's Illuminated Books*, 5: *William Blake. Milton a Poem and the Final Illuminated Works*. Ed. D. Bindman, London, 1993.

Etter, R., and Schneider, S. *Comet Halley. Memories of 1910*. New York, 1985.

Evans, D.S. "John Herschel." *Dictionary of Scientific Biography*. Ed. C.C. Gillispie. New York, 1974, 323-328.

Evans, D.S., Deeming, T.J., Evans, B.H., and Goldfarb, S., eds. *Herschel at the Cape: Diaries and correspondence of Sir John Herschel 1834-1838*. Austin/London, 1969.

Evans, H., and Evans, M. *The Life and Art of George Cruikshank. The Man who Drew the Drunkard's Daughter*. New York, 1978.

Ewing, E. *History of Children's Costume*. New York, 1977.

Faigan, J. *Paul Sandby Drawings*. Exh. cat. Sydney, 1981.

Feaver, W. *The Art of John Martin*. Oxford, 1975.

Feaver, W. *Masters of Caricature from Hogarth and Gillray to Scarfe and Levine*. Ed. A. Gould. New York, 1981.

Fennell, F. L. *Dante Gabriel Rossetti. An Annotated Bibliography*. New York/ London, 1982.

Fernandez, R., Pasachoff, J.M., and Volz, R. "*Urania Observed*, an Exhibition of Works of Art with Astronomical Content." In *The Teaching of Astronomy*. Eds. J.M. Pasachoff and J.R. Percy. Cambridge, 1989, 430-436.

Finburg, A.J. *A Complete Inventory of the Drawings of the Turner Bequest*. 2 vols. London, 1909.

Finburg, A.J. *The Life of J.M.W. Turner, R.A.* Oxford, 1961.

Finley, G. *Landscapes of Memory. Turner as Illustrator to Scott*. Berkeley/ Los Angeles, 1980.

Fischer, W.J., and Olmsted, M. "Catalogue of the Harvard Observatory Meteor Photographs." *Annals Harvard College Observatory*, 87, 1931, 233-254.

Fitzgerald, P. *Edward Burne-Jones. A Biography*. London, 1975.

Flamsteed, J. *Atlas Coelestis Britannica*. London, 1729.

Fontaney, P.J. de. "Observations of the comet of 1680 anad 1681, made at the College of Clermon." *Philosophical Collections*, 4, 1681/92, 106-113.

Forbes, E.G. *Greenwich Observatory: the Royal Observatory at Greenwich and Herstmonceaux, 1875-1975*, 1: *Origins and Early History (1675-1975)*. London, 1975. (Published as a series, Forbes, E.G., Meadows, A.J., and Howse, D. *Greenwich Observatory: the Royal Observatory at Greenwich and Herstmonceaux, 1875-1975*. 3 vols. London, 1975.)

Forbes, G. *David Gill: Man and Astronomer*. London, 1916.

Fox, C. "The engravers' battle for professional recognition in early nineteenth-century London." *London Journal*, 1, 1976, 2-32.

Fox, C., ed. *London – World City 1800-1840*. Exh. cat. New Haven/London, 1992.

Foxon, A.C. *Dante Gabriel Rossetti*. New York, 1989.

Freitag, R. *Halley's Comet: A Bibliography*. Washington, D.C., 1984.

Funnell, P. "The London Art World and Its Institutions." In *London – World City – 1800-1840*. Ed. C. Fox. Exh. cat. New Haven/London, 1992.

Gadbury, J. *De cometis....* London, 1665.

Gage, J. *Color in Turner. Poetry and Truth*. New York/Washington, D.C., 1969.

Gage, J. *J.M.W. Turner. "A Wonderful Range of Mind."* New Haven/London, 1987.

Gage, J., ed. *Collected Correspondence of J.M.W. Turner*. Oxford, 1980.

Geneva, Musée Rath. *Deux Artistes Genevois en Angleterre. Alfred-Edouard Chalon (1781-1860) et Jean-Jacques dit John Chalon (1778-1854)*. Exh. cat. Geneva, 1971.

Genuth, S.S. "Blazing Stars, Open Minds, and Loosened Purse Strings: Astronomical Research and Its Early Cambridge Audience." *Journal for the History of Astronomy*, 21, 1990, 9-20.

Genuth, S.S. "Comets, Teleology, and the Relationship of Chemistry to Cosmology in Newton's Thought." *Annali dell'Istituto e Museo di Storia della Scienza di Firenze*, 10:2, 1985, 31-65.

Genuth, S.S. "From Heaven's Alarm to Public Appeal." In *Comets and the Rise of Astronomy at Harvard. Science at Harvard University. Historical Perspectives*. Eds. C.A. Elliott and M.W. Rossier. Bethlehem/ London/Toronto, 1992, 28-54.

Genuth, S.S. "Newton and the Ongoing Teleological Role of Comets." In *Standing on the Shoulders of Giants. A Longer View of Newton and Halley*. Ed. N.J.W. Thrower. Berkeley/Los Angeles/Oxford, 1990, 298-311.

George, M.D. *Catalogue of Political and Personal Satires Preserved in the Department of Prints and Drawings in the British Museum*. 11 vols. London, 1870-1954.

George, M.D. *Hogarth to Cruikshank: Social Change in Graphic Satire*. London, 1967.

Gere, C., Rudoe, J., et al. *The Art of the Jeweller. A Catalogue of the Hull Grundy Gift to the British Museum*. Ed. H. Tait. 2 vols. London, 1984.

Getty Art History Information Program. Documents for the History of Collecting on CD-ROM of the Provenance Documentation Collaborative. Malibu, 1995.

Gill, D. "The Applications of Photography in Astronomy." *The Observatory*, 10, 1887, 267-272, 283-294.

Gill, D. "Observations of the Great Comet of 1882 made at the Royal Observatory, Cape of Good Hope." *Annals of the Royal Observatory, Cape of Good Hope*, 2:1, 1886, 2-21.

Gill, D. "On Photographs of the Great comet (b) 1882." *Monthly Notices*, 43, 1882-83, 53-55.

Gingerich, M. "Great Comets, Novae, and Lady Luck." *Sky and Telescope*, 89:6, 1995, 86-89.

Gingerich, O. "The Naming of Uranus and Neptune." *Astronomical Society of the Pacific, Leaflet*, no. 352, October 1958, 1-7.

Gizzi, C. *Dante Gabriel Rossetti*. Exh. cat. Milan, 1984.

Godfrey, R.T. "Introduction." In London, Victoria and Albert Museum, *English Caricature 1620 to the Present*. Exh. cat. London, 1984, 10-25.

Godfrey, R.T. *Wenceslaus Hollar. A Bohemian Artist in England*. Exh. cat. New Haven, 1994.

Goldschmidt, L. "Tangible fact, poetic interpretation." In *The truthful lens: A survey of the photographically illustrated book, 1844-1914*. Eds. L. Goldschmidt and W.J. Naef. New York, 1980, 3-7.

Goldsmith, D., ed. *Scientists confront Velikovsky*. Ithaca, 1977.

Graves, A. "The British Institution." *The Art Journal*, no vol. no., 1910, 336-341, 377-381.

Graves, A. *The British Institution 1806-1867. A Complete Dictionary of Contributors and Their Work from the Foundation of the Institution.* London, 1908.

Gray, J.M. *David Scott, R.S.A., and His Works. Catalogue of His Paintings, Engravings, and Drawings.* Edinburgh/London, 1884.

Greenacre, F. *Francis Danby 1793-1861.* Exh. cat. London, 1988.

Grego, J. *Rowlandson the Caricaturist.* 2 vols. New York, 1880.

Griffiths, A. "Notes on Early Aquatint in England and France." *Print Quarterly*, 4:3, 1987, 256-70.

Grigson, G. *Samuel Palmer the Visionary Years.* London, 1947.

Grimwade, A.G. *London Goldsmiths, 1697-1837: Their Marks and Lives.* London, 1990.

Grove-Hills, E.H. "The Decade 1850-1860." In *History of the Royal Astronomical Society 1820-1920.* Eds. J.L.E. Dreyer and H.H. Turner. London, 1923, 110-128. (Reprint, Oxford, 1987.)

G.T.P. "Ragnarök." *Encyclopaedia Britannica*, vol. 18. Chicago, 1972, 1103.

Guillemin, A. *The Heavens, An Illustrated Handbook of Popular Astronomy.* Ed. J.N. Lockyer. London, 1867.

Guillemin, A. *The World of Comets.* Trans. and ed. J. Glaisher. London, 1877.

Guratzsch, H., and Hasse, M. *James Gillray 1757-1815. Meisterwerke der Karikatur.* Exh. cat. Stuttgart, 1986.

Hall, A.R. *Isaac Newton: Adventurer in Thought.* Oxford/Cambridge, MA, 1992.

Hall. J. *Dictionary of Subjects and Symbols in Art.* New York, 1979.

Hall, M.B. *Promoting Experimental Learning: Experiment and the Royal Society, 1660-1727.* Cambridge, 1991.

Halley, E. "A Synopsis of the Astronomy of Comets." *Miscellanea Curiosa*, 2, London, 1723-26, 321-344.

Halsby, J. *Scottish Watercolours 1740-1940.* London, 1986.

Hammond, J.H., and Austin, J. *The Camera Lucida in Art and Science.* Bristol, 1987.

Hardie, M. *Watercolour Painting in Britain.* 3 vols. Eds. D. Snelgrove, J. Mayne, and B. Taylor. London, 1967.

Hardie, W. *Scottish Painting 1837-1939.* London, 1976.

Harrison, M., and Waters, B. *Burne-Jones.* New York, 1973.

Hart, K.W. *James Gillray. Prints by the Eighteenth-Century Master of Caricature.* Exh. cat. Hanover, NH, 1994.

Hayes, J. *The Art of Thomas Rowlandson.* Exh. cat. Alexandria, VA, 1990.

Hayes, J. *Thomas Rowlandson.* London, 1960.

Hayes, J. *Rowlandson. Watercolours and Drawings.* London, 1972.

Hayward, J.F. *The Art of the Gunmaker.* 2 vols. London, 1962.

Hayward, J.F. *European Firearms.* London, 1969.

Henshall, N. "Pig Iron and Happiness." *History Today*, 41, February 1991, 58-59.

Herman, J.K., and Barger, S. "The Moon on a Silver Plate." *Astronomy*, 15:10, 1987, 98-103.

Hermann, D.B. *The History of Astronomy from Herschel to Hertzsprung.* Trans. K. Krisciunas. Cambridge, 1984.

Herrmann, L. *Paul and Thomas Sandby.* Exh. cat. London, 1986.

Herschel, A.S. "Progress in Meteoric Astronomy during the Year 1863-64." *Monthly Notices*, 25, 1864, 158-162.

Herschel, J.F.W. *Outlines of Astronomy.* London, 1849.

Herschel, J.F.W. *Results of Astronomical Observations made during the years 1834, 5, 6, 7, 8, at the Cape of Good Hope, being the completion of a telescopic Survey of the Whole surface of the visible Heavens, commenced in 1825.* London, 1847.

Herschel, W. "A Letter [on the name of the Georgian Planet]." *Philosophical Transactions of the Royal Society*, 73, 1783, 100-101.

Herschel, W. *The Scientific Papers of Sir William Herschel.* 2 vols. London, 1912.

Hevelius, J. "The First Astronomical Observations made by Mr. Hevelius at Dantzick, Jan. 1, 1681. St. N. being an Occultation of the Bulls-Eye by the Moon. As also a short account of his Observations of the Late Comet." *Philosophical Collections*, 3, 1681, 65-67.

H.H.T. [H.H. Turner]. "Obituary: John Brett." *Monthly Notices*, 62, 1902, 238-241.

Hill, D. *Fashionable Contrasts. Caricatures by James Gillray.* London, 1966.

Hill, D. *Mr. Gillray. The Caricaturist.* London, 1965.

Hind, J.R. *The Comet of 1556; being popular replies to every-day questions....* London, 1857.

Hind, J.R. *The Comets: A Descriptive Treatise Upon those Bodies. With ...a Table of all the Calculated Comets from the Earliest Ages to the Present Times.* London, 1852.

Hind, J.R. *On the Expected Return of the Great Comet of 1264 and 1556 with a History of Former Appearances, compiled from various authors and ephemerides, for facilitating its Re-discovery.* London, 1848.

Hind, J.R. *Sweeping ephemerides for the comet of 1556.* London, 1857.

Hingley, P.D. "*Urania's Mirror* – a 170-year old mystery solved?" *Journal of the British Astronomical Association*, 104:5, 1994, 238-240.

Hoffleit, D. *Some Firsts in Astronomical Photography.* Cambridge, MA, 1950.

Hoffleit, D. "Stones From Heaven." *Memoirs Connecticut Academy of Sciences*, 23, 1992, 15-19.

Hoffleit, D. "The Thousandth Fireball-Portrait, a Harvard Antique." *The Telescope*, 8, 1941, 6-11.

Holiday, H. *Reminiscences of My Life*. London, 1914.

Honour, H. *The Age of Neoclassicism*. Exh. cat. London, 1972.

Ho Peng Yoke. "Ancient and Mediaeval Observations of Comets and Novae in Chinese Sources." *Vistas in Astronomy*, 5, 1962, 127-225.

Hoskin, M.A. "The Astronomy of the Universe." In *Cambridge Illustrated History of Astronomy*. Ed. M. Hoskin. New York and Cambridge, 1996, 198-255.

Hoskin, M.A. "Caroline Herschel." *Dictionary of Scientific Biography*. Ed. C.C. Gillispie. New York, 1974, 322-323.

Hoskin, M.A. "Rosse, Robinson, and the Resolution of the Nebulae." *Journal for the History of Astronomy*, 21, 1990, 331-344.

Hoskin, M.A. "The Telescope War." *Sky and Telescope*, 89:4, 1995, 26-28.

Hoskin, M.A. "William Herschel." *Dictionary of Scientific Biography*. Ed. C.C. Gillispie. New York, 1974, 328-336.

Hoskin, M.A. *William Herschel, Pioneer of Sidereal Astronomy*. London, 1953.

Howard, E. "Experiments and Observations on certain stony and metallic Substances at different Times said to have fallen on Earth; also on various kinds of native Iron." *Philosophical Transactions*, 92, 1802, 168-212.

Howard, H. *A Course of Lectures on Painting delivered at the Royal Academy of Fine Arts*. Ed. F. Howard. London, 1948.

Howse, D. *Greenwich Observatory: the Royal Observatory at Greenwich and Herstmonceaux, 1875-1975*, 3: *The Buildings and Instruments*. London, 1975.

Howse, D. *Nevil Maskelyne: The Seaman's Astronomer*. Cambridge, 1989.

Hudson, D. *Arthur Rackham. His Life and Work*. London, 1960.

Huggins, W. "On Comets." In *Astronomy: The Royal Institution Library of Science*, 1. Ed. B. Lowell. New York/Barking, Essex/Amsterdam, 1970, 191-202.

Hughes, D.W. "The history of Halley's Comet." *Philosophical Transactions*, 323, 1987, 349-367.

Hughes, D.W. "The History of Meteors and Meteor Showers." *Vistas in Astronomy*, 26, 1982, 325-345.

Hughes, D.W. "James Joule and Meteors." *Astronomy*, 33, 1990, 143-148.

Hughes, D.W. "A Mysterious Woodcut." *Sky and Telescope*, 74, 1987, 225.

Hughes, D.W. "The World's Most Famous Meteor Shower Picture." *Earth, Moon, and Planets*, 68, 1995, 311-322.

Hughes, P. "Paul Sandby's Tour of Wales with Joseph Banks." *The Burlington Magazine*, 117, 1975, 452-457.

Hunter, M. *Establishing the New Science: The Experience of the Early Royal Society*. Woodbridge, 1989.

Hunter, M. *The Royal Society and Its Fellows, 1660-1700: The Morphology of an Early Scientific Institution*. Oxford, 1994.

Hutchison, S.C. *The Royal Academy 1768-1986*. London, 1986. (2nd edition)

Hutchinson, John. *James Arthur O'Connor*. Exh. cat. Dublin, 1985.

Illustrated London News [The]. Various citations for articles and engravings.

Ingram, W., and Swain, K. *A Concordance to Milton's English Poetry*. Oxford, 1972.

Irwin, D. *English Neoclassical Art: Studies in Inspiration and Taste*. London. 1966.

Ivins, W. *Prints and visual communication*. Cambridge, MA, 1973.

Janaway, J. *Surrey – A Photographic Record 1850-1920*. Newbury, Berkshire, 1984.

Jerrold, B. *The life of George Cruikshank*. 2 vols. London, 1882.

Johnson, E.D.H. *Paintings of the British Social Scene from Hogarth to Sickert*. New York, 1986.

Johnson, E.D.H. *The Poetry of Earth: A Collection of English Nature Writings*. New York, 1966.

Johnston, G.H. *Scottish Heraldry Made Easy*. Edinburgh/London, 1904.

Jones, B.Z., and Boyd, J.G. *The Harvard College Observatory: The First Four Directorships, 1839-1919*. Cambridge, MA, 1971.

Jones, K.G., ed. *Webb Society deep-sky observer's handbook*. Short Hills, NJ, 1979.

Keller, H.U., et al. "Comet P/Halley's nucleus and its activity." *Astronomy and Astrophysics*, 187, 1987, 807-823 (reprinted in M. Grewing, F. Praderie, and R. Reinhard, eds. *Exploration of Halley's Comet*. Berlin, 1988, 807-823).

Keynes, G. *Blake Studies. Essays on his life and work*. Oxford, 1971. (1971a)

Keynes, G. "Blake's Visionary Heads and the Ghost of a Flea." *Bulletin of the New York Public Library*, 64, 1960, 567-572.

Keynes, G. *William Blake's Water-Colour Designs for the Poems of Thomas Gray*. Jura, 1971. (1971b)

Kidger, M.R. "The Frequency of Appearances of Bright Naked Eye Comets since 1750: Are bright comets less frequent in the 20th Century?" *Earth, Moon, and Planets*, 66, 1994, 231-238.

King, E. *Remarks Concerning Stones Said to Have Fallen from the Clouds, both in These and in Ancient Times*. London, 1796.

King, H.C., and Millburn, J.R. *Geared to the Stars. The Evolution of Planetariums, Orreries and Astronomical Clocks*. Toronto/Buffalo/London, 1978.

Kingzett, R. "A Catalogue of the Works of Samuel Scott." *The Walpole Society*, 48, 1980-82, 1-134.

Klonk, C. *Science and the Perception of Nature: British Landscape Art in the Late Eighteenth and Early Nineteenth Centuries.* New Haven and London, 1996.

Klonsky, M. *William Blake. The Seer and His Visions.* New York, 1977.

Kronk, G.W. *Comets: A Descriptive Catalog.* Hillside, NJ, 1984.

Kronk, G.W. *Cometography. A Catalogue of Comets.* 4 vols. Cambridge/New York, forthcoming.

Krumbhaar, E.B. *Isaac Cruikshank. A Catalogue Raisonné with a Sketch of his Life at Work.* Philadelphia, 1966.

Krupp, E.C. *In Search of Ancient Astronomers.* New York, 1978.

Krupp, E.C. "Recreating the Past: Powerful Pyramids, Lost Continents, and Ancient Astronauts." In *Science and the Paranormal.* Eds. G.O. Abell and B. Singer. New York, 1981, 259-264.

Lacey, N.J. *The Arthurian Encyclopedia.* New York and London, 1986.

Lambert, S. *Reading Drawings, a selection from the Victoria and Albert Museum.* Exh. cat. New York, 1984.

Lambourne, L., and Hamilton, J. *British Watercolours in the Victoria and Albert Museum.* London, 1980.

Lancaster-Brown, P. *Comets, Meteorites and Men.* London, 1973.

Lancaster-Brown, P. *Halley's Comet and the Principia.* Aldeburgh, Suffolk, 1986.

Lancaster-Brown, P. *Halley and His Comet.* Poole, Dorset, 1985.

Lankford, J. "How Not to Run an International Halley Watch." *Sky and Telescope*, 70, 1985, 530-533.

Lankford, J. "The impact of photography on astronomy." In *General History of Astronomy*, 4: *Astrophysics and Twentieth-Century Astronomy to 1950.* Ed. O. Gingerich. Cambridge/New York, 1984, 16-39.

Lassell, W., et al. "Observations on Donati's Comet." *Memoirs of the Royal Astronomical Society*, 30, 1862, 58-84.

Lee, E.M. *Translation: Turner and Printmaking.* Exh. cat. New Haven, 1993.

Lee, R.W. *A History of Valentines.* New York, 1952.

Leech, J.H., et al. *John Herschel and Victorian Science.* Exh. cat. Austin, 1966.

Leeds, Art Gallery. *Great Victorian Pictures: Their Paths to Fame.* Exh. cat. London, 1978.

Leslie, C.R. *Memoirs of the Life of John Constable, R.A.* Ed. A. Shirley. London, 1937.

Levy, D.H. *Impact Jupiter.* New York, 1995.

Levy, D.H. "More Things in Heaven and Earth: Poets and Astronomers read the Night Sky." Wolfville, Nova Scotia, 1997.

Levy, D.H. "Poet and Observer: Gerard Manley Hopkins and Some Mid-19th Century Comets." *Journal of the Royal Astronomical Society of Canada*, 75:3, 1981, 139-150.

Levy, D.H. *The Quest for Comets*, New York, 1995.

Linnell, D. *Blake, Palmer, Linnell and Co.* Lewes, East Sussex, 1994.

Lippincott, K. *The Story of Time and Space: The Old Royal Observatory at Greenwich, Past, Present and Future.* Tonbridge, Kent, 1994.

Lister, R. *Catalogue Raisonné of the Works of Samuel Palmer.* Cambridge, 1988.

Lister, R., ed. *The Letters of Samuel Palmer.* 2 vols. Cambridge, 1974. (1974a)

Lister, R. *The Paintings of Samuel Palmer.* Cambridge, 1985.

Lister, R. *The Paintings of William Blake.* Cambridge, 1986.

Lister, R. *Samuel Palmer. A Biography.* London, 1974. (1974b)

Lister, R. *Samuel Palmer, his life and art.* Cambridge, 1987.

Littmann, M., and Yeomans, D.K. *Comet Halley: Once in a Lifetime.* Washington, D.C., 1985.

Lloyd, C. *The Queen's Pictures. Royal Collections Through the Centuries.* London, 1991.

Lockhart, J.G. *Memoirs of the Life of Sir Walter Scott.* 10 vols. Edinburgh/London, 1839.

Lockyer, J.N. *The Meteoric Hypothesis. A Statement of the Result of a Spectroscopic Inquiry into the Origins of Cosmical Systems.* London, 1890.

London, Arts Council. *George Frederick Watts 1817-1904.* Exh. cat. London, 1954.

London, Arts Council of Great Britain. *James Gillray 1756-1815.* Exh. cat. London, 1967.

London, the British Astronomical Association. "The British Astronomical Association: The First Fifty Years." *Memoirs of the British Astronomical Association*, 36:2, 1948, 1-132. (Reprinted in *Memoirs of the British Astronomical Association*, 42:1, 1989, 1-132.)

London, Burlington Fine Arts Club. *Catalogue of the Collection of Early Drawings and Pictures of London.* Exh. cat. London, 1920.

London, Fine Arts Society. *Homage to Maxwell Armfield.* Exh. cat. London, 1970.

London, Guildhall Art Gallery. *Samuel Scott Bicentenary.* Exh. cat. London/Sheffield, 1972.

London, J. and S. Maas Ltd. *John Ruskin and His Circle.* Sale cat. London, 1991.

London, Royal Society. *The record of the Royal Society of London for the Promotion of Natural Knowledge.* London, 1940.

London, Science Museum. "Astronomical Recreation." Exh. checklist, 1993.

London, Victoria and Albert Museum. *English Caricature 1620 to the Present.* Exh. cat. London, 1984.

London, The Whitechapel Art Gallery. *G.F. Watts. A Nineteenth Century Phenomenon.* Exh. cat. London, 1974.

Loomis, E. *The Recent Progress of Astronomy; especially in the United States.* New York, 1850.

Lowell, B., ed. *Astronomy: The Royal Institution Library of Science.* 2 vols. New York/Barking, Essex/Amsterdam, 1970.

Lusk, L. "Milton and Art." *The Art Journal*, no vol. no., 1911-12, 129-134.

MacDonald, A., and Morrison-Low, A.D., eds. *A Heavenly Library. Treasures from the Royal Observatory's Crawford Collection.* Exh. cat. Edinburgh, 1994.

Maclear, T. "Observations of the Comet of 1843, made at the Cape of Good Hope, re-examined and reduced." *Memoirs of the Royal Astronomical Society*, 20, 1851, 62-69.

Maclear, T. "Observations of Donati's Comet, made at the Royal Observatory, Cape of Good Hope, between October 11th, 1858, and March 4th, 1859." *Memoirs of the Royal Astronomical Society*, 29, 1861, 59-83.

Maclear, T. "Observations of Halley's Comet, made at the Royal Observatory, Cape of Good Hope, in the years 1835 and 1836." *Memoirs of the Royal Astronomical Society*, 10, 1838, 91-155.

MacPherson, H., Jr. *A Century's Progress in Astronomy.* Edinburgh and London, 1906.

MacPike, E.F. *Hevelius, Flamsteed, and Halley.* London, 1937.

Manchester City Art Galleries. *Concise Catalogue of British Watercolours and Drawings.* 2 vols. Manchester, 1984.

Manchester City Art Galleries. *Masterpieces From the Manchester City Art Galleries.* Exh. cat. Manchester, 1993.

Mancoff, D.N. *The Arthurian Revival in Victorian Britain.* New York and London, 1990.

Mancoff, D.N. "'Because my Heart is Pure': Sir Edward Burne-Jones and Tennyson's *Sir Galahad*." *Avalon to Camelot*, 1:1, 1983, 24-26.

Maris Multimedia. RedShift. London, 1994. CD-ROM.

Mark, K. *Meteorite Craters.* Tucson, AZ, 1987.

Marsden, B. "Elements of Cometary Orbits." *Quarterly Journal of the Royal Astronomical Society*, 13, 1971, 428-434.

Marsden, B. "New Designations for Old." *International Comet Quarterly*, 17, 1995. 3-6.

Marsden, B. "Resolution of International Astronomical Union Commission 20." *Minor Planet Circular*, nos. 23803-23904.

Marsden, B. "Search Program for Comets." In *Asteroids, Comets, and Meteors.* Eds. A. Milani et al. Dordrecht, Holland, 1994, 1-16.

Marsden, B. "The Sungrazing Comet Group. II." *The Astronomical Journal*, 98:4, 1989, 2306-2321.

Marsden, B., and Williams, G.V. *Catalogue of Cometary Orbits 1996.* Cambridge, MA, 1996 (eleventh edition).

Martin, B. *The Young Gentleman and Lady's Philosophy, in a continued Survey of the Works of Nature and Art; By Way of Dialogue*, 1. London, 1759.

Martin, R.B. *Tennyson: The Unquiet Heart.* Oxford and New York, 1980.

Marvin, U.B. "Meteorites, the Moon and the History of Geology." *Journal of Geological Education*, 34, 1986, 140-165.

Maskelyne, N. "A plan for observing Meteors called Fireballs." *The Gentleman's Magazine*, 53, 1783, 991-992.

Matteson, L.R. "John Martin's 'The Deluge.'" *Pantheon*, 39, 1981, 220-228.

McGeachie, J. "Eighteenth Century." *ISIS*, 83:3, 1992, 496-497.

Meadows, A.J. *Greenwich Observatory: the Royal Observatory at Greenwich and Herstmonceaux, 1875-1975, 2: Recent History (1836-1975).* London 1975.

Mee, J. *Dangerous Enthusiasm. William Blake and the Cult of Radicalism in the 1790's.* Oxford, 1992.

Melosh, H.J. *Impact Cratering: A Geologic Process.* New York and Oxford, 1989.

Metz, J. *Halley's Comet, 1910: Fire in the Sky.* St. Louis, MO, 1985.

Millar, Oliver. *The Tudor, Stuart and Early Georgian Pictures in the Collection of Her Majesty the Queen.* 2 vols. Greenwich, Conn, 1963.

Millar, Oliver. *The Victorian Pictures in the Collection of Her Majesty the Queen.* 2 vols. Cambridge/New York/Oakleigh, 1992.

Millburn, J.R. "Martin's Magazine: the General Magazine of Arts and Sciences, 1755-65." *The Library (Transactions of the Bibliographic Society)*, fifth series, 28:3, London, 1973, 221-239.

Millburn, J.R., and Nuttell, J.E. "The Elliptical Pulley Cometarium and its Accuracy." Unpublished manuscript, 1985.

Mills, E. *The Life and Letters of Frederic Shields.* London, 1912.

Minneapolis, The Minneapolis Institute of Arts. *Victorian High Renaissance.* Exh. cat. Minneapolis, 1978.

Mitchison, R. "Scotland 1750-1850." In *The Cambridge Social History of Britain,* 1. Ed. F.M.L. Thompson. Cambridge, 1990, 155-207.

Montgomery, G., ed. *Concordance to the Poetical Works of John Dryden.* Berkeley and Los Angeles, 1957.

Morton, A.Q., and Wess, J.A. *Public and Private Science. The King George III Collection.* Oxford, 1993.

Morus, I., Schaffer, S., and Secord, J. "Scientific London." In *London-World City – 1800-1840.* Ed. C. Fox. Exh. Cat. New Haven/London, 129-143.

Neal, W.K., and Back, D.H.L. *Great British Gunmakers 1740-1790.* London, 1975.

New Haven, Yale Center for British Art. *The Art of Paul Sandby.* Exh. cat. New Haven, 1985.

New York, H. Schickman Gallery. *Exhibition of Old Master Drawings at the H. Schickman Gallery.* Sale cat. New York, 1968.

New York, Pierpont Morgan Library. *Icelandic Sagas, Eddas, and Art.* Exh. cat. New York, 1982.

Newby, E. *William Hoare of Bath, R.A. 1707-1792.* Exh. cat. Bath, 1990.

Nicollett, J.N. "On the Comet discovered in the Constellation Pegasus in 1821. . . ." *Memoirs of the Royal Astronomical Society*, 1:1, 1822, 154-155.

Nicollett, J.N. "The Right Ascension and the Declination of the Comet of January 1821." *Memoirs of the Royal Astronomical Society*, 1:2, 1825, 271-272.

Nonody, P.G. *The Art of Walter Crane.* London, 1902.

Noon, P. *The Human Form Divine.* Exh. cat. New Haven and London, 1997.

Norman, D. "The Development of Astronomical Photography." *Osiris*, 5, 1938, 560-594.

Nygren, J., ed. *Isaac Cruikshank and the Politics of Parody.* Exh. cat. Los Angeles/San Marino, CA, 1994.

"Obituary: Charles Piazzi Smyth." *The Observatory*, 23, 1900, 145-147.

"Obituary: Mr. Matthew Cotes Wyatt." *The Art Journal*, no vol. no, 1862, 86.

"Obituary: John Russell Hind." *Monthly Notices*, 56, 1895-96, 200-205.

"Obituary: Thomas Maclear." *Monthly Notices*, 40, 1880, 202-204.

Olbers, W. "On the Comet discovered in the Constellation Pegasus in 1821. . . ." *Memoirs of the Royal Astronomical Society*, 1:1, 1822, 156-158.

Oliver, C.P. "Obituary: William Federick Denning." *The Observatory*, 54, 1931, 276-283.

Olsen, D.J. *The Growth of Victorian London.* London, 1976.

Olson, D.W. and Olson, M.S. "William Blake and August's Fiery Meteors." *Sky and Telescope*, 78, 1989, 192-199.

Olson, R.J.M. "The Draftsman's Comet." *Drawing*, 7:3, 1985, 49-55. (1985a)

Olson, R.J.M. *Fire and Ice: A History of Comets in Art.* New York, 1985. (1985b)

Olson, R.J.M. "Giotto's Portrait of Halley's Comet." *Scientific American*, 240:5, 1979, 160-170.

Olson, R.J.M. "Much Ado About Giotto's Comet." *Quarterly Journal of the Royal Astronomical Society*, 35, 1994, 145-148.

Olson, R.J.M. *Ottocento. Romanticism and Revolution in 19th-Century Italian Painting.* Exh. cat. New York and Florence, 1992.

Olson, R.J.M. "Representations of Pope Pius VII: The First Risorgimento Hero." *Art Bulletin*, 68:1, 1986, 77-93.

Olson, R.J.M. "A water-colour by Samuel Palmer of Donati's Comet." *The Burlington Magazine*, 132, 1990, 95-96.

Olson, R.J.M., and Pasachoff, J.M. "Historical Comets Over Bavaria: The *Nuremberg Chronicle* and Broadsides." In *Comets in the Post-Halley Era*, 2. Eds. R.C. Newburn, Jr., et al. Dordrecht/Boston/London, 1991, 1309-1341.

Olson, R.J.M., and Pasachoff, J.M. "New Information on comet P/Halley as depicted by Giotto di Bondone and other Western artists." *Astronomy and Astrophysics*, 187, no. 1/2, November (II) 1987, 1-11 (reprinted in M. Grewing, F. Praderie, and R. Reinhard, eds., *Exploration of Halley's Comet.* Berlin and Heidelberg, 1988).

Olson, R.J.M., and Pasachoff, J.M. "Recording the 'wonderful meteor' of 18 August 1783: The Sandbys, 'Samuel Scott,' and heavenly bodies." *Apollo*, 146:430, 1997, forthcoming.

Olson, R.J.M., and Pasachoff, J.M. "The 1816 Solar Eclipse and Comet 1811 I in John Linnell's Astronomical Album." *Journal for the History of Astronomy*, 12, 1992, 121-133.

Omer, M. Turner. *Die Landschaften der Bibel.* Vienna, 1985.

Oppé, A.P. *The Drawings of Paul and Thomas Sandby in the Collection of His Majesty the King at Windsor Castle.* Oxford and London, 1947.

Oppé, A.P. *Thomas Rowlandson. His Drawings and Water-colours.* Ed. G. Holme. London, 1923.

Orloff, S. "Images de comètes: analogies et analyses iconographiques." *Bulletin de la Classe des Beaux-Arts*, 70, 1988, 190-244.

Paley, M.D. *The Apocalyptic Sublime.* New Haven/London, 1986.

Paley, M.D. *The Continuing City. William Blake's "Jerusalem."* Oxford, 1983.

Paley, M.D., ed. *Blake's Illuminated Books*, 1: *William Blake. "Jerusalem."* Princeton, 1991.

Palmer, A.H. *The Life and Letters of Samuel Palmer.* London, 1892.

Pang, A. S.-K. "'Stars should henceforth register themselves': astrophotography at the early Lick Observatory," *The British Journal for the History of Science*, 30, 1997, 177-202.

Pang, A. S.-K. "Victorian Observing Practices, Printing Technology, and Representations of the Solar Corona, (1): The 1860s and 1870s." *Journal for the History of Astronomy*, 25, 1994, 249-274.

Pang, A. S.-K. "Victorian Observing Practices, Printing Technology, and Representations of the Solar Corona, (2): The Age of Photomechanical Reproduction." *Journal for the History of Astronomy*, 26, 1995, 63-75.

Parris, L. *Landscape in Britain c. 1750-1850.* Exh. cat. London, 1973.

Parris, L., ed. *The Pre-Raphaelites.* London, 1994.

Parris, L., and Fleming-Williams, I. *Constable.* Exh. cat. London, 1991.

Parris, L., Shields, C., and Fleming-Williams, I. *John Constable: Further Documents and Correspondence* (*Suffolk Records Society*, 18). Ipswich, 1975.

Pasachoff, J.M. *Astronomy: From the Earth to the Universe.* Philadelphia, 1998 (fifth edition).

Pasachoff, J.M. "Not so common." *New Scientist*, 147, 1993, 51.

Pasachoff, J.M. *Saunders Videodisc on Astronomy, 1: "The Solar System."* Philadelphia, 1994.

Pasachoff, J.M., and Olson, R.J.M. "Early Comet Photography: Usherwood and Bond for Donati." *Abstract: Bulletin American Astronomical Society*, 27:4, 1995, 1331.

Pasachoff, J.M., Olson, R.J.M., and Hazen, M.L. "The Earliest Comet Photographs: Usherwood, Bond, and Donati 1858." *Journal for the History of Astronomy*, 27, 1996, 129-145.

Pasachoff, J.M., and Percy, J.R., eds. *The Teaching of Astronomy*. Cambridge, 1989.

Patten, R.L., ed. *George Cruikshank: A Revaluation*. New Brunswick, NJ, 1974.

Patten, R.L. *George Cruikshank's Life, Times, and Art*. 2 vols. New Brunswick, NJ, 1992-96.

Paulson, R. *Hogarth*. 3 vols. New Brunswick, NJ, 1993.

Paulson, R. *Hogarth: His Life, Art, and Times*. 2 vols. New Haven/London, 1971.

Paulson, R. *Hogarth's Graphic Works*. 2 vols. New Haven/London, 1970.

Payne-Gaposchkin, C. "Worlds in Collision." *Popular Astronomy*, 58, 1950, 278-286.

Pendered, M.L. *John Martin Painter*. London, 1923.

Pennington, R. *Wenzel Hollar: Beschreibendes Verzeichnis seiner Kupferstiche*. Berlin, 1982.

Peterson, H.L., ed. *Encyclopedia of Firearms*. London, 1964.

Pevsner, N. *Academies of Art Past and Present*. London, 1940.

Pidgley, M. "Cornelius Varley, Cotman, and the graphic telescope." *The Burlington Magazine*, 114, 1972, 781-786.

Piggott, Jan. *Turner's Vignettes*. Exh. cat. London, 1993.

Pillinger, C.T. "Light element stable isotopes in meteorites – from grams to picograms." *Geochimica et Cosmochim a Acta*, 48, 1984, 2739-2766.

Pillinger, C.T., and Pillinger, J.M. "The Wold Cottage meteorite: Not just any ordinary chondrite." *Meteoritics and Planetary Science*, 31:5, 1996, 589-605.

Piper, D. *Artists' London*. New York, 1982.

Piper, D. *The English Face*. London, 1957.

Piper, D., ed. *The Genius of British Painting*. New York, 1975.

Pointon, M. "The representation of time in painting: A study of William Dyce's *Pegwell Bay: A Recollection of October 5th, 1858*." *Art History*, 1, 1978, 99-103.

Pointon, M. *William Dyce 1806-1864*. Oxford, 1979.

Powell, C. *Turner's Rivers of Europe. The Rhine, Meuse, and Mosel*. Exh. cat. London, 1991.

Pressly, N.L. *The Fuseli Circle in Rome. Early Romantic Art of the 1770s*. Exh. cat. New Haven, 1979.

Punch, or the London Charivari. Various articles and prints.

Raine, K. *Blake and Tradition*. 2 vols. Princeton, 1968.

Raleigh, North Carolina Museum of Art. *Catalogue of Paintings, 2: British Paintings to 1900*. Raleigh, 1969.

Ramsey, J.T., and Licht, A.L. *The Comet of 44 B.C. and Caesar's Funeral Games*. Atlanta, 1997.

Ranyard, A.C., ed. "Observations made during total Solar Eclipses." *Memoirs of the Royal Astronomical Society*, 41, 1879, 1-792.

Rawlinson, W.G. *The engraved work of J.M.W. Turner, R.A.* 2 vols. London, 1908, 1913.

Rees, A., ed. *Cyclopaedia: or Universal Dictionary of Arts, Sciences and Literature*. 45 vols. London, 1819-20.

Reid, G.W. *A Descriptive Catalogue of the Works of George Cruikshank*. 2 vols. London, 1871.

Reilly, R., and Savage, G. *Wedgwood, the Portrait Medallions*. London, 1973.

Reynolds, G. *The Later Paintings and Drawings of John Constable*. 2 vols. New Haven/London, 1984.

Reynolds, G. *Victorian Painting*. New York, 1966.

Rinder, F. *The Royal Scottish Academy 1826-1916*. Glasgow, 1917.

Roberts, J. *Views of Windsor. Watercolours by Thomas and Paul Sandby*. Exh. cat. London, 1995.

Robertson, B. "Paul Sandby and the early development of English watercolor." 2 vols. Ph.D. diss., Yale University, 1987.

Robinson, L.J. "Book review of *Celestial Objects for Common Telescopes*." *Sky and Telescope*, 26, 1963, 95-96.

Roget, J.L. *History of the Old Water-Colour Society, now the Royal Society of Painters in Water-Colour*. 2 vols. London, 1891.

Rosen, D., and Rosen, S. *London Science*. London, 1994.

Rosenblum, N. *A History of Photography*. New York, 1989.

Rosenblum, R. "British Art and the Continent, 1760-1860." In F. Cummings and A. Staley. *Romantic Art in Britain: Paintings and Drawings 1760-1860*. Exh. cat. Philadelphia, 1968, 11-16.

Rosenblum, R. *Transformations in Late Eighteenth Century Art*. Princeton, 1967.

Rosse, W. Parsons, Earl of. *The Scientific Papers of William Parsons, Third Earl of Rosse, 1800-1867*. Ed. Sir C. Parsons. London, 1926.

Rothermel, H. "Images of the Sun: Warren De la Rue, George Biddell Airy and celestial photography." *The British Journal for the History of Science*, 26, 1993, 137-169.

Ruffner, J.A. "The Curved and the Straight: Cometary Theory from Kepler to Hevelius." *Journal for the History of Astronomy*, 2, 1971, 178-194.

Ryan, W.F. "John Russell, R.A., And Early Lunar Mapping." *The Smithsonian Journal of History*, 1, 1966, 27-48.

Sadler, D.H. "Place in Space." In *Greenwich Observatory: 300 years of astronomy*. Ed. C.A. Ronan. London, 1975, 17-22.

Sandby, W. *Thomas and Paul Sandby Royal Academicians*. London, 1892.

Schaaf, F. *Comet of the Century: From Halley to Hale-Bopp*, New York, 1997.

Schaffer, S. "Comets and the world's end." In *Predicting the Future*. Eds. L. Howe and A. Wain. Cambridge, 1993, 52-76.

Schaffer, S. "'The Great Laboratories of the Universe': William Herschel on matter theory and planetary life." *Journal for the History of Astronomy*, 11, 1980, 81-111.

Schiaparelli, G.V. "Sur la relation qui existe entre les comètes et les ètoiles filantes." *Astronomisches Nachrichten*, 68, 1873, 331-334.

Schiff, G. *Johann Heinrich Füssli, 1741-1825*. 2 vols. Zurich/Munich, 1973.

Schleinitz, O. von. *Walter Crane*. Bielefeld/Leipzig, 1902.

Schmadel, L.D. *Dictionary of Minor Planet Names*, Berlin, 1993 (second edition).

Schmidt, R.E. "Samuel Scott's Comet Halley of 1759." *Abstract: Bulletin of the American Astronomical Society*, 17, 1985, 844.

Schofield, R.F. "The Industrial Orientation of Science in the Lunar Society of Birmingham." In *Science, Technology and Economic Growth in the Eighteenth Century*. Ed. A.E. Musson. London, 1972, 136-147.

Schofield, R.F. *The Lunar Society of Birmingham*. Oxford, 1963.

Scott, W.B. *Autobiographical Notes of the Life of William Bell Scott*. 2 vols. New York, 1892.

Scott, W.B. *Memoir of David Scott R.S.A.* Edinburgh, 1850.

Shanes, E. *Turner's Human Landscape*. London, 1990.

Shanklin, J. "Estimation of DC – an Experiment." *International Comet Quarterly*, 17, 1995, 85-87.

Simpson, R. *Camelot Regained. The Arthurian Revival and Tennyson 1800-1849*. Woodbridge, Suffolk, 1990.

Sims, C., and Sims, A. *Picture Making. Technique and Inspiration*. London, 1934.

Singer, C. "The First English Microscopist: Robert Hooke." *Endeavour*, 14, 1955, 12-18.

Smetham, J. *Letters of James Smetham*. Eds. S. Smetham and W. Davies. London, 1892.

Smith, R.W. "The Cambridge Network in Action: The Discovery of Neptune." *Isis*, 80, 1989, 395-422.

Smout, C. "Scotland 1850-1950." In *The Cambridge Social History of Britain*, 1. Ed. F.M.L. Thompson. Cambridge, 1990, 209-280.

Smyth, C.P. "On Astronomical Drawing." *Memoirs of the Royal Astronomical Society*, 15, 1843, 71-83.

Smyth, C.P. "On photographic illustrations for books." *Transactions of the Scottish Society for the Arts*, 5, 1861, 87-92.

Smyth, C.P. "Remarks to Accompany the Foregoing Observations." *Memoirs of the Royal Astronomical Association*, 20, 1851, 68-69.

Smyth, C.P. "Some remarks on theories of cometary physics." *Transactions of the Royal Society of Edinburgh*, 20, 1853, 134-145.

Smyth, G., and Hyde, S. *Walter Crane. Artist, Designer and Socialist*. Exh. cat. London, 1989.

Smyth, W.H. *A Cycle of Celestial Objects*. 2 vols. London, 1844.

Smyth, W.H. "Observations on Halley's Comet." *Memoirs of the Royal Astronomical Society*, 9, 1836, 229-246.

Somerville, M. *Mechanism of the Heavens*. London, 1831.

Somerville, M. *Personal Recollections, From Early Life to Old Age, of Mary Somerville*. London, 1873.

Southey, R. *Mr. Rowlandson's England*. Ed. J. Steel. Woodbridge, Suffolk, 1985.

Spadafora, D. *The Idea of Progress in Eighteenth-Century Britain*. New Haven/London, 1990.

Speed, H. "Charles Sims, R.A." In *The Old Water-colour Society's Club 1928-1929*. London, 1929, 45-66.

Spencer, B.M. *Don Africana Library. Select Acquisitions and Bibliographical Notes*. Cape Town, 1988.

Spencer, I. *Walter Crane*. London, 1975.

Stafford, B.M. *Artful Science: Enlightenment Entertainment and the Eclipse of Visual Education*. Cambridge, MA/London, 1994.

Stainton, L. *British Landscape Watercolours 1600-1860*. Exh. cat. London, 1985.

Staley, A. "British Landscape Painting, 1760-1860." In F. Cummings and F. Staley. *Romantic Art in Britain. Paintings and Drawings 1760-1860*. Exh. cat. Philadelphia, 1968, 25-30.

Staley, A. *The Pre-Raphaelite Landscape*. Oxford, 1973.

Stephens, F.G. *A Memoir of George Cruikshank*. New York, 1891.

Sterne, L., and Kollmeier, H.H. *A Concordance to the English Prose of John Milton*. Binghampton, 1985.

Stewart, P. "Science and Superstition: Comets and the French Public in the 18th century." *American Journal of Physics*, 54, 1, 1986, 16-24.

Stone, E.J. "Address delivered by the President on presenting the Gold Medal of the Society to Mr. Common." *Monthly Notices*, 44, 1883-84, 221-222.

Stone, E.J. "Note on a Crayon Drawing of the Moon, by John Russell R.A. at the Radcliffe Observatory, Oxford." *Monthly Notices*, 55, 1896, 88-95.

Stoney, G.J. *On the connexion between Comets and Meteors*. London, 1867

Story, A.T. *The Life of John Linnell*. 2 vols. London, 1892.

Stott, C., and Hughes, D.W. "Two Piazzi Smyth Comet Paintings." *Annals of Science*, 46, 1989, 165-172.

Struve, O., and Winnecke, A. "Pulkowaer Beobachtungen des Grossen Cometen von 1858." *Mémoires de L'Académie Impériale des Sciences de St.-Pétersbourg*, 2:1, 1859, 1-69.

Struve, O., and Zebergs, V. *Astronomy of the 20th Century*. New York, 1962.

Sugden, K. "An Eclectic Astronomer." *Sky and Telescope*, 63, 1982, 27-29.

Surtees, V. *The Paintings and Drawings of Dante Gabriel Rossetti (1828-1882)*. 2 vols. Oxford, 1971.

Tennant, C. *The Box of Stars*. Boston/Toronto/London, 1993.

Tennyson, A. *The Poems of Tennyson*. Ed. C. Ricks. London/New York, 1969.

Thompson, E. *William Blake and the Moral Law*. New York, 1994.

Thompson, E. *Witness Against the Beast. William Blake and Moral Law*. Cambridge, 1993.

Thompson, F.M.L., ed. *The Cambridge Social History of Britain 1750-1950*. 3 vols. Cambridge, 1990.

Thrower, N.J.W., ed. *Standing on the Shoulders of Giants: A Longer View of Newton and Halley*. Berkeley/Los Angeles/Oxford, 1990.

Tolstoy, J. *James Clerk Maxwell: A Biography*. Chicago, 1981.

Trombino, D.F. "Dr. John William Draper." *Journal of the British Astronomical Association*, 90, 1980, 565-571.

Viscomi, J . *Blake and the Idea of the Book*. Princeton, 1993.

Vsekhsvyatskii, S.K. *Physical Characteristics of Comets*. Jerusalem, 1964.

Walker, A. *Beauty Illustrated by an Analysis and Classification of Beauty in Women*. London, 1852.

Wark, R.R. *Drawings by Thomas Rowlandson in the Huntington Collection*. San Marino, CA, 1975.

Warner, B. *Charles Piazzi Smyth, Astronomer-Artist; His Cape Years 1835-1845*. Cape Town/Rotterdam, 1983.

Warner, B. "The Great Comet of 1843." *Monthly Notices of the Astronomical Society of South Africa*, 29, 1980, 69-75. (1980a)

Warner, B. "Charles Piazzi Smyth at the Cape of Good Hope." *Sky and Telescope*, 59, 1980, 4-5. (1980b)

Warner, D.J. "Smyth, Charles Piazzi." *Dictionary of Scientific Biography*. New York, 1974, 498-499.

Wasson, J.T. *Meteorites*. New York, 1985.

Waterfield, R.L. *A Hundred Years of Astronomy*. London/New York, 1938.

Waterhouse, E. *Painting in Britain 1530 to 1790*. Harmondsworth, Middlesex, 1962.

Watt, C.B. "Predicting the mid-eighteenth-century return of Halley's Comet." In R. Taton and C. Wilson. *General History of Astronomy, 2: Planetary astronomy from the Renaissance to the rise of astrophysics, pt. B: The eighteenth century*. Cambridge and New York, 1995, 281.

Webb, T.W. *Celestial Objects for Common Telescopes*. London, 1859.

Webb, T.W. "On the Great Comet of 1861." *Monthly Notices*, 22, 1861-62, 305-314.

Webb Weekend 1990 – Nov. 24th-25th. Hereford, 1990.

Wedgwood, B., and Wedgwood, H. *The Wedgwood Circle 1730-1897. Four Generations of a Family and Their Friends*. Westfield, NJ, 1980.

Wees, D.J. "*Darkness Visible.*" *The Prints of John Martin*. Exh. cat. Williamstown, MA, 1986.

Weinek, L. "Der grosse Sternschnuppenfall am 27 Nov. 1885." *Astronomische Beobachtungen an der K.K. Sternwarte zu Prag in den Jahren 1885, 1886 und 1887*. Prague, 1890, 44-50.

Weinglass, D.H. *Prints and Engraved Illustrations By and After Henry Fuseli*. Aldershot, Hants, 1994.

Westfall, R.S. *Never at Rest: A Biography of Isaac Newton*. Cambridge, 1980.

Westrum, R. "Science and Social Intelligence about Anomalies: The Case of Meteorites." *Social Studies of Science*, 8. London and Beverly Hills, 1978, 461-493.

Wheatland, D.P. *The Apparatus of Science at Harvard 1765-1800*. Cambridge, MA, 1968.

Whipple, F.L. "The Rotation Period of Comet Donati." *Nature*, 273, 1978, 134-135.

Whiston, W. *An Account of a Surprising Meteor Seen in the Air March 19. 1718/19*. London, 1719.

White, C. *English Landscape 1630-1850. Drawings, Prints and Books from the Paul Mellon Collection*. Exh. cat. New Haven, 1977.

Wilcox, S. *British Watercolours: Drawings of the 18th and 19th Centuries from the Yale Center for British Art*. Exh. cat. New York, 1985.

Wilkening, L. *Comets*. Tucson, AZ, 1982.

Williams, H.S. *The Great Astronomers*. New York, 1930.

Williams, I.A. *Early English Watercolours*. Bath, 1952.

Williamson, G.C. *John Russell, R.A.* London, 1894.

Willsdon, C. "Dyce 'in camera': new evidence of his working methods." *The Burlington Magazine*, 132, 1990, 760-765.

Wilton, A. *J.M.W. Turner. France Italy Germany Switzerland*. New York, 1982.

Wilton, A. *J.M.W. Turner. His Art and Life*. New York, 1979.

Wilton, A. *Painting and Poetry. Turner's "Verse Book" and his Work of 1804-1812*. Exh. cat. London, 1990.

Wilton, A. "Painting in London in the Early Nineteenth Century." In *London – World City – 1800-1840*. Ed. C. Fox. Exh. cat. New Haven/London, 1992, 167-186.

Wilton, A. *Turner and the Sublime*. Exh. cat. London, 1980.

With, G.H. "Observations of Coggia's Comet." *Monthly Notices*, 36: 6, 1876, 277-278.

Wood, C. *The Dictionary of Victorian Painters*. Woodbridge, Suffolk, 1978.

Wright, T., ed. *The Works of James Gillray, The Caricaturist*. London, n.d. (1970 reprint).

Wright, T., and Evans, R.H. *Historical and Descriptive Account of the Caricatures of James Gillray*. London, 1831.

Yau, K.K., Weissman, P.R., and Yeomans, D.K. "Meteorite falls in China and some related human casualty events." *Meteoritics*, 29, 1994, 864-871.

Yeomans, D.K. *Comets: A Chronological History of Observation: Science, Myth, and Folklore*. New York, 1991.

Young, I.B. *A Concordance to the Poetry of Byron*. 4 vols. Austin, Texas, 1965.

Index